Biochips as Pathways to Drug Discovery

Drug Discovery Series

Series Editor

Andrew Carmen

Johnson & Johnson PRD, LLC
San Diego, California, U.S.A.

Drug Discovery Series/7

Biochips as Pathways to Drug Discovery

Edited by
Andrew Carmen
Gary Hardiman

CRC Press
Taylor & Francis Group
Boca Raton   London   New York

CRC Press is an imprint of the
Taylor & Francis Group, an informa business

CRC Press
Taylor & Francis Group
6000 Broken Sound Parkway NW, Suite 300
Boca Raton, FL 33487-2742

International Standard Book Number-10: 1-57444-450-6 (Hardcover)
International Standard Book Number-13: 978-1-57444-450-6 (Hardcover)

Library of Congress Cataloging-in-Publication Data

Biochips as pathways to drug discovery / edited by Andrew Carmen and Gary Hardiman.
    p. cm.
    Includes bibliographical references.
    ISBN 1-57444-450-6 (alk. paper)
    1. Biochips. 2. Drug development. I. Carmen, Andrew. II. Hardiman, Gary, 1966-

R857.B5B54 2006
615'.19--dc22   2006045577

Visit the Taylor & Francis Web site at
http://www.taylorandfrancis.com

and the CRC Press Web site at
http://www.crcpress.com

Preface

In the summer of 1982 as a biochemistry undergraduate at Cornell University I had the good fortune to pursue a research project in a prestigious fly (*Drosophila melanogaster*) laboratory. The then very young John Lis took a chance on me, offering the opportunity to do undergraduate research in his laboratory. It was an exciting time even though I was on the periphery, gaining experience with *Drosophila*, molecular biology technique, and the excitement of seeing the first "blue" flies from the heterologous-controlled expression of β-galactosidase driven by the HSP70 promoter, a cover photo in *Cell* back in 1983. But the upshot was that I wanted to know what controlled genes, what caused the chromosomal puffing, and what were the chromatin (and epigenetic) factors.

As a graduate student and McKnight Scholar at the University of California at Davis, I continued my interest in transcription. Through interactions with Peter Yau (now at University of Illinois, Champaign-Urbana) and Morton Bradbury, an interest was spawned in me in the acetylation of histones and other epigenetic factors. Naturally, I moved on to Mike Grunstein's laboratory at the University of California at Los Angeles (UCLA). I knew that epigenetic factors, and certainly histones, were key elements in transcription. My personal "holy grail" at the time was the elusive histone acetyltransferase. Being somewhat naive in regard to biochemistry at the time, I thought purification would be no problem. To my shock and dismay, I could purify it and follow its activity, but it fell apart, no matter how fast I worked or what I tried. Fortunately for me, I decided to see if I could find a histone deacetylase activity in my extracts. Fortuitously, again, I had done a great job at purifying a fairly stable histone deacetylase complex in some of my "acetyltransferase" extracts. Not much of it, but it was relatively stable; some quick math determined it was doable for protein sequence, if purified from approximately 5 kg of yeast. As luck would have it, Thomas Sutherland at UCLA ran a fermentor facility that would allow this to happen. Thus, I was enabled to purify the relatively scarce yeast HDA complex.

Of course, this was only the setup for a bigger problem: Now you have the enzyme, and good lord, the yeast genome was just sequenced, but you have four other genes similar to HDA1, including RPD3, HOS1, HOS2, and HOS3. We had some work to interpret this complexity. The fractionated enzyme activities would disappear with their corresponding deletion, but how do you tell what they were actually doing in the cell? Single deletions didn't seem to have much effect. Fortunately for me, I was in the laboratory concurrent to Andreas Hecht, (now at Max Planck Freiburg) and Stephen Rundlett, who were developing chromatin IP cross-linking techniques coupled with PCR for determination of targeting of protein factors involved in gene-silencing. We decided to take two approaches: see if the deacetylase proteins could be found associated directly or if a telltale trail of histone acetylation could be seen in a gene-specific manner. In order to do this we developed a full set of specific antibodies directed toward every possible histone acetylation site and every deacetylase. A large team of graduate students and "postdocs" took on the daunting task of multiplex PCR, with limited automation. The deacetylases proved elusive, but their trail of action could be followed. Eventually, we found site-specific targeting of the enzyme but clearly, though, there was a better way.

It was the dawn of microarray. As Mike Grunstein said to me at the time, "Pat Brown has it on a Web site. You can build it." No more of that tedious multiplex PCR stuff. Not knowing what I was getting into, I agreed to the project, along with the help of the very skilled Rick Klufas and the late Mike Eng, both highly motivated UCLA instrumentation facility staff. They were key to the success of the project. To me, I got my feet wet in microarray and found it attractive enough to use as a base for my next position. No more linear science, do the whole genome, or any subset, in one go. Answer questions faster in days that previously took years.

Having spent, arguably, too much time at UCLA "having fun" as a postdoc and assistant research scientist, I wanted a new challenge. I had spent a number of years collaborating with Merck scientists and saw that I had more opportunity than I had been led to believe in pharmaceutical R&D. An actual application and direction, rather than just pure science, was compelling. The perfect position was found in La Jolla at the new R.W. Johnson facility, now Johnson & Johnson Pharmaceutical Research & Development, managing a small genomic operation. There I wouldn't just be building the technology, but I could drive it. If it didn't exist, I had the resources to define the new direction.

So, it was strive for the new chip to answer the questions that just 10 years ago were daunting. Find the targets, bind the drugs, optimize them, put them in rats, and test in people. Sounds simple. The truth is that there are many problems and inefficiencies in drug discovery. In a fiercely competitive marketplace, pharmaceutical companies can not afford to spend excess dollars on developing drugs that will fail to get FDA approval or will have some profoundly poor characteristics. In this book we present a comprehensive look at how the industry faces these challenges, in many cases with new technologies such as biochips to reduce the cost of drug discovery and improve drug safety. The industry is getting smarter, finding the targets and weeding out potentially problematic drugs sooner, thus cutting costs. In short order, we may also find that the one drug for everyone may not be the norm. Pharmacogenomics presents a hope not only to get better drugs, but to fit the right drugs to the right people. This might also have implications that we may improve selection for clinical trials. Here, we look at how these trends will affect the industry and what the outcomes might be on the science and long-term prospects of these technologies and the companies utilizing them.

Andrew Carmen, Ph.D.
San Diego, California

In 1989 I received my first introduction to molecular biology in the laboratory of Frank Gannon, then at the National University of Galway in Ireland. As a B.Sc. honors student in microbiology, I had, like my classmates, to complete a 6-month laboratory project; mine was on the examination of strains of *E. coli* for the production of lambda phage extracts. This I dove into with the help of the late

Riche Powell, and after many late nights, the lack of success in producing successful extracts was attributed to the *E. coli* strains not being what they were supposed to be. Nevertheless, my interest was piqued, and when Frank offered me a postgrad position in his lab, working on the effects of saltwater adaptation on gene expression in the Atlantic salmon, this seemed like a good way to spend the next few years. At that time in late 1989, little of any genome, and in particular the salmon, had been sequenced by today's standards. Much of 1989 and 1990 consisted of sampling fish, at local fish farms and at the National Diagnostics Center in Galway, extracting RNA and building up a repository of salmon at different developmental stages. By today's standards the approach I took, although sound, seems quaint. I generated first-strand cDNA from these salmon and used this material to screen salmon liver and kidney cDNA libraries for clones that exhibited differences in their hybridization patterns. Not surprisingly, the majority of the cDNA clones appeared the same in fresh and saltwater salmon. I have a great memory, though, of looking at a series of autorads on a long Irish summer evening and finding a series of cDNAs that clearly had elevated levels in the saltwater fish compared to the freshwater fish. The next two years was spent capitalizing on this find and characterizing these cDNAs. Not surprisingly, many of the cDNAs I had uncovered were what one would expect, namely genes encoding proteins involved in aerobic metabolism and growth.

In 1993, I decided that after 8 years in the same university, in a location well-known for lots of rain, I wanted to live in a sunny climate for a couple of years. I had the opportunity to spend a brief stint in the lab of Frank Talamantes at the University of California, Santa Cruz in 1992 and, really liking California, I decided that would be my next move. As luck would have it, I found a postdoc position at DNAX Research Institute in Palo Alto in October 1993. At that time DNAX was ramping up its in-house sequencing efforts and applying the high-throughput approach to novel factor discovery under the direction of Gerard Zurawski and the late Jacques Chiller. I joined the lab of Fernando Bazan and Rob Kastelein and, with Fernando Rock, became part of a structural biology group involved in bioinformatics-based gene discovery, with a strong emphasis on comparative genomics, in particular, the characterization of novel signaling molecules and pathways in both human, fly, and nematode systems. This allowed me to first work with DNA microarrays in the mid 1990s as DNAX had a key interest in the technology. My earliest memory of DNA microarray data is a 4 mB file Excel file (reasonably large by my mid 1990s standards) that someone had aptly named "the complete enchilada."

In 1998, I joined the ill-fated Axys Pharmaceuticals. This, at the time, seemed like a unique opportunity, the chance to do interesting science in a biotech setting. Working at Axys afforded me the chance to work with microarrays, in the context of both oncology and nematode projects, but more importantly it allowed me to participate in the Molecular Dynamics Early Technology Access Program, which among other opportunities, introduced me to Andrew, ultimately leading to this publication. In 1999, after living in four cities in an 18-month period and my Chrysler Le Baron convertible having traversed the U.S. twice on two occasions, I slowly began to think hard about what to do next. Another Boston winter was not an option. Most of my colleagues seemed to be working for or starting dot-com companies.

Around this time, when Axys closed its doors in La Jolla in the fall of 1999, my colleague and friend Antonio Tugores made me aware that the University of California at San Diego (UCSD) wanted to hire a director to oversee the running of the UCSD Biomedical Genomics Microarray (BIOGEM) Core Facility. This seemed like the perfect opportunity to help establish a new genomics facility and assist diverse researchers in applying this technology to a variety of biological questions. Being able to optimize the technology, build microarrays that were not commercially available, and help bring in emerging technologies has been enormously satisfying. Working on the challenges of biochip technology, particularly dealing with small sample sizes and applying the technology to the clinical setting, are current interests of mine.

In this edition, we provide a comprehensive overview of the current state of biochip technology and the effect biochips are having on biomedical research, in particular the pharmaceutical industry. Technology platforms are presented and covered in detail. The clinical and pharmacogenomic relevance of biochips, ChIP-chip assays, and high-throughput approaches are all reviewed in depth. Chapters are presented detailing the application of biochips to the study of malaria, toxicogenomics, and SNPs. Intellectual property and market overviews are presented as current and forward-looking perspectives. The DNA microarray field will thrive in the coming years, an expansion that will encompass robotics, nucleic acid chemistries, and informatics. Multidisciplinary approaches will help this field mature and find its niche in the clinical arena. I trust that you find this book a valuable reference.

Gary Hardiman, Ph.D.
La Jolla, California

Note: This preface expresses the views of the authors and is not intended to express any views of their respective employers.

Acknowledgments

I would like to thank all my colleagues who contributed to this work, for without these innovators we could not have been able to put together this innovative work. I also thank my employer for allowing me to pursue this endeavor. I thank my wife, Jeanette, for her support, love, and patience. — **A.C.**

This publication contains chapters from world expert scientists, academics, programmers, and engineers skilled in the different facets of microarray technology, and I am grateful to all those who contributed material. Many people behind the scenes have contributed to the success of this project. Thanks go out to the UCSD Biomedical Genomics Facility (BIOGEM) and my colleagues at UCSD, in particular professors Chris Glass, Geoff Rosenfeld, Scott Emr, Bill McGinnis, and Tony Wynshaw-Borris. I would also like to thank Cyndy Illeman and Deborah Seidle at UCSD Core Bio Services and the members of BIOGEM past and present, particularly Jennifer Lapira, Colleen Eckhardt, Ivan Wick, Kristin Stubben, Karin Bacon, Allen Lee, Barbara Ruggeri, and Roman Sasik. A special thank you goes to Ivan Wick for providing original images for the book's cover. I thank my colleagues, friends, and family for their constant support. Finally, I thank my wife Patricia and daughter Elena for their love and affection, and their patience every time I disappeared with my laptop to write or edit. — **G.H.**

The Editors

Andrew Carmen, Ph.D., is the associate director, Global Pharma R&D, Johnson & Johnson Pharmaceutical Research & Development, San Diego, and the former manager of Genomic Operations, where he focused on establishing automation technologies for microarrays. He received his B.S. (biochemistry) from Cornell University in 1984 and his Ph.D. (biochemistry) as a McKnight Fellow from the University of California, Davis, in 1991. He is also concurrently pursuing an M.B.A. at the UCSD Rady School of Management. As a postdoctoral fellow and assistant research scientist at UCLA he purified and identified the first yeast histone deacetylase complex, HDA, a global transcriptional regulator. Currently, he seeks opportunities in technology leading to breakthrough discoveries in pharmaceutical research.

Gary Hardiman, Ph.D., is the director of BIOGEM (an organized research unit and core facility) specializing in high throughput genomic approaches, including DNA microarray technology and bioinformatics, at the University of California, San Diego. He is an assistant professor in the Department of Medicine at UCSD and teaches classes on microarray technology. His Ph.D. is in microbiology from the National University of Ireland (Galway). He completed postdoctoral research fellowships at DNAX Research Institute, Palo Alto, and has worked as a senior scientist at Axys Pharmaceuticals (La Jolla and South San Francisco). His current research interests include the application of genomic approaches to understand the molecular mechanisms of disease. He is the editor of the popular *Microarray Methods and Applications Book* (Nuts and Bolts Series) published by DNA Press, Inc.

Contributors

Francis Barany
Department of Microbiology and
 Immunology
Weill Medical College of Cornell
 University
New York, New York

Arindam Bhattacharjee
Agilent Technologies Inc.
Andover, Massachusetts

Anton Bittner
Johnson & Johnson Pharmaceutical
 Research & Development
San Diego, California

Jacobus Burggraaf
Centre for Human Drug Research
Leiden, The Netherlands

Andrew Carmen
Johnson & Johnson Pharmaceutical
 Research & Development
San Diego, California

Pengchin Chen
NuGEN Technologies
San Carlos, California

Roberto Ciccocioppo
Department of Pharmacological
 Science and Experimental Medicine
University of Camerino
Camerino, Italy

Adam F. Cohen
Centre for Human Drug Research
Leiden, The Netherlands

Nadine Cohen
Johnson & Johnson Pharmaceutical
 Research & Development
Raritan, New Jersey

Jacques Corbeil
Université Laval
Quèbec, Canada

Alan Dafforn
NuGEN Technologies
San Carlos, California

Heng Dai
Johnson & Johnson Pharmaceutical
 Research & Development
San Diego, California

Marieke de Kam
Centre for Human Drug Research
Leiden, The Netherlands

Glenn Y. Deng
NuGEN Technologies
San Carlos, California

Roberto Fagnani
Biocept Inc.
San Diego, California

Reyna Favis
Johnson & Johnson Pharmaceutical
 Research & Development
Raritan, New Jersey

J. Bruce German
Nestlé Research Center
Lausanne, Switzerland

Gary Hardiman
Department of Medicine
University of California-San Diego
La Jolla, California

Michael Herrler
NuGEN Technologies
San Carlos, California

Ewoud van Hoogdalem
Johnson & Johnson Pharmaceutical
  Research & Development
Beerse, Belgium

Dawn M. Iglehart
NuGEN Technologies
San Carlos, California

Sergey E. Ilyin
Johnson & Johnson Pharmaceutical
  Research & Development
Spring House, Pennsylvania

C. Richard Jones
Biopharma
Basel, Switzerland

Michiel Kemme
Centre for Human Drug Research
Leiden, The Netherlands

Yung-Kang Ken Fu
AutoGenomics Inc.
Carlsbad, California

Phillip Kim
AutoGenomics Inc.
Carlsbad, California

Richard Kliman
Department of Biological Sciences
Cedar Crest College
Allentown, Pennsylvania

Sriveda Koritala
NuGEN Technologies
San Carlos, California

Fareed Kureshy
AutoGenomics Inc.
Carlsbad, California

Nurith Kurn
NuGEN Technologies
San Carlos, California

Susan M. Lato
NuGEN Technologies
San Carlos, California

Karine Le Roch
Genomics Institute of the Novartis
  Research Foundation, San Diego,
and Department of Cell Biology
The Scripps Research Institute
La Jolla, California

Albert Leung
Johnson & Johnson Pharmaceutical
  Research & Development
San Diego, California

Qingqin Li
Johnson & Johnson Pharmaceutical
  Research & Development
Raritan, New Jersey

Xuejeun Liu
Johnson & Johnson Pharmaceutical
  Research & Development
San Diego, California

Peter G. Lord
Johnson & Johnson Pharmaceutical
  Research & Development
Raritan, New Jersey

Vijay Mahant
AutoGenomics Inc.
Carlsbad, California

Michael McMillian
Johnson & Johnson Pharmaceutical
  Research & Development
Raritan, New Jersey

Lisa Milne
Neotrove Informatics
West Chester, Pennsylvania

Peter Morrison
Biodiscovery Inc.
El Segundo, California

David M. Mutch
Nestlé Research Center
  and Center for Integrative Genomics
University of Lausanne
Lausanne, Switzerland

Alex Y. Nie
Johnson & Johnson Pharmaceutical
  Research & Development
Raritan, New Jersey

Vicki G. Norton
Wilson, Sonsini, Goodrich & Rosati
San Diego, California

Margriet Ouwens
Department of Molecular Biology
Leiden University Medical Center
Leiden, The Netherlands

Anton Petrov
BioDiscovery Inc.
El Segundo, California

Susheela Pillarisetty
NuGEN Technologies
San Carlos, California

Tony Pircher
Biocept Inc.
San Diego, California

Prankaj Prakash
BioDiscovery Inc.
El Segundo, California

Reshma Purohit
NuGEN Technologies
San Carlos, California

Matthew-Alan Roberts
Nestlé Purina Pet Care
St. Louis, Missouri

Daniel Robyr
Department of Genetic Medicine and
  Development
University of Geneva Medical School
Geneva, Switzerland

Barbara Ruggeri
Department of Medicine, BIOGEM
University of California-San Diego
La Jolla, California

Soheil Shams
BioDiscovery Inc.
El Segundo, California

Sejal Sheth
Affymetrix Inc.
Santa Clara, California

Laura Soverchia
Department of Pharmacological
  Science and Experimental
  Medicine
University of Camerino
Camerino, Italy

Phillip Stafford
Biomining.com
Tempe, Arizona

Pavel Tsinberg
Biocept Inc.
San Diego, California

Leah Turner
NuGEN Technologies
San Carlos, California

Massimo Ubaldi
Department of Pharmacological
  Science and Experimental Medicine
University of Camerino
Camerino, Italy

M.B.A. van Doorn
Centre for Human Drug Research
Leiden, The Netherlands

Martin Wang
NuGEN Technologies
San Carlos, California

Shenglong Wang
NuGEN Technologies
San Carlos, California

Gary Williamson
Nestlé Research Center
Lausanne, Switzerland

Elizabeth Winzeler
Genomics Institute of the
  Novartis Research Foundation
San Diego, and
Department of Cell Biology
The Scripps Research Institute
La Jolla, California

Huinian Xiao
Johnson & Johnson Pharmaceutical
  Research & Development
San Diego, California

Jing Xu
BioMinerva Group
San Diego, California

Lynn Yieh
Johnson & Johnson Pharmaceutical
  Research & Development
San Diego, California

Table of Contents

1 DNA Biochips — Past, Present, and Future: An Overview

Gary Hardiman and Andrew Carmen

CONTENTS

INTRODUCTION

Drug discovery is a complex and costly process, with greater than 99% of the investigated experimental compounds discarded as failures. Only a handful of the molecules evaluated as part of the discovery and preclinical phases reach the marketplace [1]. In the current economic climate, the pharmaceutical industry is faced with the double-edged dilemma of increased research and development costs and a decline in the number of novel therapeutics dispensed to the public. The main consequence of this is that the industry has been forced to devise and adapt methodologies that increase the number of new drug candidates in the pipeline, within a much shorter time frame [2]. In the drug discovery process, the identification of viable drug target for a therapeutic area of interest is of key importance [3]. This target is invariably a protein whose function or dysfunction is implicated in the pathology or progression of the disease, for example, a growth factor. A well-characterized example is the epidermal growth factor (EGF) receptor family. Interaction of the extracellular EGF ligand with its receptor results in a signal transduction cascade, ultimately leading to cell division, the synthesis of new proteins, and tumor progression.

Once a target has been identified or validated, the subsequent step is the design of a drug that will interact with the target and deliver the desired therapeutic effect. Knowledge of the ligand–receptor interaction is an important element of the design

1

process, as the drug molecules typically insert at the functional or critical site of the target protein, analogous to a lock and key scenario. Recent advances in protein structure elucidation methods and improvements in three-dimensional modeling techniques have yielded sophisticated approaches to the generation of drug candidates, which are collectively termed *rational drug design*. Candidates are tailored to the three-dimensional structure of target binding and active sites. Medicinal and combinatorial chemistry techniques are employed to generate large compound libraries whose structures correspond to the target's strategic site. These libraries are subsequently screened using high-throughput approaches to identify compounds that reflect the activity of the target protein. Screening assays reveal those compounds that achieve optimal *in vitro* effects.

Drugs can be classified according to their chemical composition into small molecule drugs (SMDs) and biologics (therapeutic hormones, enzymes, monoclonal antibodies, cytokines, and antisense drugs). The tyrosine kinase inhibitor ST1571 (imatinib mesylate, Gleevac; Novartis Pharmaceuticals Corp., East Hanover, NJ) is an excellent example of an SMD that has had a huge impact on the treatment of chronic myelogenous leukemia and gastrointestinal stromal tumors. Monoclonal antibodies (MAbs) have found application in the treatment of cancer, autoimmune disease, viral infection, and myocardial infarction, and as diagnostic agents. Rixtuximab (Rituxan; Genentech, Inc., South San Francisco, CA) was the first MAb approved for the treatment of cancer, specifically for non-Hodgkin's B cell lymphoma. Trastuzumab (Herceptin; Genentech, Inc.), a humanized MAb that targets the extracellular portion of the human epidermal growth factor receptor 2 (HER2)/Neu receptor, overexpressed in many breast cancers, is another success story. Remicade (Infliximab; Centocor, Inc., Horsham, PA) and Enbrel (Entanercept; Immunex, Thousand Oaks, CA) both target tumor necrosis factor alpha TNFalpha and block its inflammatory response, with indications for Crohn's disease (Remicade) and various forms of rheumatoid and psoriatic arthritis.

In recent years, high-density DNA microarrays or biochips have revolutionized biomedical research and greatly accelerated target validation and drug discovery efforts [4]. The utility of microarray technology is that it permits highly parallel gene expression profiling, providing snapshots of the transcriptome in both healthy and diseased states. This knowledge obtained from such comparisons is highly valuable as it identifies gene families and more importantly pathways that are affected by the disease, in addition to those that remain unaffected [5]. Similar expression profiles may imply that genes are coregulated, and this allows researchers to formulate hypotheses about genes with hitherto unknown functions by comparing their expression to those with well-defined functions. Biochips thus can be used to identify and prioritize drug targets, based on their ability to confirm a massive number of gene expression measurements in parallel.

Microarrays are still predominantly used for gene expression analyses, but they are also finding utility in genotyping and resequencing applications, in addition to comparative genomic hybridization and genomewide (epigenetic) localization studies, as covered later in this book. Biochips have been utilized to address *in vitro* pharmacology and toxicology issues, and are being widely applied to improve the processes of disease diagnosis, pharmacogenomics, and toxicogenomics [6–9].

The DNA microarray market is expected to thrive in the coming years, an expansion that will encompass robotic devices, biochip chemistries, nucleic acid labeling and detection strategies, and the significant informatics and data management systems required to store, maintain, and tease the meaning from the voluminous data generated from such studies. This growth is not surprising, given the incredibly powerful nature of this technology and its application to understanding the genomic basis of disease.

This chapter examines the brief history of the biochip field and tracks the evolution of the major platforms in use today by the pharmaceutical industry.

EVOLUTION AND DEVELOPMENT OF BIOCHIPS

Gene expression analysis has rapidly progressed from a classical "single gene" analytical approach to a series of robust technologies that allow highly detailed surveys of complete genomes in a variety of organisms. The origin of the microarray or biochip has its roots in the seminal discovery by Edwin Southern 30 years ago that DNA could be attached to a solid support and interrogated for sequences of interest [10]. Southern described a process whereby a DNA sequence termed a *probe* could easily be labeled with a radioactive or fluorescent marker and hybridized to a mobilized DNA target sample on a filter membrane. The DNA probe hybridized to complementary DNA sequences in the respective sample and the association of the probe, and target sequences could be visualized via autoradiography or chemiluminescent detection mechanisms.

In the 1980s, a decade noted for the increasing use and application of recombinant cloning methodology, researchers adapted the Southern Blot method to facilitate screening of genomic and cDNA libraries. DNA filters were generated from these libraries consisting of bacterial colonies grown on standard laboratory petri dishes. These crude bacterial lysates represent the earliest arrays whereby cDNA or gene banks could be routinely screened for DNA sequences of interest [10–12]. Advances in laboratory automation facilitated the creation of complex high-density filters with a very large number of DNA sequences immobilized in a two-dimensional addressable grid or array format. Differential screening techniques utilizing total cellular RNA and membrane arrays have been widely applied to these macroarrays to study differences in gene expression in different tissues and cells undergoing differentiation [13].

The biochips widely in use today, however, owe their existence to innovation in miniaturization in both the private and academic sectors. Innovators in the development of this technology include Hyseq (Sunnyvale, CA), Affymetrix (formerly Affymax) (Santa Clara, CA), Oxford Gene Technologies (Oxford, UK), and Stanford University (Stanford, CA). Scientists working at these respective organizations developed and advanced technologies related to manufacturing, experimental processing, and genomic profiling. Hyseq developed a method for sequencing DNA by hybridization on an array, using oligonucleotide probes with lengths between 11 and 20 nucleotides that are hybridized to the target nucleic acid sequence. The complementary oligonucleotide probe sequences were overlapping in length and thereby enabled identification of the target nucleic acid sequence. The Hyseq technology permitted the discrimination of perfect match hybrids from hybrids that contained a single nucleotide mismatch. This allowed highly accurate DNA sequencing in a high-throughput array format [14]. Affymetrix developed a technology to manufacture polymers on solid supports, using light-directed spatially parallel chemical synthesis. Their approach to chip manufacturing utilized Very Large Scale Immobilized Polymer Synthesis (VLSIPS™) substrate technologies that could be applied for the synthesis of both peptides and oligonucleotides. The process utilized a series of photo-labile groups attached to solid supports that upon exposure to light activation could react with monomers such as nucleotides and amino acids. Affymetrix has successfully applied this technology to DNA sequencing, DNA fingerprinting, chromosomal mapping, and specific interaction screening [15]. Oxford Gene Technologies (OGT), a company established by Edwin Southern, also developed technology utilizing a solid support containing an array of oligonucleotides to identify DNA sequences, under hybridization conditions where discriminations can be made between matched and mismatched oligonucleotide probes.

The spotted microarray represents an important format and widely utilized application of this technology that was developed at Stanford University by Patrick Brown and colleagues. This experimental paradigm compares mRNA abundance in two different samples via a competitive hybridization. Fluorescent targets are prepared separately from control and test mRNA species, and both are mixed and hybridized together on the same microarray slide. The target gene sequences are allowed to hybridize to their complementary sequences present in the array features. The relative intensities of the resulting signals on the individual features are proportional to the amounts of specific mRNA transcripts in each sample, thereby enabling an estimation of the relative expression levels of the genes in the test and control populations [16]. The DNA arrays are fabricated using a capillary dispenser, which deposits DNA at specific array positions. Spotted microarray production remains a highly automated process, utilizing either capillary-pin-based or inkjet microdispensing liquid handling systems [17]. This particular genre of biochip has been widely adapted by the academic community due to the open source nature of the approach. Many protocols, software tools, and detailed blueprints for robotic printing devices have been freely disseminated. Nevertheless, spotted arrays found commercial utility. Synteni, a company founded in 1994, commercialized this technology and eventually became the microarray division of Incyte upon its acquisition. Agilent (Palo Alto, CA) has utilized inkjet technology to fabricate spotted cDNA arrays from PCR amplicons, although this array format has been largely retired.

BIOCHIPS PLATFORMS: COMPARISONS AND CONTRASTS

Many competing technologies have been adapted by the pharmaceutical industry including oligonucleotide and full-length cDNA arrays [18,19]. These platforms enable the comparison of mRNA abundance in two different biological samples, on identical or replicate microarrays. Affymetrix (Santa Clara, CA) has been a leader in the field for many years, applying photolithographic technologies derived from the semiconductor industry to the fabrication of high-density biochips. The GeneChip™

rapidly became a pharmaceutical industry standard owing to its extensive genome coverage, high levels of reproducibility, and relative ease of use. It is comprised of short single-stranded oligonucleotides. GeneChip fabrication is achieved via a combination of photolithography and solid-phase DNA synthesis. Arbitrary polynucleotides are synthesized in a highly specific manner at defined locations. Initially, a series of synthetic linkers containing photolabile groups are attached to a silicon substrate. Ultraviolet light is subsequently targeted to specific areas on the chip surface using a photolithographic mask. This has the effect of causing localized photodeprotection. The DNA chemical building blocks, hydroxyl-protected deoxynucleosides, are added to the surface, and coupling occurs at the sites that have been illuminated. Additional steps involve directing light to alternate areas of the substrate by using a different photolithographic mask, followed by DNA synthesis. A major advantage of GeneChips is its *in silico* design, which eradicates the requirement of cDNA or oligonucleotide libraries and the potential likelihood of mislabeled features [20]. Additionally the small feature size allows the fabrication of very dense arrays. A disadvantage of this platform is that it demands commitment to GeneChip-specific hardware. Furthermore, it utilizes short 25-mer oligonucleotides, which are inherently less sensitive than the longer 60-mers utilized in other technologies. Affymetrix overcomes this shortcoming, perhaps, via a sophisticated multiple match and mismatch strategy, and powerful algorithms to deconvolute the data.

Alternative platforms have recently emerged. Illumina (San Diego, CA) has developed a bead-based technology for SNP genotyping and gene expression profiling applications on two distinct substrates, the Sentrix LD BeadChip and the Sentrix Array Matrix (which multiplex up to 8 and 96 samples, respectively). Both formats employ an "array of arrays," which increases throughput by enabling the processing of multiple samples simultaneously. Each array contains thousands of tiny etched wells, into which thousands to hundreds of thousands of 3-micron beads self-assemble in a random fashion. Then, 50-mer gene-specific probes concatenated with "address or zip-code" sequences are immobilized on the bead surface. Once bead assembly has occurred, the array is "decoded," using a proprietary process, to determine which bead type containing a particular sequence is present in each well of the substrate. The advantages of this platform are its sensitivity and reproducibility, and small feature size. The oligonucleotide probes can be validated off-line. The Illumina technology offers major increases in throughput to the pharmaceutical industry, but similar to the Affymetrix technology, it demands a commitment to dedicated hardware and software.

The Applied Biosystems Expression Array System (Foster City, CA) employs standard phosphoramidite chemistry to synthesize 60-mer oligonucleotides that are validated off-line by mass spectrometry and are subsequently deposited onto a derivatized nylon substrate. The 3' end of the oligonucleotide is covalently coupled to the nylon via a carbon spacer, thereby elevating the oligonucleotide off the surface and avoiding steric hindrance. The use of chemiluminescence rather than fluorescence distinguishes this platform from others. The advantage of the chemiluminescent scheme is lower background signal intensities. Additionally, once gene targets of interest have been discovered, validation can be carried out rapidly using prevalidated, real-time PCR probes designed from the same genomic region as the microarray probe.

A disadvantage with this platform at present is that it is not readily amenable to customization, offering custom arrays for a limited number of species. In a recurring theme to other commercial platforms, dedicated hardware is required.

Another platform that has been used by the pharmaceutical industry is the CodeLink™ Bioarray from GE Healthcare (Piscataway, NJ), in which 30-mer oligonucleotides are synthesized *ex situ* using standard phosphoramidite chemistry in a similar manner to the ABI biochips. Using piezoelectric deposition technology, the probes are spotted on a proprietary three-dimensional gel matrix. Covalent attachment of the probes is accomplished via covalent interactions between 5' amine groups on the oligonucleotide probes and functional groups on the slide surface. The three-dimensional nature of the slide surface supports an aqueous biological environment and solution-phase kinetics, which improve the limit of detection [21]. The platform is relatively open and in principle the arrays can be utilized with most microarray scanners. The limitations of this platform are a much larger feature size than Illumina or Affymetrix. Consequently, smaller numbers of features can be packed into similar biochip real estate. Additionally, the nature of printed biochips, not just CodeLink™, is such that imperfections will exist with certain probes, albeit in a very small number. Such features are identified as MSR, where the probe was masked after printing, because it represented a suboptimal probe. In such a case, data cannot be obtained from these features. With proper experimental design though, this need not be a major detractor for these arrays.

Agilent Technologies (Palo Alto, CA) relies on the *in situ* synthesis of 60-mer probes by inkjet printing using phosphoramidite chemistry. The 60 mers provide enhancements in sensitivity over 25 mers in part to the larger area available for hybridization. Another advantage is that only one 60 mer per gene or transcript is required [22]. Although short oligonucleotides should in theory provide the greatest discrimination between related sequences, they often have poor hybridization properties. Hughes et al. [22] carried out a detailed study on the effects of oligonucleotide probe length to examine the effects of hybridization specificity and concluded that the beneficial effects of long oligonucleotides were due to both steric and nonsteric effects. The Agilent platform is both reproducible and sensitive. Furthermore, considerable cost savings are realized with this biochip platform, as it is a two-color assay, unlike the others described previously. However, the two-color approach has the potential disadvantage of different fluorescently labeled nucleotides incorporating into nucleic acid targets with different frequencies, thereby altering ratios due to enzymatic parameters rather than actual transcript abundance. Additionally, multiple experiment comparisons are not possible without replicating the reference sample (which, in the case of certain samples such as biopsy material, may be impractical to obtain).

Combimatrix (Mukilteo, WA) has established a solid-phase oligonucleotide synthesis system by using a method that electrochemically places monomers to specific locations on substrates [23]. Alternatives to conventional photolithography with chromium masks are being utilized to fabricate biochips. Nimblegen has synthesized microarrays containing 380,000 features using a digital light processor that creates digital masks to synthesize specific polymers [24]. Microarray devices containing microfluidic structures have also been developed. Microfluidic-based systems have been established that detect hybridization events using electrochemical methods such as voltammetry, amperometry, and conductivity, which will alleviate the need for target sample labeling [25].

PROTEIN BIOCHIPS

Biochemical studies have traditionally focused on the analyses of single-protein species [26]. Two approaches have been widely used to characterize multiple proteins in biological samples. One approach has utilized two-dimensional gels and permitted the separation and visualization of up to 10,000 proteins at once. Upon separation, proteins of interest have been excised from the gel matrix and characterized by mass spectrometry, a time-consuming endeavor applicable only to abundant proteins. Furthermore, limitations exist with current two-dimensional gel separation technology. Although mass spectrometric methods remain unquestionably an excellent means of uncovering potential targets and novel biomarkers, they are not suitable for validation studies, where the initial finding needs to be subjected to rigorous follow-up. The changes observed in the levels of proteins of interest often need to be retested multiple times in a variety of tissues under different conditions and time points. The pharmaceutical industry needs a lower-cost alternative screening technology, amenable to high throughput, such as a protein biochip.

Protein biochips were worth an estimated $122 million in 2002 and have a predicted value of $545 million in 2008 [27]. In November 2004, the first commercially available high-density protein biochip was released by Invitrogen (Carlsbad, CA), containing 1800 unique human proteins, encompassing a wide cross-section of proteins, including kinase, membrane-associated, cell-signaling, and metabolic proteins. Increasing attention is at present being focused on the development of protein microarrays [28–30]. Protein arrays, analogous to their DNA counterparts, are comprised of a library of proteins, immobilized in a two-dimensional, addressable grid. Different chip formats currently exist, including glass and matrix slides and nanowells, with a typical array containing 10^3 to 10^4 features within a total area of 1 cm^2 [31].

In contrast to nucleic acid biochips where miniaturization has increased throughput considerably with decreased bioreagent costs, the inherent structural diversity and complexity in proteins has made the development of protein arrays technically very difficult. Nucleic acid analysis is relatively straightforward in comparison, as both DNA and mRNA molecules are relatively homogenous, and possess high affinity and high-specificity binding partners. Proteins in comparison do not possess straightforward binding partners. The rapid production of proteins is hindered by the lack of a Polymerase Chain Reaction (PCR) equivalent. Major technical hurdles exist with protein biochips, primarily related to acquisition, arraying, and stable attachment of proteins to chip surfaces, and subsequently the detection of interacting proteins.

Proteins are highly sensitive to the physiochemical properties of the chip support material. Polar arrays, for example, are chemically treated to bind hydrophilic proteins. Such surfaces are unsuitable for cell membrane proteins such as G-protein-coupled receptors possessing exposed hydrophobic moieties. Membrane proteins represent the majority of all potential drug targets and are very difficult to stabilize. As opposed to nucleic acids, proteins do not all behave in a similar fashion when exposed to the same surface chemistry. Surface chemistries may promote retention of certain proteins and cause denaturation or loss in activity of others. Often, proteins that are soluble in their native environments may precipitate on chip surfaces. One of the major difficulties is selecting a surface chemistry that permits diverse proteins to retain their native folded conformation and biological activity. Affinity tags are utilized to offset these problems by providing gentle immobilization conditions that maintain protein stability and function. They permit a common immobilization strategy that can be applied to a variety of proteins.

Antibodies represent an all-purpose high-affinity, high-selectivity, protein-binding reagent that has found utility in the generation of protein arrays. However, at present, antibodies are available for a mere fraction of the proteome, and the specificity of many of these antibodies remains poorly documented. Many antibodies are glycosylated and contain large protein-based supporting structures, and cross-reaction with other proteins is, consequently, not uncommon, resulting in large numbers of false positives and questionable data. Yet another layer of complexity with protein detection is that the range in cells of protein concentrations is several orders of magnitude greater than that for mRNAs. Protein microarray detectors require a greater dynamic range of operation, up to a factor of 10^8, as compared to 10^4 for mRNAs. This presents difficulties in the design of a global protein array as separate chips are needed for the detection of rare and abundant proteins, respectively. Because of these limitations, chip-based protein biochips may never obtain the same level of penetration as the nucleic-acid-based chips. It is likely that alternative nonchip strategies such as chromatography-coupled mass spectrometry may prove more useful in this field.

DNA Biochip Limitations and Challenges

Each of the respective microarray platforms has a demonstrated efficiency with respect to signal dynamic range, the ability to discriminate mRNA species, the reproducibility of the raw data, and the fold change and expression level values. Nevertheless, technological, standardization, and patent use limitations exist with biochip technologies. Biochips currently permit the analysis of the relative levels of mRNA species in one tissue sample compared to another. Although a measure of abundance is obtained, biochips do not permit the absolute quantification of specific transcript. This poses a greater challenge requiring in-depth knowledge of the hybridization of each probe to its cognate mRNA species. As each probe–target interaction represents a unique interaction, this is not a trivial pursuit [32]. Additionally, microarrays are limited by the fact that the data obtained merely indicate whether a certain messenger RNA is above the system's threshold level of detection. If the signal is significantly above the background intensity, one can say with a high degree of confidence that it is present. However, the absence of signal does not indicate that the particular mRNA is not expressed. There is a very strong possibility that the mRNA is expressed, albeit at low levels and, further, this low-level expression may be of importance.

DNA microarrays are limited in their ability to detect gene transcription; it should also be noted that mRNA abundance in a cell often correlates poorly with the amount of protein synthesized [33]. Expression analysis using DNA microarrays measures only the transcriptome. Important regulation takes place at the levels of translation and enzymatic activities. The only effect of signal transduction that is observed in a gene expression experiment lies downstream and may be at the end point of a given pathway. Furthermore, DNA microarrays currently have little utility in determining biologically relevant posttranslational modifications, which influence the diversity, affinity, function, cellular abundance, and transport of proteins. DNA microarrays, for example, are not applicable to samples lacking mRNA, such as bodily fluids like urine.

Yet another limitation is that alternative splicing is virtually ignored in many of the current array iterations. Therefore, it is difficult to address definitively whether changes in signal from a particular message are because of alternative splicing events rather than a change in transcript abundance. Current knowledge of alternative splicing in the transcriptome is limited, but this deficiency will likely be addressed in future generations of biochips. A great difficulty for biochips is that mRNA is an unstable molecule. Messenger RNAs are programmed for enzymatic degradation, and the half-lives of different species vary considerably. Those transcripts with short half-lives may be difficult to extract in reproducible quantities. Thus, regulation in expression of a gene with a very short half-life may be impossible to detect with any degree of statistical significance. Poor experimental practices can also lead to differential degradation in samples, making comparisons dubious.

Microarray experimentation is a complex process, and significant time and effort are required to design biologically sound and statistically robust experiments. Once target genes are identified, additional time and expense are required to validate their selection and relevance. Drug discovery programs utilizing microarray technologies must, therefore, consider all available technologies before allocating precious resources. Extensive platform evaluations are impractical for the majority of researchers as this involves considerable expenditure and often a commitment to dedicated hardware and software [18,19,34–37]. The choice of platform utilized by the pharmaceutical industry will continue to be guided by the content on that platform and the amount of RNA available for experimentation.

The existence of multiple technologies has raised the possibility of cross-platform comparison and integration of data. Carefully designed studies have been performed to evaluate the interchangeability of data from different platforms. The outlook for cross-platform integration of data to date is more encouraging than the initial studies suggested. Nevertheless, it poses formidable challenges. The cross-platform discordance observed is attributable to the differences inherent in each of the respective platforms. The probes utilized in different platforms may cause inaccurate expression measurements owing to overlap with related gene family members and the inability to discriminate between splice variants. In view of these issues, cross-platform data from microarray analysis needs to be interpreted cautiously and preferably using sequence-matched probes. As commercial manufacturers make probe data more readily available, we can expect to see improvements in data integration and better standards.

Moreover, when commercial manufacturers adopt standard DNA chip manufacturing practices, and arrays begin to be implemented as clinical diagnostic tools, many of the quality control methods currently employed in the semiconductor industry will appear. This will result in higher-quality, higher-density arrays with greater sensitivity and reproducibility, facilitating a more robust analysis of cellular gene expression. The critical issues that remain are standardization, reproducibility, development of appropriate controls, reference standards, and regulatory compliances. Reference standards will be available in the near future from institutes such as the National Institute for Biological Standards and Control (NIBSC).

MICROARRAYS AND FUTURE DIRECTIONS

The boundaries of technology will be continually challenged as this technology progresses, and novel applications are devised [34]. In addition to the evolving technical approaches of DNA microarray systems, new applications for microarrays are being developed. Recent progress in combining the use of chromatin (ChIP) assays with DNA microarrays has allowed genomewide analysis of transcription factor localization to specific regulatory sequences in living cells [38]. Higher-density arrays, tiling the entire genome, will permit high-resolution maps and global views of the functional relationships amongst transcriptional machinery, chromatin structure, and gene expression in human cells [39].

Additionally, microarrays are being employed as gene delivery vectors that transfect cell monolayers cultured on the array surfaces. The widespread use of RNA interference (RNAi) has prompted several groups to fabricate RNAi cell microarrays permitting discrete, in-parallel transfection with thousands of RNAi reagents on a microarray slide [40]. Though still in their infancy, RNAi cell microarrays promise to increase the efficiency, economy, and ease of genomewide RNAi screens. Tissue microarrays are permitting histological analyses in a high-throughput, parallel manner [41]. Microarrays have entered the clinical arena and hold much promise for molecular diagnostics and clinical medicine. Automated chip platforms permitting multiplexed assays such as the INFINITI™ System from Autogenomics (Carlsbad, CA) (discussed later in this book) should improve the throughput and quality of genetic testing [42]. Affymetrix has developed the GeneChip System 3000Dx, which will enable clinical laboratories to analyze microarray diagnostics, such as the Roche AmpliChip CYP450 Test. This test can be used to identify certain naturally occurring variations in the drug-metabolism genes *CYP2D6* and *CYP2C19* that affect the rate at which a person metabolizes many commonly used drugs. Affymetrix has also introduced an automation system discussed in this text, which offers the ability to run low-cost dense arrays in plate-based formats. Perhaps, as the cost of arrays is reduced via economy of scale, array technologies may see new applications in plate-based arrays, such as in toxicogenomic, preclinical and clinical proof-of-concept studies. SNP-based arrays, not a focus of this chapter, are also emerging as powerful tools in drug discovery and are discussed in subsequent chapters.

REFERENCES

1. Cunningham, M.J. Genomics and proteomics: the new millennium of drug discovery and development. *J Pharmacol Toxicol Methods* 44(1): 291–300, 2000.
2. Kennedy, T. Managing the drug discovery/development interface. *Drug Discovery Today* 2(10): 436–444, 1997.

3. Avidor, Y., Mabjeesh, N.J., and Matzkin, H. Biotechnology and drug discovery: from bench to bedside. *South Med J.* 96(12): 1174–1186, 2003.

4. Marton, M.J., DeRisi, J.L., Bennett, H.A., Iyer, V.R., Meyer, M.R., Roberts, C., Stoughton, R., Burchard, J., Slade, D., Dai, H., Bassett, D.E., Jr., Hartwell, L.H., Brown, P.O., and Friend, S.H. Drug target validation and identification of secondary drug target effects using DNA microarrays. *Nat. Med.* 4: 1293–1301, 1998.

5. Vilo, J. and Kivinen, K. Regulatory sequence analysis: application to the interpretation of gene expression. *Eur Neuropsychopharmacol* 11: 399–411, 2001.

6. Waring, J.F., Ciurlionis, R., Jolly, R.A., Heindel, M., and Ulrich, R.G. Microarray analysis of hepatotoxins in vitro reveals a correlation between gene expression profiles and mechanisms of toxicity. *Toxicol Lett.* 120: 359–368, 2001.

7. Hamadeh, H.K., Amin, R.P., Paules, R.S., and Afshari, CA: An overview of toxicogenomics. *Curr. Issues Mol. Biol.* 4(2): 45–56, 2002.

8. Johnson, J.A. Drug target pharmacogenomics: an overview. *Am. J. Pharmacogenomics* 1(4): 271–281, 2001.

9. Kruglyak, L. and Nickerson, D.A. Variation is the spice of life. *Nat. Genet.* 27: 234–236, 2001.

10. Southern, E.M. Blotting at 25. *Trends Biochem Sci.* 25: 585–588, 2000.

11. Grunstein, M. and Hogness, D.S. Colony hybridization: a method for the isolation of cloned DNAs that contain a specific gene. *Proc Natl Acad Sci U S A* 72(10): 3961–3965, 1975.

12. Williams, J.G. and Patient, R.K. *Genetic Engineering.* IRL Press, Washington, D.C., 1988.

13. Hardiman, G. and Gannon, F. Differential transferrin gene expression in Atlantic salmon *(Salmo salar L.)* freshwater parr and seawater smolts. *J. Appl. Ichthyol.* 12: 43–47, 1996.

14. Wallace, R.B., Shaffer, J., Murphy, R.F., Bonner, J., Hirose, T., Itakura, K. Hybridization of synthetic oligodeoxyribonucleotides to phi chi 174 DNA: the effect of single base pair mismatch. *Nucl. Acids Res* 6: 3543–3557, 1979.

15. Chee, M., Yang, R., Hubbell, E., Berno, A., Huang, X.C., Stern, D., Winkler, J., Lockhart, D.J., Morris, M.S., Fodor, S.P.A. Accessing genetic information with high-density DNA arrays. *Science* 274: 610–614, 1996.

16. Schena, M., Shalon, D., Davis, R.W., and Brown, P.O. Quantitative monitoring of gene expression patterns with a complementary DNA microarray. *Science* 270: 467–470, 1995.

17. Bowtell, D.D.L. Options available — from start to finish for obtaining expression data by microarray. *Nat. Genet.,* 21: 25–32, 1999.

18. Hardiman, G. Microarray platforms — comparisons and contrasts. *Pharmacogenomics* 5(5): 487–502, 2004.

19. Wick, I. and Hardiman, G. Biochip platforms as functional genomics tools for drug discovery. *Curr Opin Drug Discov Dev.* 8(3): 347–354, 2005.

20. Knight, J. When the chips are down. *Nature* 410: 6831, 2001.

21. Ramakrishnan, R., Dorris, D., Lublinsky, A., Nguyen, A., Domanus, M., Prokhorova, A., Gieser, L., Touma, E., Lockner, R., Tata, M., Zhu, X., Patterson, M., Shippy, R., Sendera, T.J., and Mazumder, A. An assessment of Motorola CodeLink™ microarray performance for gene expression profiling applications. *Nucl. Acids Res.* 30: e30, 2002.

22. Hughes, T.R., Mao, M., Jones, A.R., Burchard, J., Marton, M.J., Shannon, K.W., Lefkowitz, S.M., Ziman, M., Schelter, J.M., Meyer, M.R., Kobayashi, S., Davis, C., Dai, H., He, Y.D., Stephaniants, S.B., Cavet, G., Walker, W.L., West, A., Coffey, E.,

Shoemaker, D.D., Stoughton, R., Blanchard, A.P., Friend, S.H., and Linsley, P.S. Expression profiling using microarrays fabricated by an ink-jet oligonucleotide synthesizer. *Nat. Biotechnol* 19: 342–347, 2001.

23. Nuwaysir, E.F., Huang, W., Albert, T.J., Singh, J., Nuwaysir, K., Pitas, A., Richmond, T., Gorski, T., Berg, J.P., Ballin, J., McCormick, M., Norton, J., Pollock, T., Sumwalt, T., Butcher, L., Porter, D., Molla, M., Hall, C., Blattner, F., Sussman, M.R., Wallace, R.L., Cerrina, F., and Green, R.D. Gene expression analysis using oligonucleotide arrays produced by maskless photolithography. *Genome Res* 12: 1749–1755, 2002.

24. Singh-Gasson, S., Green, R.D., Yue, Y., Nelson, C., Blattner, F., Sussman, M.R., and Cerrina, F. Maskless fabrication of light-directed oligonucleotide microarrays using a digital micromirror array. *Nat. Biotechnol.* 17(10): 974–978, 1999.

25. Gao, X., LeProust, E., Zhang, H., Srivannavit, O., Gulari, E., Yu, P., Nishiguchi, C., Xiang Q, and Zhou X. A flexible light-directed DNA chip synthesis gated by deprotection using solution photogenerated acids. *Nucl. Acids Res* 29, 4744–4750, 2001.

26. Pandey, A. and Mann, M. Proteomics to study genes and genomes. *Nature* (Insight), 405: 837–846, 2000.

27. Shaw, G. Cheaper chips find a good fit with hit validation. *Drug Discovery Dev.* 2, 2005.

28. Mitchell, P. A perspective on protein microarrays. *Nat Biotechnol* 20: 225–229, 2002.

29. Service, R.F. Searching for recipes for protein arrays. *Science* 294: 2080–2082, 2001.

30. Cahill, D. Protein arrays: a high throughput solution for proteomics research? *Proteomics: A Trends Guide,* Elsevier, London, 2000, pp. 47–51.

31. Zhu, H., Bilgin, M., Bangham, R., Hall, D., Casamayor, A., Bertone, P., Lan, N., Jansen, R., Bidlingmaier, S., Houfel, T., Mitchell, T., Miller, P., Dean, R.A., Gerstein, M., and Snyder, M. Global analysis of protein activities using proteome arrays. *Science* 293: 2101–2105, 2001.

32. Rouse, R.J., Espinoza, C.R., Niedner, R.H., and Hardiman, G. Development of a microarray assay that measures hybridization stoichiometry in moles. *Biotechniques* 3: 464–470, 2004.

33. Gygi, S.P., Rochon, Y., Franza, B., and Abersold, R. Correlation between protein and mRNA abundance in yeast. *Mol. Cell Biol.* 19: 1720–1730, 1999.

34. Rouse, R. and Hardiman, G. Microarray technology: an intellectual property retrospective. *Pharmacogenomics* 4(5): 623–632, 2003.

35. Hardiman, G. Microarray platforms and drug discovery. *Trends in Drug Discovery — Screening* 1: 1–4, 2005.

36. Hardiman, G. Microarray platforms — comparisons and contrasts. *Pharmacogenomics* 5(5): 487–502, 2004.

37. Stafford, P. and Liu, P. Microarray technology comparison, statistical analysis, and experimental design. *Microarray Methods and Applications — Nuts and Bolts,* DNA Press, Eagleville, PA, 2003, pp. 273–324.

38. Ren, B., Robert, F., Wyrick, J. et al. Genome-wide location and function of DNA-associated proteins. *Science* 290: 2306–2309, 2000. [Description of the Genome-wide location process, which identified all of the known genes for the yeast transcription factor Gal4, as well as several new genes that were confirmed by conventional ChIP assays and additional functional assays.]

39. Kim, T.H., Barrera, L.O., Zheng, M., Qu, C., Singer, M.A., Richmond, T.A., Wu, Y., Green, R.D., and Ren, B. A high-resolution map of active promoters in the human genome. *Nature* 436(7052): 876–880, 2005.

40. Wheeler, D.B., Carpenter, A.E., and Sabatini, D.M. Cell microarrays and RNA interference chip away at gene function. *Nat. Genet.* 37(Suppl.): S25–30, 2005.

41. Espineda, C., Seligson, D.B., James Ball, Jr., W., Rao, J., Palotie, A., Horvath, S., Huang, Y., Shi, T., and Rajasekaran, A.K. Analysis of the Na,K-ATPase alpha- and beta-subunit expression profiles of bladder cancer using tissue microarrays. *Cancer* 97: 1859–1868, 2003.
42. Mahant, V., Kureshy, F., Vairavan, R., and Hardiman, G. The *INFINITI™* system — an automated multiplexing microarray platform for clinical laboratories. *Microarray Methods and Applications — Nuts and Bolts*, DNA Press, Eagleville, PA, 2003, pp. 325–338.

2 Three-Dimensional HydroArrays: Novel Microarrays for Genomic and Proteomic Studies

Roberto Fagnani, Pavel Tsinberg, and Tony J. Pircher

CONTENTS

INTRODUCTION

The research unveiled in 1995 by Schena et al. started a new era of monitoring gene transcription [1]. The paper by Schena described the simultaneous measurement of the expression of thousands of genes using DNA microarrays. Although techniques such as Northern blot, RT-PCR, and RNase protection are typically and reproducibly used to measure levels of gene expression, they do not allow for simultaneous assessment of thousands of genes in a parallel fashion. As high-density microarrays became more prevalent, the role of the older techniques shifted from discovery to validation; now assays like Northern blots are used to verify data collected through microarray technology.

Although initial screening of genomes is best done using high-density microarrays, the identification of genes or pathways of interest as well as the study of specific genomic events and pathways may be best performed using arrays containing smaller subsets of selected genes, which we called *focus arrays*. In many cases, when only a small group of genes is sampled, having thousands of gene probes on one array may provide too much information. Analysis of such massive data sets is time consuming, requires complex bioinformatics, and may obscure relevant data. We, therefore, propose smaller, more precise, highly reproducible "focus" (or "pathway") arrays as an alternative method for validating, verifying, and studying data initially collected by high-density arrays. In this chapter, we describe one such array, including its unique three-dimensional features and its applications to genomics and proteomics studies.

THREE-DIMENSIONAL HYDROARRAYS

In a standard oligonucleotide or cDNA microarray, probes are arranged in a monolayer fashion directly on a solid surface [1,2]. To print such arrays, the probes are suspended in a printing buffer and transferred onto a solid substrate where they react with the chemically derived surface, after which excess probes are washed away. For the purpose of this chapter, we define such microarrays as two-dimensional arrays.

We have developed three-dimensional polyurethane-urea-hydrogel-based arrays (three-dimensional HydroArrays). With HydroArrays, probes are not bound directly to the solid substrate, but are rather derived to and suspended within swellable hydrogel. The hydrogel is comprised mainly of "soft," hydrophilic polyethylene glycol (PEG) chains, with the probe moieties bound directly into the gel matrix.

A comparison between a two-dimensional and three-dimensional array is illustrated in Figure 2.1. In the two-dimensional array, the probes are bound in a two-dimensional fashion directly to the surface. In contrast, the three-dimensional HydroArray probes are bound in a multilayer fashion within a three-dimensional microdroplet.

The three-dimensional structures of three-dimensional HydroArrays allow for a much greater number of possible probe attachment points than a standard two-dimensional substrate. Each microdroplet effectively consists of many layers of probes covalently tethered to and uniformly dispersed throughout the gel matrix.

FIGURE 2.1 Two-dimensional vs. three-dimensional microarray.

With this configuration, each microdroplet contains 10^{11} to 10^{12} probes. This is an order of magnitude greater than the number of probes available in conventional two-dimensional arrays.

The three-dimensional distribution of capture probes allows for greater spatial separation between individual probes. This overcomes steric hindrance arising from probes being too close to one another. In addition, biomolecules covalently tethered in the aqueous environment of the hydrogel retain their native conformation and activity. The probes are now free to rotate fully in three dimensions, allowing for better hybridization.

Because of the relatively large molecular weight, length, and hydrophilic nature of PEG chains, the gel, in its fully swollen state, is made up of over 95% water, therefore permitting free diffusion of target molecules in and out of the matrix. Much like the more commonly used polyacrylamide gel, hydrogel microdroplets are permeable to both macromolecules (up to at least 160,000 Da) as well as small molecules.

HYDROARRAY CHEMISTRY

Trimeric PEG polyol, which is end-derived with an aliphatic isocyanate (NCO), serves as the starting point. Trimeric PEG polyol is dissolved in anhydrous organic solvents to yield a PEG-NCO or solvent mixture prepolymer. This prepolymer is then mixed with buffered aqueous solutions containing the biological probes. Essentially, the initiation of polymerization is triggered by water.

Other gel-based arrays have been reported in the literature [3]. These arrays are typically made of polyacrylamide and require either free radicals, UV, or harsh organic solvents to initiate polymerization. In contrast, the polymerization of PEG-NCO begins as soon as the prepolymer is mixed with the aqueous buffer containing the biological probes. This avoids the exposure of sensitive biological probes, such as proteins, to harsh polymerization conditions.

During polymerization, a series of amine–isocyanate reactions simultaneously perform three different reactions, as indicated in Figure 2.2: (1) cross-link the

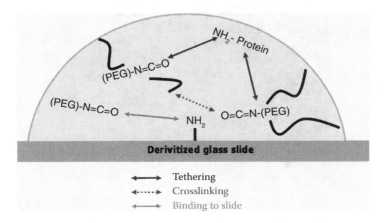

FIGURE 2.2 PEG-Hydrogel chemistry: amine-isocyanate reaction.

polymer; (2) tether the amine-derived bimolecular probes to the emerging polymer matrix, and (3) bind the hydrogel microdroplets to primary-amine-derived glass slides, firmly linking the microdroplet to the solid support.

In the case of genomic HydroArrays, the probes are 45-mer oligonucleotides derived via their 5' NHS end. Later in this chapter, we will also discuss the use of protein as probes for proteomic applications. Although not discussed in this chapter, we have also produced arrays whereby the capture probes are made of small biological molecules, such as biotin or metal chelators, to immobilize proteins as part of our proteomic arrays program. We have also succeeded in encapsulating functional, living bacterial and mammalian cells within hydrogel microdroplets in an array format to generate cell-based microarrays.

In the printing step, the prepolymer is mixed with the probes in aqueous buffer in a microtiter plate, which is then used as a source plate for printing. Next, a robotically controlled matrix of solid printing pins (specifically designed to optimize transfer of hydrogel) picks up small amounts (approximately one nanoliter) of probe or hydrogel mixtures from the source plate and deposits or "prints" an array of microdroplets on a glass slide. After printing, arrays are allowed to polymerize to completion, a process that takes 3 h. No further processing of the hydrogel is necessary, and the HydroArrays are ready for use.

The size of each microdroplet is typically 300 μM in diameter, with a height of 40 μM. The estimated volume varies between 1.9 and 5.7 nL, using a slightly rounded cylinder as the model. Due to its hydrophilic properties, the hydrogel readily absorbs water, and is therefore fully hydrated during hybridization. When dried, it rapidly loses water and, with it, most of its volume. In its fully dried state, the hydrogel collapses to 5% of its maximum volume, with a height of just 5 μM. Hydrogel is optically clear in both its hydrated and dried states, as shown in Figure 2.3. Both features were designed for fluorescent detection of hybridization, using standard confocal laser scanners such as ScanArray or Axon Gene-Pix.

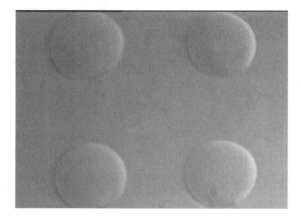

FIGURE 2.3 Phase-contrast image of a set of four dried spots. (From Gurevitch, D., et al., A novel three-dimensional hydrogel-based microarray platform, *JALA*, 6(4): 87–91, 2001. With permission.)

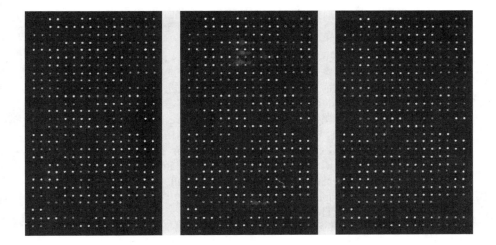

FIGURE 2.4 Three identical hybridized three-dimensional HydroArrays.

The high number of capture probes and the low steric hindrance provided by the aqueous microenvironment of hydrogel microdroplets create hybridization conditions that result in a high signal-to-noise ratios, low spot-to-spot and print-to-print variation, and very tight coefficient of variations (CV). The performance of a typical three-dimensional HydroArray is visually illustrated in Figure 2.4. Table 2.1 shows the CVs and the reproducibility of three separate print-to-print hybridizations. Note the tight CVs, varying between 1 and 6% (4% average).

TABLE 2.1
Coefficient of Variations and Fluorescence Signal Intensity of Three Separate Hybridizations

	Three Different Hybridizations with Cy3-Labeled Targets				
Probe Id	Name	Slide 1	Slide 2	Slide 3	Avg	CV
2732	ACTB	28453	31709	29889	30017	5%
2736	RPL13A	39686	40660	40148	40165	1%
2742	RPL19	6141	6401	6220	6254	2%
2745	GAPD	19606	21063	21678	20782	5%
2746	PDHB	31631	31534	30126	31097	3%
2753	RPS9	30751	29008	27224	28994	6%
2796	PDHA1	25189	25673	24427	25096	3%
2797	UBC	35247	37414	36071	36244	3%
2798	HPRT1	8804	8878	8158	8613	5%
2799	HLA-C	27284	26614	25673	26524	3%
2827	ARA1	33456	35672	34778	34635	3%
2836	ARA4	21602	23204	22474	22427	4%
					Avg cv	4%

EXAMPLE OF FOCUS THREE-DIMENSIONAL HYDROARRAYS

The analysis of the genes involved in apoptosis provides an example of focus three-dimensional HydroArrays. Antihuman CD95 (FAS), a widely studied inducer of apoptosis [4], was chosen as the model system. Human T-cell leukemia cell line, Jurkat clone E6-1 cells (American Cell Culture Collection, Rockville, MD), were cultured in RPMI-1640 media and 10% FBS. After reaching a concentration of 7×10^5 cells/mL, cells were treated with 100 ng/mL antihuman CD95 (eBioscience, San Diego, CA) for 6 h. Poly-A RNA was extracted, and Cy3-labeled targets were prepared by incubating 2-μg poly-A RNA with Cy3-labeled random hexamers (IDT, Skokie, IL) and reverse transcriptase (SuperScript II, Invitrogen, Carlsbad, CA) for 2 h. In this procedure, targets were not amplified, and each synthesized target contained one Cy3 molecule on the 5' end. The focus HydroArrays are also compatible with target processing systems that utilize amplification, e.g., T7 RNA amplification.

In this example, the targets were hybridized to an apoptosis pathway HydroArray, consisting of 256 apoptosis-related genes and control probes, for 16 h. The distribution of gene induction, normalized against housekeeping genes, is shown in Figure 2.5A. Figure 2.5B shows the genes that are induced threefold or more under the experimental conditions described previously. These genes include DAPK1, BIRC4, CRADD, and TNFSF12, all of which have previously been described to be involved in FAS signaling and apoptosis. DAPK1 is a proapoptotic kinase, reportedly induced by FAS [5,6]. BIRC4 has been described as an inhibitor of FAS-mediated apoptosis [7]. CRADD interacts with the intracellular death domain of CD95 (FAS) [8]. Finally, TNFSF12 (TWEAK) has also been described as an inducer of apoptosis [9]. Few of the housekeeping genes were affected by FAS treatment. These results demonstrate that the apoptosis HydroArray effectively identifies alterations in gene transcription in cells subjected to apoptotic events.

To demonstrate the responsiveness of the capture probes used in the preceding example, we used the same apoptosis three-dimensional HydroArray to compare

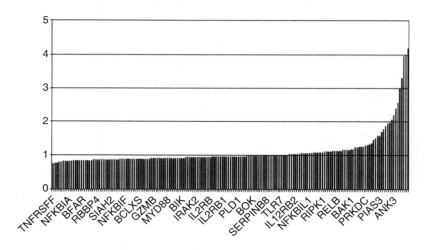

FIGURE 2.5 Effect of antihuman CD95 (FAS) antibody on gene expression.

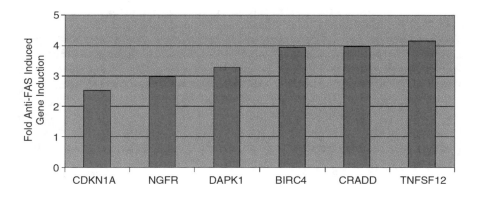

Genes induced in Jurkat T-cells by anti-human CD95 (FAS) antibody.

FIGURE 2.5. (continued).

two different tissue samples with high differential expression to visualize alterations in gene transcription. Here, we synthesized Cy3-labeled targets from equal amounts (2 μg) of commercially obtained poly-A RNA (Ambion) from human liver and brain, using Cy3-labeled random hexamers in a reverse transcriptase reaction, and hybridized the prepared single-stranded targets to the apoptosis three-dimensional HydroArray described in the previous paragraph. Figure 2.6 illustrates visible differences in gene expression levels between liver and brain tissues.

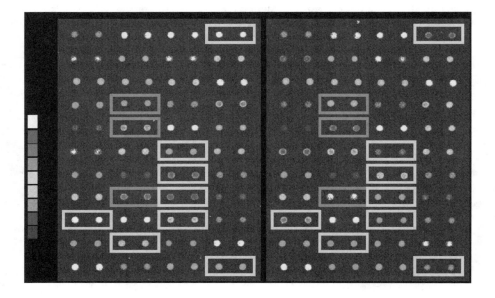

FIGURE 2.6 Differential expression between liver and brain genes.

IN VITRO DIAGNOSTIC APPLICATIONS

The examples of focus three-dimensional HydroArrays provided in this chapter indicate that HydroArrays can be used as life-science research tools. However, HydroArrays can also provide effective diagnostic assays. The recent launch of the Roche Amplichip CYP450 array, jointly developed by Affymetrix and Roche Diagnostics to monitor drug toxicity and metabolism, illustrates this new application of microarrays. Consequently, we have applied our focus, low-density array platform to the prenatal diagnosis of fetal abnormalities and have developed the Chromosomal Disorder array (CD HydroArray). In this section, we will highlight the key features of such array.

Approximately 4 million deliveries take place every year in the U.S. The aging population and the trend to delay pregnancies until later in life has resulted in an increase in the number of prenatal diagnostic procedures, since the incidence of fetal chromosomal disorders is known to increase with the age of the mother. A list of common chromosomal disorders, including aneuploidies and microdeletions, and their incidence at birth is listed in Table 2.2.

About 300,000 amniocentesis and chorionic villous sampling (CVS) procedures take place every year in the U.S. For amniocentesis, a needle is inserted through the mother's abdomen into the uterus. Amniotic fluid containing fetal cells is drawn and sent to a lab for karyotyping analysis. CVS involves the insertion of a catheter through the vagina and the cervix and into the uterus to the developing placenta under ultrasound guidance to remove sample of placental cells from the placental chorionic villi. Alternative approaches of CVS are the transvaginal and transabdominal routes.

Both procedures are invasive and carry a small risk of injury to the fetus, which may cause spontaneous abortions. CVS in particular can cause morbidity to the fetus at a rate that is about 0.5 to 1% higher than that of amniocentesis. Moreover,

TABLE 2.2

Common Aneuploidies and Microdeletions Tested by CD HydroArray and Their Incidence at Birth

Type/Key Examples	Disorder	Incidence at Birth
Aneuploid		
Kleinfelter, XXY	Sex chromosome	1/800 male birth
Turner, XO	Sex chromosome	1/4,000 female birth
Triple X, XXX	Sex chromosome	1/700 female birth
XYY	Sex chromosome	1/1,000 male birth
Patau syndrome	Chromosome 13	1/10,000 birth
Edwards syndrome	Chromosome 18	1/6,000 birth
Down syndrome	Chromosome 21	1/800 birth
Microdeletion		
Di George syndrome	22q11	1/4,000–8,000 birth
Cri-du-Chat syndrome	5p15.2	1/20,000–50,000 birth
Williams-Beuren syndrome	7q11.2	1/10,000 birth

karyotyping analysis of amniocentesis samples requires culturing these cells for about 2 weeks to allow enough cell propagation for chromosomal analysis. Diagnostic assays capable of detecting genetic abnormalities in a noninvasive manner and in a more rapid time frame are required.

FISH (fluorescent *in situ* hybridization) is a tool commonly used to diagnose chromosomal disorders to a fairly high degree of accuracy. The technique, however, though time proven, is not only time-consuming but also requires allocating large amounts of a skilled technician's time. Furthermore, though FISH can accurately detect aneuploidies, it cannot detect microdeletions.

Building on the advantages of reproducibility and sensitivity of our three-dimensional HydroArrays, we have designed a diagnostic array with both sensitivity and specificity greater than 99.5% for detecting both aneuploidies and microdeletions in fetal DNA samples. Specifically designed 45-mer probes to genes located on nine different chromosomes are printed using six spots each on a single array, providing a statistically significant number of data points. Each group of six spots consists of two groups of three spots, representing two separate transfer pins used to manufacture each cluster. A cluster of blanks, which do not contain any probes, is also included as a control. Two identical slides are used per patient, and the final result is only computed if the data obtained from both slides is identical to within 5%.

The ability of CD HydroArrays to correctly identify chromosomal disorders is visually illustrated in Figure 2.7 to Figure 2.9. The quantification of signal intensity, and the statistical correlations leading to diagnostic values are indicated in Table 2.3 and Table 2.4.

The detection of male and female sex chromosomes in normal subjects and the identification of genetic disorders caused by sex chromosome aneuploidism is illustrated in Figure 2.7. Control probes are hybridized in the first lane of each of the three panels.

Figure 2.8 illustrates the detection of trisomy at chromosomes 13, 18, and 21. The first panel represents normal male values (46 chromosomes XY). The second panel illustrates a male patient's with Patau syndrome, characterized by trisomy at chromosome 13. In this case, the signal generated by the corresponding probe is greater than that generated by the control sample in the first panel. Panel 3 indicates elevated values of chromosome 18 probes, characteristic of trisomy 18, or Edward syndrome. Panel 4 illustrates the greater intensity generated by the chromosome 21 probes, indicative of the chromosome 21 trisomy typical of Down syndrome.

The detection of chromosome microdeletions is visually illustrated in Figure 2.9. The first panel shows normal values. The second panel shows the values of a patient with Cri-du-chat syndrome, characterized by a microdeletion at chromosome 5. Relative to the signal generated by a normal sample, the Cri-du-chat patient sample had virtually no signal at the chromosome 5 probe.

The third panel shows the values of a patient with Di George syndrome, characterized by a microdeletion at chromosome 22. Relative to the signal generated by the chromosome 22 probe of a normal sample, Di George patients generated a signal at the corresponding probe that was significantly lower than that of the control probes.

The relative fluorescence intensities generated by the CD HydroArrays indicated in the previous paragraphs were quantified, their intensity ratios computed against

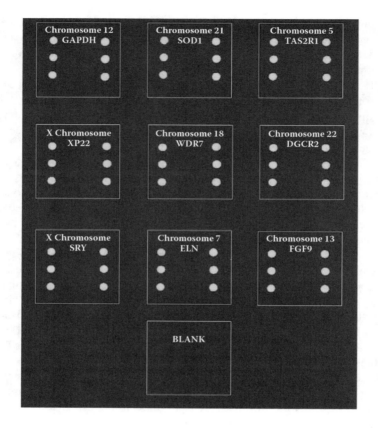

FIGURE 2.7   Detection of sex-chromosome aneuploid.

GAPDH for each of the eight genes being tested. These data are summarized in Table 2.3 and Table 2.4.

As previously stated, the CD HydroArray contains nine genes, of which eight are specific for genes at risk. The remaining probe is for the internal control GAPDH, a "housekeeping" gene, known to be unaffected by any of the disorders studied in this test. To normalize the data for each patient, intensity ratios are computed against GAPDH for each of the eight genes being tested. The normalization procedure marginalizes differences in starting material. Ratios to GAPDH are the same for all normal patients, regardless of the absolute fluorescent intensities. Sample data were accumulated over time, averaged, normalized against GAPDH gene, and used as the acceptance criteria by which all incoming patient data were evaluated. These data are summarized in the first two columns of Table 2.3.

Quantifying the signal generated by the male Y chromosome on the CD HydroArray can easily identify blind normal male and female samples. Because females lack the Y chromosome, the difference between males and females is determined by the presence of signal in the SRY (Y chromosome) position in male samples. After normalization against the GAPDH gene, male samples gave rise to

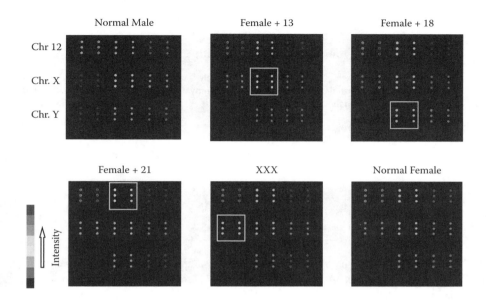

FIGURE 2.8 Detection of trisomy 13, 18, and 21.

ratio values of 0.166, whereas female samples did not produce detectable signals owing to lack of the Y chromosome (hence, the blank SRY value in the Normal Female column). Similarly, female samples, by virtue of possessing two X chromosomes vs. only one for males, can be accurately identified by the signal produced by XP22 position of the X chromosome. Indeed, these values are almost double those of males (0.35 for females vs. 0.19 for males).

Similar trends in signal ratios are utilized to detect sexual aneuploidies. Kleinfelter syndrome is characterized by the XXY genotype, i.e., patients have two

TABLE 2.3
Detection of Sex Chromosome Disorders with CD HydroArrays

	Normal Male	Normal Female	XO	XXX	XXY
XP22	0.19	0.35	0.18	0.51	0.38
SRY	0.17	—	—	—	0.15
SOD1	0.68	0.65	0.65	0.65	0.66
ATP7B	0.59	0.56	0.55	0.52	0.53
WDR7	0.26	0.25	0.24	0.22	0.24
ELN	0.35	0.33	0.33	0.32	0.32
TAS2R1	0.25	0.25	0.24	0.25	0.25
DGCR2	0.35	0.32	0.34	0.38	0.31

TABLE 2.4
Detection of Trisomy and Microdeletions with CD HydroArrays

	Normal Male	Normal Female	Female, Trisomy 18	Female, Trisomy 21	Male, Trisomy 18	Male, Trisomy 21	Di George (Female)	Cri-du-Chat (Male)
XP22	0.19	0.35	0.33	0.36	0.20	0.19	0.35	0.19
SRY	0.17	—	—	—	0.16	0.15	—	—
SOD1	0.68	0.65	0.64	0.88	0.64	0.90	0.70	0.72
ATP7B	0.59	0.56	0.59	0.60	0.62	0.63	0.57	0.57
WDR7	0.26	0.25	0.39	0.26	0.37	0.26	0.22	0.23
ELN	0.35	0.33	0.37	0.34	0.36	0.35	0.36	0.38
TAS2R1	0.25	0.25	0.23	0.22	0.25	0.27	0.22	0.12
DGCR2	0.35	0.32	0.32	0.29	0.34	0.34	0.22	—

X chromosomes and one Y chromosome. When samples of these patients were analyzed with our CD HydroArray, they generated XP22 ratios (typical of X chromosome) of 0.38, similar to the 0.35 value of normal females. Their SRY value (characteristic of Y chromosome) was 0.15, consistent with normal male ratio of 0.17 (Table 2.3). Therefore, a fetal sample with XP22 value of 0.38 and an SRY value of 0.15 can correctly diagnose a male fetus carrying Kleinfelter syndrome.

Patients affected by Turner syndrome, characterized by XO, lack the Y chromosome. When Turner's samples patients were analyzed with our CD HydroArray, XP22 values were 0.18, comparable to those of normal males. Because these patients lack the Y chromosome, their SRY values were not detectable (Table 2.3). Similarly, Triple X (or XXX) patients gave rise to XP22 values of 0.51, or 50% higher than that of normal females due to the presence of three X chromosomes, but gave rise to no SRY value, due to the lack of the Y chromosomes.

Chromosome microdeletions, which are characterized by selective loss of gene parts, give rise to signals that are lower than the normal values. Samples of such patients were tested with our CD HydroArray, and the data are summarized in Table 2.4, where green highlights indicate the affected gene.

The patient in column three has XP22 values of 0.33 and undetectable SRY values: This patient therefore is female. The other ratios are very similar to normal female ratios, except for the WDR7 value, which is 0.36, about 50% greater than the normal value of 0.25. The WDR7 gene is located on Chromosome 18; the value reported indicates the presence of an extra Chromosome 18. This patient is therefore diagnosed as a female with Trisomy 18.

PROTEIN THREE-DIMENSIONAL HYDROARRAYS

The design of protein chip microarrays faces two major challenges: (1) the need for easy microfabrication methods, and (2) a microenvironment on the surface of the chip capable of maintaining proteins in hydrated conditions. This is essential for proteins to retain their native three-dimensional configuration and consequently their biological activity.

Because of these challenges, only a select number of protein arrays have been microfabricated thus far. Zhu H. et al. [10] reported the fabrication of an array of microwells made of the disposable silicone elastomere poly(dimethylsiloxane). With this platform, the authors have been able to fabricate a number of densely packed microwells on a small chip, each well providing the physical segregation between analytes necessary to enable the screening of multiple analytes on the same chip. Proteins are covalently attached to each microwell using the cross-linker 3-glycidoxypropyltrimethoxysilane. The authors reported the attachment of up to 8×10^9 $\mu g/\mu m^2$ of protein on the surface of each microwell. These wells, however, are exposed to ambient air, with potential fluid evaporation and drying of the captured proteins, and consequent loss of their biological activity.

The same authors reported the analysis of 6566 yeast proteins, representing 5800 different yeast gene expression products, with a microarray printed on a glass slide. Proteins were immobilized to aldehyde-treated glass slides; their free amino group binding via a Shiff base formation [11]. However, Shiff bases are unstable binding events that require stabilization with reducing agents. In the absence of such treatment, dissociation would occur, resulting in detachment of the proteins from the solid support. Alternatively, the authors linked proteins onto slides treated with nickel, to which proteins would bind through the expression of a histidine tag. This paper is one of the first reports to describe a massive array with a large number of proteins for rapid parallel screening. However, in these arrays, the immobilized proteins are exposed to ambient air and thus are liable to dry on the surface of the chip, which results in loss of biological activity. With such a chip, maintaining the proteins in a constantly hydrated state to retain their folded configuration is a challenging task.

MacBeath G. and Schreiber S.L. [12] printed a protein array on a glass slide. The proteins were in 40% glycerol to prevent evaporation of the nanodroplets. Slides were previously treated with aldehyde-containing silane to form Shiff-base linkages with the amino groups of the proteins. When slides are washed and prepared for interaction with test proteins at the end of the immobilization phase, glycerol is equally washed away and, as before, unless proteins are constantly maintained in a hydrated state, they can easily dry and denature.

In the first reported example of a three-dimensional chip, Zlatanova and Mirzabiekov [3] developed an array made of micropads of polyacrylamide. DNA was deposited on the top of each micropad and allowed to passively diffuse into the gel. However, the fabrication of a microarray of polyacrylamide micropads is cumbersome, requiring photomasks and equipment commonly used in photolithographic processes. In addition, the diffusion of DNA into the polyacrylamide micropads is a slow process, requiring up to 48 h.

To immobilize protein on a solid support in a way that preserves their folded configuration, Arenkov P. et al. [13] arrayed functionally active proteins within microfabricated polyacrylamide pads and microelectrophoresed proteins to accelerate diffusion. However, the polymerization of acrylamide requires free radicals, which may damage proteins as well as produce an unstable morpholino derivative that can negatively affect the stability and shelf life of the chip itself.

Polyurethane hydrogels provide a viable alternative to the methods mentioned previously. Unlike acrylamide, the polymerization of polyurethane hydrogels is initiated by water. This bypasses the need to use free radicals or organic solvents, creating a microenvironment into which proteins can be directly encapsulated during polymerization. In addition, (1) the initiation of polymerization, (2) the binding of proteins to the hydrogel backbone, and (3) the binding of the hydrogel microdroplets to the glass slide are mediated by the same isocyanate reactive groups in conditions that can be meticulously controlled. Consequently, the molar concentration of proteins immobilized within each microdroplet can be carefully quantified. Moreover, the fact that the hydrogel droplets contain up to 96% water results in the encapsulated proteins being maintained in a constant state of hydration, thereby retaining their three-dimensional native conformation.

In the following paragraphs, we will describe a series of protein chips that utilize our PEG-NCO hydrogel platform. In particular, we will show examples of arrays suitable for antigen–antibody recognition, protein–protein interactions, protein–DNA interactions, and enzymatic reactions within hydrogel microdroplets.

PROTEIN DIFFUSION AND MOLECULAR RECOGNITION: AN EXAMPLE OF ANTIBODY–ANTIGEN BINDING

To demonstrate that proteins can diffuse into the hydrogel droplets in a functionally viable state, we chose the FITC–Anti-FITC antigen–antibody system. In this experiment, FITC was first encapsulated within the microdroplet. Then, anti-FITC antibodies labeled with Alexa Fluor594 were allowed to diffuse through the microdroplets to bind to FITC, as indicated schematically in Figure 2.9A.

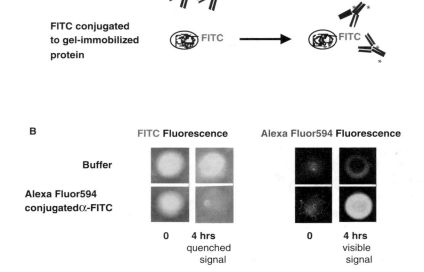

FIGURE 2.9 Detection of chromosome microdeletions. (From Gurevitch, D., et al., A novel three-dimensional hydrogel-based microarray platform, *JALA*, 6(4): 87–91, 2001. With permission.)

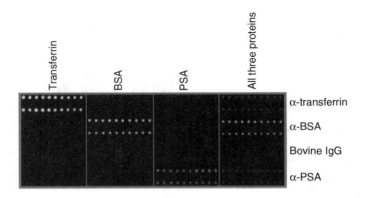

FIGURE 2.10 Example of complex protein–protein interactions.

In the absence of anti-FITC antibodies, the fluorescence of FITC embedded within the hydrogel can be readily detected (Figure 2.9B). Incubation of anti-FITC antibody leads to diffusion of the antibody into the hydrogel microdroplet, where the binding of the antibody with its cognate antigen (FITC) results in the quenching and subsequent loss of FITCs fluorescence (Figure 2.9B). The presence of the anti-FITC antibody in the hydrogel microdroplet is further confirmed by the red fluorescence emitted by Alexa Fluor594 (the fluorophor used to label the antibodies) when excited at 590 nm (Figure 2.9B).

To further demonstrate that antibodies can specifically bind to their respective antigens within hydrogel microdroplets, the reverse experiment was also executed. In this case, antibodies raised against transferrin PSA or BSA were first tethered within the gel matrix of microdroplets. Fluorescent-labeled transferrin PSA or BSA were then allowed to diffuse through the microdroplets, where they bound specifically to their respective antibodies tethered to the hydrogel (Figure 2.10). The bovine IgG antibody control did not interact with any of the targets.

These experiments demonstrate that antibodies and other large proteins are capable of diffusing through hydrogel microdroplets and binding to their cognate antigens. This not only provides a method to detect the presence of various antigens within the hydrogel microdroplets but also opens many possibilities for carrying out biochemical reactions in a microarray format. Therefore, we explored whether other protein assays can be miniaturized in a microarray format. The following sections describe our results in developing assays aimed at performing protein–protein interactions, protein–DNA interactions, and enzymatic activities within hydrogel microdroplets.

PROTEIN–PROTEIN INTERACTIONS

The ability to study protein–protein interactions is important for drug discovery and structural proteomics. We studied the interaction between calmodulin and calcineurin as a model for complex protein–protein interactions. For these experiments, we used antibody capture to support a selective calcium-mediated interaction between calcineurin (a heterodimer) and fluorescently labeled calmodulin (Figure 2.11). The calcium-independent calmodulin or anticalmodulin antibody interaction served as positive control.

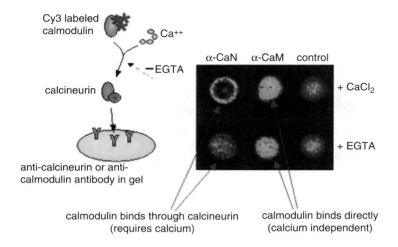

Cy3 labeled calmodulin

Ca++

α-CaN   α-CaM   control

—EGTA

calcineurin

+ CaCl₂

+ EGTA

anti-calcineurin or anti-
calmodulin antibody in gel

calmodulin binds through calcineurin
(requires calcium)

calmodulin binds directly
(calcium independent)

FIGURE 2.11 (*A color version follows page 204*) Sequence-specific protein–DNA recognition in three-dimensional HydroArray. (From Gurevitch, D., et al., A novel three-dimensional hydrogel-based microarray platform, *JALA*, 6(4): 87–91, 2001. With permission.)

Protein–DNA Interactions

Transcription factor binding to a recognition DNA sequence is an essential event in gene expression. Building upon our DNA microarrays, we developed a transcription factor DNA array as a model system for specific DNA–protein interactions. This provided a novel tool to help further our understanding of gene regulation and control.

Bacterial λ repressor binding sequence O_R2O_R1 and its mutant, carrying a single base mutation at the binding site, are printed and hybridized to their corresponding complementary sequences, as described in Figure 2.12A and Figure 2.12B. Binding of the Cy3-labeled λ repressor to its native operon sequence results in the gain of fluorescent signal in the corresponding spots. The absence of a strong fluorescence in the mutant spots indicates that the interaction is sequence specific (Figure 2.12C). Comparison of the SYBR Gold (a double-stranded DNA strain) stained fluorescence of the printed slides with the Cy3 fluorescence from λ repressor confirms that it is the sequence-specific λ repressor–λ operon interaction, rather than nonspecific protein binding to unevenly printed DNA that gives rise to the Cy3 signal associated with the wild type O_R2O_R1 sequence.

A second example of protein–DNA interaction with our three-dimensional HydroArray platform is provided by the binding of the Estrogen Receptor (ER) (UBI, Lake Placid, NY) to its consensus estrogen response element (ERE). The sequence of the wild-type ERE is as follows: 5′-acggtagAGGTCActgTGACCTctac-ccg-3′, and the two ER binding sites are highlighted in capital letters [14]. A mutant ERE sequence was used as the negative control. The wild-type ERE sequence differs from the mutant sequence by four nucleotides in a region known to be critical for binding by the receptor, as described by Vanacker et al. [14]. The binding of the estrogen receptor to the estrogen response element forms the ER-ERE complex,

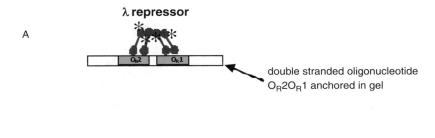

λ repressor

double stranded oligonucleotide
O_R2O_R1 anchored in gel

B **wt** tc**taacaccgtgcgtgtt**gactattt**tacctctggcggtgata**atgg

mutant tc**ttacaccgtgcgtgtt**gactattt**tacctctggcggtgaaa**atgg

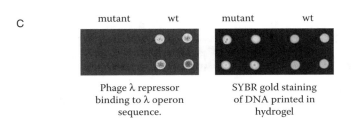

C

| mutant | wt | mutant | wt |

Phage λ repressor
binding to λ operon
sequence.

SYBR gold staining
of DNA printed in
hydrogel

FIGURE 2.12 Transcription factor binding to target DNA.

which was then detected with anti-ER antibodies coupled to a fluorescent signal display. These data are shown in Figure 2.13.

Transcription factor–DNA interactions with short half-lives (i.e., less than 1 min) can be better detected with electrophoresis gel shift assays than with nitrocellulose filter assays. This is due in part to the "caging effect" provided by the gel matrix [15]. In these conditions, transcription factor molecules that dissociate from their cognate DNA sequences cannot diffuse away because they are trapped in close proximity to one another by the gel matrix. Thus, they can rapidly recombine with each other, increasing the stability of the complex. Because of these caging effects, gel shift assays are more sensitive than nitrocellulose filter assays, where kinetically labile complexes are prone to dissociate during washing of the filter. Our experiments demonstrate that protein–DNA binding studies can be performed within hydrogel microdroplets, possibly due to a cage-effect-like activity of the hydrogel scaffold.

The ability to analyze DNA–protein interactions on microarrays offers a route to more efficient screening of agents, focused on transcriptional regulation as possible therapeutic targets. We anticipate that double-stranded DNA microarrays designed in various formats may have broad applications in studying protein–DNA interactions, including inhibitors and activators of sequence-specific transcription factors, agonists and antagonists, transcription regulatory proteins, and synergy among transcription regulators. We also anticipate that double-stranded DNA microarrays will be very

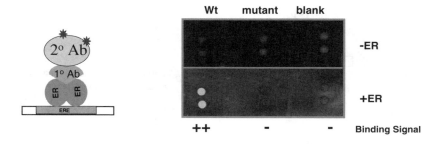

ER: estrogen receptor
ERE: estrogen response element

FIGURE 2.13 Enzymatic reactions within three-dimensional HydroArrays.

effective substitutes for cumbersome protein–DNA interaction assays currently used in the field, such as gel mobility shift assays or filter binding assays.

ENZYMATIC REACTIONS WITHIN THREE-DIMENSIONAL HYDROARRAYS

We have further extended the sequence-specific protein–DNA interaction study to protein–DNA interactions associated with enzymatic activity. Figure 2.14 shows the results of a tyrosine phosphatase model enzymatic system. Nine unique peptides were deposited as probes, with six of the nine being tyrosine phosphopeptides. The color of the spots is proportional to the amount of fluorescent dye bound to the microarray. The lower intensity (weakest signal — least binding) is blue in color, while the highest intensity (strongest signal — most binding) is red. There is no signal on either array for replicates of the three peptides that are not phosphorylated. The control panel shows the signals generated by the tyrosine phosphopeptides.

The tyrosine phosphatase, *Yersinia enterocolitica* YOP phosphatase (NEB, Beverly, MA) was applied to the array on the right. Following incubation for 10 min, the residual phosphopeptide on the treated and control arrays was then detected by

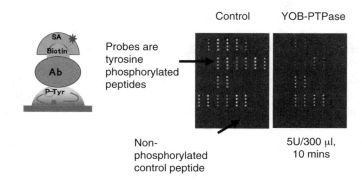

FIGURE 2.14

biotinylated antiphosphotyrosine antibody, followed by Cy3-labeled streptavidin for signal display. Each polypeptide was printed as two quadruplet spots: the first eight spots represent blank hydrogel controls for nonspecific binding and the next eight spots contain a negative control, the serine and threonine phosphate substrate. When the phosphatase cleaves a phosphate from the peptide, one expects a corresponding loss of signal.

The results shown in Figure 2.14 demonstrate the activity of tyrosine phosphatase on phosphopeptides tethered in the hydrogel. Not all the peptides are dephosphorylated at the same rate. Peptides with higher acidity (containing aspartic acid or glutamic acid) in positions 4, 3, 2, or 1 form the phosphotyrosine in 0 position relative to the aminoterminal and are known to be efficiently dephosphorylated [16]. In our study, peptides containing ENDY(P), EDNEY(P), and DADEY(P) were efficiently dephosphorylated. Peptides containing ENAEY(P) were dephosphorylated more slowly, whereas DRVY(P), TRNIY(P), and VVPLY(P) were dephosphorylated at the slowest rate. The relative dephosphorylation rates are consistent with the literature. An important application of this technology is for drug screening, including evaluating the activity of potential enzyme inhibitors.

CONCLUSIONS

The experiments described in this chapter indicate that our new, three-dimensional HydroArrays provide a diverse novel platform with increased sensitivity and specificity for genomic studies. By virtue of immobilizing a greater number of capture probes in a three-dimensional microdroplet format, the arrays exhibited greater sensitivity and very low (6%) coefficient variations. The greater accuracy of three-dimensional HydroArrays enabled the detection of 1.5-fold differences in gene expression. This has allowed the development of a diagnostic three-dimensional HydroArray, which correctly identifies a number of chromosomal abnormalities.

The high water content (95%) provides a microenvironment wherein the proteins are fully hydrated and retain their natural three-dimensional configurations. This has allowed the microfabrication of a number of protein chip prototypes, such as those suitable for antigen–antibody binding, protein–protein interactions, protein–DNA interactions. In addition, we demonstrated that enzymatic reactions can also be miniaturized and carried out in a microarray format.

REFERENCES

1. Schena, M., Shalon, D., Davis, R.W., and Brown, P.O. Quantitative monitoring of gene expression patterns with a complementary DNA microarray. *Science* 270: 467–470, 1995.
2. DeRisi, J., Penland, L., Brown, P.O., Bittner, M.L., Meltzer, P.S., Ray, M., Chen, Y., Su, Y.A., and Trent, J.M. Use of a cDNA microarray to analyse gene expression patterns in human cancer. *Nat Genet.* 14: 457–460, 1996.
3. Zlatanova, J. and Mirzabiekov, A. Gel-immobilized microarrays of nucleic acids and proteins: production and applications for macromolecular research. *Methods Mol Biol.* 170: 17, 2001.

4. Nagata, S. and Goldstein, P. The Fas death factor. *Science* 88: 355–365, 1995.

5. Cohen, O. and Kimchi, A. A DAP-kinase: from functional gene cloning to establishment of its role in apoptosis and cancer. *Cell Death Differ.* 8(1): 6–15, 2001.

6. Cohen, O., Inbal, B., Kissil, J.L., Rayeh, T., Berissi, H., Spivak-Kroizaman, T., Feinstain, E., and Kimchi, A. DAP-kinase participates in TNF-alpha- and Fas-induced apoptosis and its function requires the death domain. *J Cell Biol.* 146(1): 141–148, 1999.

7. Deveraux, Q.L., Leo, E., Stennicke, H.R., Welsh, K., Salvesen, G.S., and Reed, J.C. Cleavage of human inhibitor of apoptosis protein XIAP results in fragments with distinct specificities for caspases. *EMBO J.* 18(19): 5242–5251, 1999.

8. Ahmad, M., Srinivasula, S.M., Wang, L., Talanian, R.V., Litwack, G., Fernandes-Alnemri, T., and Alnemri, E.S. CRADD, a novel human apoptotic adaptor molecule for caspase-2, and FasL/tumor necrosis factor receptor-interacting protein RIP. *Cancer Res.* 57: 615–619, 1997.

9. Kaplan, M.J., Ray, D., Mo, R.R., Yung, R.L., and Rickardson, B.C. TRAIL (Apo2 ligand) and TWEAK (Apo3 ligand) mediate CD4+ T cell killing of antigen-presenting macrophages. *J Immunol.* 164(6): 2897–2904, 2000.

10. Zhu, H., Klemic, J.F., Chang, S., Bertone, P., Casamayor, A., Klemic, K.G., Smith, D., Gerstein, M., Reed, M.A., and Snyder, M. Analysis of yeast protein kinases using protein chips. *Nat Genet.* 26(3): 283–289.

11. Zhu, H.M., Bilgin, R., Bangham, D., Hall, A., Casamayor, P., Bertone, N., Lan, R., Jansen, S., Bidlingmaier, T., Houfek, T., Mitchell, P., Miller, R., Dean, A., Gerstein, M., and Snyder, M. Global analysis of protein activities using proteome chips. *Science* 293(5537): 2101–2105, September 14, 2001.

12. MacBeath G. and Schreiber S.L. Printing proteins as microarrays for high-throughput function determination. *Science* 289: 1760, 2000.

13. Arenkov, P., Kukhtin, A., Gemmell, A., Voloshchuk, S., Chupeeva, V., and Mirzabekov, A. Protein microchips: use for immunoassay and enzymatic reactions. *Anal Biochem.* 278(2): 123–131, February 15, 2000.

14. Vanacker, J.-M., Pettersson, K., Gustafsson, J.-Å., and Laudet, V. Transcriptional targets shared by estrogen receptor-related receptors (EERs) and estrogen receptor (ER)α, but not ERB. *EMBO J.* 18: 4270, 1999.

15. Fried, M. and Crothers, D.M. Equilibrium studies of the cyclic AMP receptor protein — DNA interaction. *J. Mol Biol.* 172: 263–282, 1984.

16. Zhang, Z.-Y., Thieme-Sefler, A.M., MaClean, D., McNamara, D.J., Dobrusin, E.M., Sawyer, T.K., and Dixon, J.E. Substrate specificity of the protein tyrosine phosphatases. *Proc Natl Acad Sci U S A* 90: 4446–4450, 1993.

3 Biochips in Malaria for Antiparasitic Discovery

Karine Le Roch and Elizabeth Winzeler

CONTENTS

INTRODUCTION

Plasmodium falciparum is the causative agent of the most deadly form of human malaria, killing 1 to 3 million individuals per year. The emergence and spread of resistance to widely used antimalarials make the development of novel therapeutic approaches an urgent task. The recent publication of the entire genome of the *P. falciparum* revealed over 5400 genes of which 60% encode for hypothetical proteins with unknown function and will potentially have a high impact for malaria drug discovery. Moreover, high throughput functional genomics and especially DNA microarray technology are already opening the doors to better understanding of the malaria parasite biology. In this chapter, we describe some of the genomic approaches that have been used to increase the knowledge of the biology of this eukaryotic parasite, an essential step in the antiparasitic discovery pipeline.

MALARIA

Malaria is one of the most ancient and devastating parasitic diseases of humans with an estimated 300 to 500 million infections and 1.5 to 3 million deaths annually (Breman et al., 2001). Caused by four species of apicomplexan parasites belonging to the genus *Plasmodium,* malaria is transmitted to humans by the *Anopheles* mosquito. Despite massive efforts over the past century to eradicate the disease through mosquito control and prophylactic antimalarial drugs such as chloroquine (CQ), or sulphadoxine-pyrimethanine (SP), malaria continues to pose a severe threat to human health worldwide. The common clinical signs are fever, chills, prostration, and anemia. With severe disease progression, symptoms can include delirium, acidosis, cerebral malaria, multiorgan failure, coma, and death. Recent analyses suggest that the medical and economic impacts of malaria in endemic areas are underestimated (Breman et al., 2001; Gallup and Sachs, 2001). Moreover, the increasing emergence of multidrug-resistant *Plasmodium* spp. as well as insecticide resistant *Anopheles* spp. and the lack of an effective malaria vaccine highlight a critical need for the identification of new chemotherapeutics and vaccines.

LIFE CYCLE OF THE MALARIA PARASITE

The malaria parasite has a complex life cycle (Figure 3.1). Transmission is initiated when an infected mosquito takes a blood meal and haploid sporozoites are released from its salivary gland into the human host blood stream. These parasites rapidly invade hepatocytes where they differentiate and replicate, releasing free merozoites into the bloodstream. These merozoites rapidly infect host erythrocytes where the

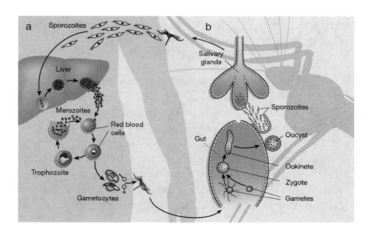

FIGURE 3.1 Life cycle of the malaria parasite. (From Wirth, D.F., Biological revelations, *Nature*, 2002, 419, 495–496. With permission.)

parasite differentiates through ring, trophozoite, and the multinucleate schizont stage over a 48- to 72-h period (depending on the particular *Plasmodium* spp.). At the end of this asexual cycle, the erythrocytes lyse, releasing a new batch of merozoites that can invade new erythrocytes. This erythrocytic cycle is responsible for the clinical symptoms of malaria. By unknown molecular mechanisms, some erythrocytic parasites do not undergo asexual multiplication and instead differentiate to form male and female gametocytes. When mature gametocytes are taken up by a feeding mosquito, gamete formation and fertilization occur in the gut; the resultant diploid zygote (ookinete) penetrates the stomach wall where it encysts (oocyst); within this oocyst, sporozoites form and, when released, invade the mosquito's salivary glands.

GENOME OF *P. FALCIPARUM*

In an international effort to accelerate the discovery of drugs and protective vaccines, the entire 22.8 Mb genome of the most lethal *Plasmodium* species, *P. falciparum*, which consists of 14 chromosomes, a linear mitochondrial genome, and a circular plastidlike genome, was published in October 2002 (Gardner et al., 2002). The *falciparum* genome is twice the size of the yeast *Schizosaccharomyces pombe*'s genome, and represents the richest A+T genome sequenced to date (80.6% overall, and 90% in the introns and intergenic regions). The malaria genome sequencing consortium estimates that there were more than 5409 predicted open reading frames (ORFs) encoded in the *P. falciparum* genome, 60% of which lack sequence similarity to genes from any other known organism (Gardner et al., 2002). However, peptides from over 2400 of these ORFs have been detected by mass spectrometry, validating the gene prediction algorithms (Florens et al., 2002; Lasonder et al., 2002). Thus, almost two thirds of the proteins appear to be unique to this organism, a proportion higher than that of any other eukaryote. This may reflect a greater evolutionary distance from other organisms, increased by the reduction of the sequence similarity due to the A+T richness of the genome (Gardner et al., 2002). Particular categories of genes appear to be overrepresented, such as those involved in immune evasion and host–parasite interactions. Other categories appear to be underrepresented, such as those associated with cell cycle, cell organization and biogenesis, enzymes, transporters, or transcription factors (Gardner et al., 2002). The presence of under-represented gene families does not necessarily mean that fewer genes are involved in these processes relative to other organisms, but highlights a lack of biological knowledge about the malaria parasite. Although defining putative roles for these ORFs in the absence of similarities to other organisms remains challenging, discovery of their roles and identification of *Plasmodium*-specific key regulatory sequences could be fundamental to develop new antimalarials.

FUNCTIONAL GENOMICS AND DISCOVERY OF ANTIPARASITIC AGENTS

Following the publication of the *P. falciparum* genome sequence, efforts were made to translate the genomic information into a better understanding of the biology of the malaria parasite to design more effective and affordable antimalarial drugs. Because of

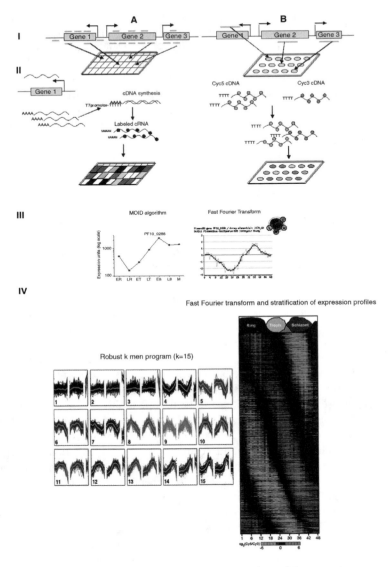

FIGURE 3.2 *(A color version follows page 204)* Comparison of the two microarray methods used for the malaria parasite gene expression's life cycle (A, short oligonucleotides by *in situ* synthesis, Affymetrix, Santa Clara, CA) or malaria cell cycle (B, long oligonucleotides by robotic deposition of nucleic acids onto a glass side). I — Probe Design: For the high-density 25-mer oligonucleotide array, multiple probes per gene are placed on the array (A). In the case of robotic deposition, a single (75-mer) probe is generally used for each gene. II — Preparation of labeled material for measurement of gene expression using a cRNA-labeled protocol (A) or a cDNA labeling protocol using the Cy3 (or Cy5) for a two-color strategy (B). III — Experimental design and expression level using different algorithms: MOID algorithm (A) and Fourier transform (B). III — Cluster analysis using the robust k-mean algorithm (A) or the fast Fourier transform (FTT) (B).

the complexity of the parasite's life cycle, as well as the fact that the organism is haploid during most of its development, the use of traditional forward and reverse genetic techniques have been challenging and time-consuming. Understanding this complex organism at the biological level requires more than just knowledge of genes and the genome, it required knowledge about the complex parasite–host interactions that occur throughout the life cycle. As a result, high-throughput functional techniques (transcriptomic and/or proteomic techniques) emerged as significant tools to elucidate gene function more rapidly and cost effectively. DNA microarray technologies represent a powerful tool for a whole genome approach to generate quantitative gene expression data. Variations in gene expression can reflect important aspects of biological function. Up- or downregulation of a gene or set of genes at a particular stage of the life cycle can provide predictive information about their function, involvement in a metabolic pathway, or their interaction with the host. Systematic characterization of expression patterns can provide information for interpreting biological significance and understanding how developmental events are controlled, facilitating effort to design better methods for interrupting these expression pattern and developmental events.

To distinguish between differences among the hundreds of potential hypothetical proteins, changes in the expression profiles for thousands of genes throughout the life cycle were analyzed. Two transcriptional analyses covering the entire genome of *P. falciparum* were published (Bozdech, Llinas et al., 2003; Le Roch et al., 2003). Le Roch et al. used a high-density, 25-mer oligonucleotide array generated by *in situ* synthesis by photochemistry and mask-based photolithography (Affymetrix, Santa Clara, CA). Bozdech et al. used a custom-made array with a robotic disposition of long 70 nt oligonucleotides. These studies were used to identify developmental events at the steady-state mRNA level throughout the life cycle or the erythrocytic cell cycle, respectively. Despite differences in technologies (short oligonucleotide vs. long oligomer microarray), both studies showed comparable expression patterns for the erythrocytic stage (Figure 3.2). This serves to emphasize that microarray technologies are reliable and powerful techniques. When well-designed, biochips can be a critical tool for answering significant biological questions. For *P. falciparum*, one such question to be answered is, what biochemical targets might serve as the basis for the rational design of antimalarials?

SHORT OLIGONUCLEOTIDE VS. LONG OLIGONUCLEOTIDE MICROARRAY

ARRAY AND PROBE DESIGN

Before the completion of the *P. falciparum* genome project, a customized high-density oligonucleotide array was designed in early 2002 at the Genomic Institute of the Novartis Foundation (GNF) U.S., using the available genome draft and annotation. The array had a distinctive design, unique in three ways: rather than a fixed number of probes per gene, the array used a variable number of probes matching the length of the predicted ORFs (from 1 to more than 200 probes); the array included probes in the intergenic and antisense coding regions; and the probe design differed from the standard Affymetrix technology, which uses perfect match (PM) and its corresponding mismatch (MM), differing by a single base in the middle. As the MM tends to overcorrect the signal by increasing the noise level and occupies half of the probe space, researchers at GNF designed a whole malaria genome array with PM only containing 260,596 25-mer single-stranded probes from predicted coding sequences (including mitochondrion and plastid genome sequences) and 106,630 probes from the noncoding sequence covering the 23-Mb genome with an average density of one probe per 150 bases on both DNA strands.

Around the same time that GNF created its Affymetrix custom-made malaria array, Bozdech et al. constructed a gene-specific microarray of the *P. falciparum* genome sequence, using the publicly available resource from the Malaria Genome Consortium. For each of the predicted ORFs, they designed 70-mer oligonucleotide array elements for the entire genome (Bozdech, Zhu et al., 2003). The DNA microarray used 7462 70-mer oligonucleotides representing 4488 of the 5409 annotated ORFs. The 70-bp oligonucleotides were synthesized in well format (Operon Technologies, California) and robotically spotted onto glass slides to detect cDNA hybridization level.

EXPERIMENTAL DESIGN AND EXPRESSION LEVEL

For the short oligonucleotide array experiment, total RNA was extracted from different stages of the malaria parasite life cycle, labeled by a strand-specific protocol and hybridized on the array (Le Roch et al., 2002; Le Roch et al., 2003). Life cycle stages used were carefully chosen: sporozoites, seven time points spanning the intraerythrocytic cell cycle (using two independent synchronization methods to obtain replicates and reveal genes that were under true cell-cycle control), and mature gametocytes. Because no mismatch probes were designed in this custom-made high-density oligonucleotide array, standard Affymetrix algorithms depending on the PM-MM values could not be utilized. To analyze the expression level within the array, the Match Only Integral Distribution (MOID) algorithm was used (Zhou and Abagyan, 2002; Zhou and Abagyan, 2003) to give an absolute expression level for each gene.

The correlation coefficient of the logarithm-transformed expression values between synchronizations and hybridizations (values range from 0.87 to 0.91) demonstrates high reproducibility and establishes that 88% of the predicted genes were expressed in at least one stage of the life cycle where expression levels throughout the life cycle varied by 5 orders of magnitude.

To profile the malaria parasite life cycle, a one-way statistical test was applied to identify differentially expressed genes, using time course data. To be considered a regulated gene, a minimum of a 1.5-fold change across the life cycle with a P value less than 0.05 was required. Forty-nine percent of the expressed genes were found to be life cycle regulated. As a result, the 51% that did not pass the statistical barrier were considered to be constitutively expressed. This group contained mainly hypothetical proteins and housekeeping genes. Although one cannot suggest a particular function for these hypothetical proteins using this transcriptome data, one can speculate about their importance as to the maintenance of the parasite through its life cycle. Moreover, those with no strong human orthologs may serve as potential drug targets.

The 70-nt long oligonucleotide array study was confined to a high-resolution analysis of the erythrocytic cell cycle. The mRNA was collected from a highly synchronized *in vitro* culture (using the sorbitol treatment technique) of *P. falciparum* strain HB3. Using a large-scale culturing technique (4.5-l bioreactor), samples were collected hourly for a 48-h period during the intraerythrocytic developmental cycle (IDC) of *P. falciparum.*

Because experimental variability is high for glass slide arrays, expression values were normalized throughout the IDC to a common pool control in a standard two-color competitive hybridization (Eisen and Brown, 1999). This normalization produced a graph of expression induction that cannot be directly compared to absolute values. The relative abundance of individual mRNAs varied continuously throughout the IDC with a single maximum and a single minimum. Experiments were reproducible. Pearson correlation (r) was greater than 0.90 for 68% and 0.75 for 86% of the transcripts represented by multiple oligonucleotides with detectable expression during the IDC. In contrast to the short oligonucleotide array, they found only 20% of the genes with a relative constant expression profile. This discrepancy may be due to differences in the criteria used.

CLUSTER ANALYSIS VS. FAST FOURIER TRANSFORM (FFT) ANALYSIS

Various methods of cluster analysis for genomewide expression data using statistical algorithms can be used to organize and group genes according to similarities in gene expression patterns. The main objective of expression profiling using the short oligonucleotide array was to demonstrate that genes that performed or that are involved in similar molecular processes have similar expression profiles. This makes it possible to assign functions to thousands of uncharacterized proteins encoded by the *P. falciparum* genome and to establish participation in specific biochemical pathways. To this end, genes whose expression was regulated were first grouped on the basis of expression time through the life cycle. Groups were assigned using a robust k mean algorithm. The cluster number, k = 15, was arbitrarily chosen as a reasonable estimate for the biological conditions analyzed. A higher number of clusters (e.g. k = 20 or 30) tended to overfractionate genes with similar molecular processes, whereas lower cluster numbers tended to assemble genes within molecular pathways that were not biologically related. Despite the fact that most of the genes in these clusters were hypothetical proteins (48 to 88%), a sufficient number of them have already been described experimentally and/or sequence homologies, and were defined to have specific cellular roles. For example, 13 of the 18 differentially expressed genes described as proteasome endopeptidases were found in a cluster of 110 genes. Our aim here is not to provide an exhaustive list of genes or descriptions of the different clusters observed throughout the parasite life cycle, but rather to convey the general idea that components in a cluster are both logical and nonrandom. By comparing gene ontology rosters with gene cluster rosters, it was observed that genes in each cluster shared particular functions by several orders of magnitude less than would be predicted by chance, allowing us to conclude that gene expression profiling can provide an insight into the gene's cellular role. Genes from cluster 4, for example, demonstrate upregulation expression at the merozoite and ring stages. Genes in this cluster exhibit gene ontology annotation, indicating their role in the establishment of the parasite into the red blood cell (i.e., early transcribed membrane proteins, erythrocyte-binding antigens, or genes involved in lipid and fatty acid metabolism). Hypothetical proteins that belong to this cluster are expected to perform similar functions. Cluster 15 groups genes that were highly expressed at the late schizont stage as well as genes involved in the invasion of the erythrocyte. Given that most of the genes considered as candidates for blood-stage vaccines reside in cluster 15, it can be reasonably assumed that the 90 hypothetical genes in this cluster could possibly be used in the development of new vaccines. Using this cluster analysis, more than 1000 genes within the malaria genome were given a hypothetical function and several numbers of them were seen as potential new drug or vaccine candidates.

Despite the usefulness of hierarchical clustering for comparing sets of expression data, Bozdech et al. used a different approach in the analysis of expression patterns. They applied the simple Fourier analysis technique to calculate the apparent phase and the frequency of expression for each gene during IDC. A score for each expression profile was then calculated based upon the period tightness in the periodicity and the amplitude of the peak in order to create a phaseogram of the IDC transcriptome of *P. falciparum*.

As with the short oligonucleotide array and the data set cluster analyses, the IDC phaseogram showed a cascade of expression from the ring to the schizont stage. Using this FFT method analysis, they demonstrated a programmed cascade of cellular processes that ensured the completion of the *P. falciparum* IDC and described that functionally related genes usually have common expression profiles (Figure 3.2).

AFFYMETRIX VS. LONG OLIGONUCLEOTIDE MICROARRAY

These two alternative formats for oligonucleotide-based microarray are commonly utilized. Both techniques allow for effective and complete genome design. Yet despite the use of dissimilar technologies, different *falciparum* strains (3D7 vs. HB3), varying methods of synchronization (sorbitol vs. thermocycling incubation), and varying sample time points throughout the erythrocytic cycle, the expression profiles were almost identical (Figure 3.2). In addition, both studies came to the same conclusion: genes with correlated temporal expression patterns often share similar functional roles. To conclude, to further biological relevance of the data sets and design best potential antimalarials, additional experiments and computational tools are needed.

BIOLOGICAL RELEVANCE OF EXPRESSION PROFILING

The advantage of using gene expression profiling for the malaria parasite can be enormous. In contrast to the analysis of human tissues, in which one often works with nonhomogeneous samples and is dependent upon access to good clinical and medical data, malaria parasites tend to be highly synchronized with homogeneous samples. Therefore, with a good bioinformatics platform and data processing, it is easier to correlate a change in gene expression to biological relevance. Although cluster analysis together with the "guilt by association" principle constituted a significant advance for a better understanding of many hypothetical proteins in *Plasmodium*, there is still a need to maximize the vast amount of gene expression data sets to fully exploit the potential of high-throughput genomic approach and to narrow the research for identifying new targets. To make the most of high-throughput techniques, Zhou et al. (2004) used the high-density oligonucleotide array gene expression data sets to describe and apply a novel data-mining algorithm: the ontology-based pattern identification (OPI). The OPI systematically discovers the expression patterns that best represent functionally related genes based on the principle of guilt by association. The OPI uses the gene ontology (GO) consortium to partially organize clusters. The OPI combined the malaria life cycle expression data sets with the *Plasmodium* gene annotation to begin with some *a priori* knowledge of the functional classification of a subset of genes and uses this functional information to dynamically create a cluster maximized for their biological information content. Using this technique, rather than the k mean algorithm, genes can be grouped into multiple functional categories according to their expression profiles and their association with multiple biological functions, a model closer to biological reality. OPI analysis for the malaria life cycle expression profile yielded 320 significant gene clusters representing 320 biological processes, cellular components, and molecular functions and allowed the functional annotation of uncharacterized genes based on existing ontology knowledge. Analogous methods have been published recently with a similar idea: to challenge *in silico* biological interpretations of microarray experiments, using the hierarchical nature of GO terms (Breitling et al., 2004a; Breitling et al., 2004b; Lee et al., 2004; Toronen, 2004). This method was used to successfully identify, within the malaria life cycle expression data sets (Le Roch et al., 2003) and additional gametocytogenesis time courses, a cluster of 271 genes, including 204 hypothetical genes, most likely to be involved in sexual differentiation and, potentially, targets for transmission-blocking vaccine and drug development (Young, 2004).

METABOLIC PATHWAY ANALYSIS

Although it is obvious that genes involved in multiprotein complexes or in similar function should be coexpressed, it is less evident that genes involved in a single metabolic pathway would cluster together. Indeed, genes involved in a single metabolic pathway may be posttranscriptionally regulated or specifically activated at the protein level only when needed. To investigate possible coexpression of genes involved in single metabolic pathways, Young et al. (2005) used the OPI for this analysis. They found, for example, that the cluster "carbohydrate metabolism" includes seven of the ten enzymes involved in the glycolysis with the exception of the hexokinase (first step) aldolase, and pyruvate kinase (last step). Overall, this result demonstrated that genes associated with similar biochemical pathways generally cluster together. One trying observation was that, in addition to genes involved in glycolysis, this cluster possesses several ribosomal proteins that represent a significant overlap between protein synthesis and carbohydrate metabolism. Although the two processes could possibly be under similar transcriptional regulation, this result may highlight the fact that the numbers of biological conditions examined thus far were insufficient to distinguish all molecular processes. This overlap might be avoided by analyzing additional conditions and time points across the malaria life cycle (e.g., several mosquito stages as well as liver stage). Although OPI attempts to separate functionally unrelated coexpressed genes by clustering according to biological knowledge, single hypothetical genes may be found in multiple clusters that contain overlapping genes. This observation is somewhat troublesome when attempting to predict functions of coclustering genes, but such data are extremely valuable for generating hypotheses about gene networks and multifunctional genes. This new computational tool will generate more complete knowledge of the parasite's metabolic pathways or *metabolome* and identify parasite-specific enzymes that can serve for the identification of potential drug targets.

Chips for New Drug

The genome, expression profiles, and functional characterization of hypothetical genes using rational databases and computational queries have allowed a glimpse of unique biochemical targets in *P. falciparum* with no human homologues and can serve as the basis for the rational design of antimalarials. This eukaryotic parasite contains a unique organelle, an apicoplast, a plastid homologous to chloroplasts acquired by the process of endosymbiosis (Foth and McFadden, 2003). The apicoplast of prokaryotic origin is semiautonomous. It possesses its own genome and expression machinery in addition to numerous proteins encoded by nuclear genes. The apicoplast has been implicated in various metabolic functions, including synthesis of lipids, heme, and isoprenoids (Ralph et al., 2004). Inhibitions of plastid-associated proteins have been shown to kill the parasite and demonstrate that the apicoplast is an essential organelle. Computational analysis predicted that 550 nuclear proteins that targeted the plastid may potentially be excellent drug targets. A large percentage of these plastid proteins have unknown functions; analysis of the life cycle data, using computational algorithms such as the OPI, will allow a functional characterization of a subset of them, thereby narrowing the research of the best candidates for such drug discovery.

Proteases have also been shown to play an important role in the metabolism of the erythrocytic cell cycle (Rosenthal, 2002). Some serve for the degradation of the host cell hemoglobin in the food vacuole to produce amino acids essential for protein synthesis. Inhibition of growth by protease inhibitors validates their importance for the parasite development (Rosenthal et al., 2002). Ninety-two proteases have been identified in the *P. falciparum* genome (Wu et al., 2003); variation in their expression profiles shows that they are involved in cellular processes in addition to hemoglobin degradation. Association with OPI clusters will allow for the identification of protease functions throughout the life cycle and to suggest protease inhibitors with efficacy at all stages of the parasite's life cycle.

Molecular mechanisms regulating cell proliferation and development in the malaria parasite are still largely unknown. Cell cycle controls and signal transduction pathways responsible for the developmental stage transitions have been difficult to analyze thus far. Identification of putative homologues for a number of eukaryotic cell cycle regulators such as cyclines, cycline-dependant kinases (CDKs), and components involved in transduction pathways (e.g., the Map kinase pathways have been limited by the fact that the plasmodial sequence homologies, because of their high A + T content, are usually weak). The importance of the cell cycle and life cycle progression in the malaria parasite and the fact that inhibition by kinase inhibitors kills the parasite validate the drug potential of these genes (Doerig, 2004). Gene expression profiling and cluster analysis will help elucidate these complex pathways and identify specific malaria targets.

Mechanisms controlling transcriptional or posttranscriptional activations are fundamental in eukaryotic cells. Previous analyses of gene expression in the malaria parasite have shown that transcription was generally monocistronic and developmentally regulated (Horrocks et al., 1998; Horrocks et al., 1996; Lanzer et al., 1993; Scherf et al., 1998). This has been widely confirmed by microarray data sets, which show that there is a good correlation in the timing between when a gene is expressed and when its product is required by the cell (Bozdech, Llinas et al., 2003; Le Roch et al., 2003). Multiple sequence alignments of protein domains using the profile-hidden Markov model (HMM) obtained from transcriptional regulators, was able to identify only 71 protein hits, a third of the transcriptional control elements expected for the *Plasmodium* genome size based on comparisons with fungi, plants, and animals (Coulson et al., 2004). On the contrary, protein motifs that have been shown to have a role in regulating mRNA stability, localization and translation seem to be prevalent in *Plasmodium*. Additional malaria sequences, whose functions have been shown to alter chromatin structure, as well as the presence of adjacent cotranscriptional genes (Le Roch et al., 2003) and the presence of chromosomal clusters of coexpressed proteins (Florens et al., 2004) suggest a potential mechanism for gene control in *Plasmodium* by chromatin structure. Analysis of the expression profiles of these control elements across the life cycle shows a high degree of stage specificity. Examination of these profiles through computational algorithms and/or OPI analysis will highlight the association of coregulated genes to produce additional hypotheses. Understanding this complex and apparently specific transcriptional regulation in the malaria parasite will undoubtedly lead to the identification of vital regulator elements that can be excellent potential drug candidates.

Lipid metabolism is also clearly of interest for drug design. *P. falciparum* intraerythrocytic growth is associated with a dramatic increase in total membrane content resulting from parasite enzymatic activities. Parasite membranes are associated with essential structure and specific processes (cell invasion, nutrient acquisition, trafficking, modulation of the host membranes, or immune evasion against the host immune system). Furthermore, it was recently shown that *Plasmodium* has unique phospholipid metabolic pathways (Vial et al., 2003). The identification of fatty acid biosynthesis, as well as the isoprenoid biosynthesis (in the apicoplast), confirmed the uniqueness of lipid metabolism in the malaria parasite. In addition, drugs that target the phosphatidylcholine biosynthesis inhibit parasite growth and validated this metabolic pathway as a potential drug target (Wengelnik et al., 2002). Complete elucidation of this metabolism will highlight possible new candidates.

Possibilities of discovering drugs against the malaria parasite are enormous. Here, we have merely presented an overview of what is currently known and researched. We have not discussed the potential inhibitors targeting the parasite's transport mechanisms, the possible inhibitors of the invasion process, or the ubiquitin regulation system pathway, largely underinvestigated and likely to have a central role in the cell cycle progression. It should also be kept in mind that genes involved in sexual development will likely be excellent transmission blocking targets. Expression profile analyses fetched substantial elucidation within these specific metabolic pathways, but functional elucidation of a significant number of hypothetical proteins will certainly bring innovative insights. There is no doubt that complete functional characterization of specific pathways or enzymes crucial to parasite survival (and not that of the host) will bring additional features to the rational design of new chemotherapeutic agents.

THE USE OF BIOCHIPS FOR ELUCIDATING THE MECHANISM OF DRUG ACTION

To this point, our analysis has focused on the identification of potential drug targets using life cycle expression profiling and elucidation of gene function. A reverse approach could provide additional insight. When micro-organisms or tissues are treated with small molecules that inhibit basic cellular processes, genes in the inhibited pathway may be transcriptionally up- or downregulated (Evans and Guy, 2004; Gunther et al., 2003; Hatzixanthis et al., 2003; Reinoso-Martin et al., 2003; Schuller et al., 2004). This observation leads to the evaluation of the global transcriptional response to drug treatment, a useful tool for identifying the cellular processes affected by the drug, as well as for finding new potential targets within the affected pathway. Today, despite many years of investigation, the mechanisms of action of the most effective antimalarials such as quinoline, antifolate, or artemisinin-derivatived compounds remain uncertain. To identify genes implicated in drug interactions and to study the drug's mechanism of action, genomewide microarray analyses were performed. To our knowledge, few analyses of transcriptional changes under drug treatments have been performed. Ganesan et al. (2003) analyzed the effect of the lethal antifolate WR99210 directed at the dihydrofolate reductase-thymidylate synthase using the long oligonucleotide microarray. RNA for *de novo* pyrimidine biosynthesis and folate biosynthesis pathway show only subtitle changes (less than 25%) in dying cells. The antimalarial choline analog, T4, a compound in preclinical studies, which targets the inhibition of the phosphatidylcholine biosynthesis, has been extensively investigated using the high-density oligonucleotide array (Le Roch et al., 2006). Transcriptome analysis using the hierarchical nature of GO terms reveals a significant induction (2- to 69-fold changes) in stress-related genes and genes involved in sexual differentiation after more than 30 h of incubation with synchronized parasites. No significant changes were observed for the enzymes involved in the lipid biosynthesis pathway. An arrest of the genes involved in the cell cycle progression was also detected, which illustrates that the parasites can detect a chemical stress and stimulate sexual development. This effect is not surprising; it has been demonstrated that gametocytogenesis can be induced by stress. Indeed, gamete formation is a toll that enables the parasite to escape the host's death by a rapid transmission to the *Anopheles* mosquito. To date, no other study has shown a significant drug response of the metabolic pathway, theoretically involved in a drug's action. However, it should be noted that the malaria parasite, as an obligate intracellular parasite, has evolved in a buffered intracellular environment in which the evolutionary forces may have induced the loss of genes involved in transcriptional feedback responses. This may explain why *Plasmodium* has a tight and specific transcriptional regulation across its life cycle, which may imply a certain susceptibility to specific new antimetabolites.

VACCINE DEVELOPMENT

Over the last 20 years, extensive research has focused on vaccine development, but so far the outlook for vaccines is less optimistic than for drug discovery. An effective malaria vaccine must induce a protective immune response equivalent to or better than that provided by natural immunity. Indeed, when an adult who acquired natural immunity returns to his or her endemic area after a few months, he has usually lost his protective immunity and become sick. For this reason, an effective malaria vaccine requires new methods of maximizing the longevity of the protective immune response. The "winning" vaccine will possess multiantigenic determinants with multistage expression. The most promising antigens under evaluation for use in vaccine development against the erythrocytic stage are targeted at the invasive step of the malaria parasite (e.g., merozoite surface proteins, erythrocyte binding antigens, and rhoptry proteins). Interestingly, all of these potential targets have similar expression profiles across the cell cycle and are extensively expressed at the late schizont stage. Identification of hypothetical genes coexpressed with cell invasion genes are great potential vaccine candidates. Approximately 100 to 200 hypothetical proteins have been identified with such profiles (Bozdech, Llinas et al., 2003; Le Roch et al., 2003).

The detection of single-feature polymorphism (SFP) within the *Plasmodium* genome will certainly bring new insights for the identification of possible vaccine candidates. Large-scale identification of SFPs is now achievable for complex genomes (Borevitz et al., 2003). Genomic DNA hybridization to high-density nucleotide arrays, together with new analytical tools, can detect an increase or decrease of the hybridization intensity level for the 25-mer probes and identify SFPs across the whole genome. This has been successfully tested for four isolates of *P. falciparum* from geographically diverse areas (Honduras, Southeast Asia, Sierra Leone, and Brazil) with the complete sequence of *P. falciparum* chromosome 2 (Volkman et al., 2002). Variations were mostly concentrated in subtelomeric regions of the chromosome end, a region known to be encoded by multigene families involved in immune evasion. Membrane-associated proteins were found to be responsible for more than 85% of all detected polymorphisms. A number of hypothetical proteins were also detected as highly polymorphic, suggesting that these genes may be under genetic selection pressures similar to those observed with antigenic and membrane protein genes. This study led to the conclusion that a whole genome analysis of malaria strains will identify new and efficient vaccine candidates under genetic selection pressures across all stages of the malaria life cycle (Kidgell, 2006).

CONCLUSIONS

Functional genomic analyses, using high-throughput studies of the whole genome have generated huge data sets. Complete reference genome sequences for *Plasmodium* species, genome annotations, transcriptional analysis, steady-state levels using microarray technologies, and cluster analysis of coexpressed genes are part of these and offer a huge advantage in understanding the parasite biology. But efficient antiparasitic discoveries need to integrate the relationship between transcriptome and proteome. Analyses of protein levels across the malaria life cycle have also been highly informative (Florens et al., 2002; Lasonder et al., 2002); analysis of the protein–protein interactions will assist in the construction of the *Plasmodium's* interactome and unquestionably yield fundamental biological information. Polymorphic diversities and comparative genomic analyses will certainly provide additional informative data. All these data need to be stored, organized, analyzed using the latest computational tools, and be widely accessible online. The *Plasmodium* genome database PlasmoDB (http://PlasmoDB.org) provides the latest and most comprehensive collection of *Plasmodium*-related data sets. Once stored, data can be integrated and linked in relational databases; depending upon the relationship between multiple data types, integrated queries can then be submitted for "*in silico* research" in order to filter and structure new hypotheses. These revolutionary technologies will undoubtedly bring success to antiparasitic discovery although it should be borne in mind that experimental and clinical validations will always be required.

REFERENCES

Borevitz, J.O., D. Liang, D. Plouffe, H.S. Chang, T. Zhu, D. Weigel, C.C. Berry, E. Winzeler, and J. Chory. 2003. Large-scale identification of single-feature polymorphisms in complex genomes. *Genome Res* **13**: 513–523.

Bozdech, Z., M. Llinas, B.L. Pulliam, E.D. Wong, J. Zhu, and J.L. DeRisi. 2003a. The transcriptome of the intraerythrocytic developmental cycle of *Plasmodium falciparum. PLoS Biol* **1**: 5.

Bozdech, Z., J. Zhu, M.P. Joachimiak, F.E. Cohen, B. Pulliam, and J.L. DeRisi. 2003b. Expression profiling of the schizont and trophozoite stages of *Plasmodium falciparum* with a long-oligonucleotide microarray. *Genome Biol* **4**: R9.

Breitling, R., A. Amtmann, and P. Herzyk. 2004a. Graph-based iterative Group Analysis enhances microarray interpretation. *BMC Bioinformatics* **5**: 100.

Breitling, R., A. Amtmann, and P. Herzyk. 2004b. Iterative group analysis (iGA): A simple tool to enhance sensitivity and facilitate interpretation of microarray experiments. *BMC Bioinformatics* **5**: 34.

Breman, J.G., A. Egan, and G.T. Keusch. 2001. The intolerable burden of malaria: a new look at the numbers. *Am J Trop Med Hyg* **64**: iv–vii.

Coulson, R.M., N. Hall, and C.A. Ouzounis. 2004. Comparative genomics of transcriptional control in the human malaria parasite *Plasmodium falciparum. Genome Res* **14**: 1548–1554.

Doerig, C. 2004. Protein kinases as targets for anti-parasitic chemotherapy. *Biochim Biophys Acta* **1697**: 155–168.

Eisen, M.B. and P.O. Brown. 1999. DNA arrays for analysis of gene expression. *Methods Enzymol* **303**: 179–205.

Evans, W.E. and R.K. Guy. 2004. Gene expression as a drug discovery tool. *Nat Genet* **36:** 214–215.

Florens, L., X. Liu, Y. Wang, S. Yang, O. Schwartz, M. Peglar, D.J. Carucci, J.R. Yates, III, and Y. Wub. 2004. Proteomics approach reveals novel proteins on the surface of malaria-infected erythrocytes. *Mol Biochem Parasitol* **135:** 1–11.

Florens, L., M.P. Washburn, J.D. Raine, R.M. Anthony, M. Grainger, J.D. Haynes, J.K. Moch, N. Muster, J.B. Sacci, D.L. Tabb, A.A. Witney, D. Wolters, Y. Wu, M.J. Gardner, A.A. Holder, R.E. Sinden, J.R. Yates, and D.J. Carucci. 2002. A proteomic view of the *Plasmodium falciparum* life cycle. *Nature* **419:** 520–526.

Foth, B.J. and G.I. McFadden. 2003. The apicoplast: a plastid in Plasmodium falciparum and other Apicomplexan parasites. *Int Rev Cytol* **224:** 57–110.

Gallup, J.L. and J.D. Sachs. 2001. The economic burden of malaria. *Am J Trop Med Hyg* **64:** 85–96.

Ganesan, K., Jiang, L., White, J, Rathod, P. 2003. Rigidity of the Plasmodium Transcriptome Revealed by a Lethal Antifolate. Molecular Parasitology Meeting (Woods Hole, MA).

Gardner, M.J., N. Hall, E. Fung, O. White, M. Berriman, R.W. Hyman, J.M. Carlton, A. Pain, K.E. Nelson, S. Bowman, I.T. Paulsen, K. James, J.A. Eisen, K. Rutherford, S.L. Salzberg, A. Craig, S. Kyes, M.S. Chan, V. Nene, S.J. Shallom, B. Suh, J. Peterson, S. Angiuoli, M. Pertea, J. Allen, J. Selengut, D. Haft, M.W. Mather, A.B. Vaidya, D.M. Martin, A.H. Fairlamb, M.J. Fraunholz, D.S. Roos, S.A. Ralph, G.I. McFadden, L.M. Cummings, G.M. Subramanian, C. Mungall, J.C. Venter, D.J. Carucci, S.L. Hoffman, C. Newbold, R.W. Davis, C.M. Fraser, and B. Barrell. 2002. Genome sequence of the human malaria parasite *Plasmodium falciparum*. *Nature* **419:** 498–511.

Gunther, E.C., D.J. Stone, R.W. Gerwien, P. Bento, and M.P. Heyes. 2003. Prediction of clinical drug efficacy by classification of drug-induced genomic expression profiles in vitro. *Proc Natl Acad Sci U S A* **100:** 9608–9613.

Hatzixanthis, K., M. Mollapour, I. Seymour, B.E. Bauer, G. Krapf, C. Schuller, K. Kuchler, and P.W. Piper. 2003. Moderately lipophilic carboxylate compounds are the selective inducers of the *Saccharomyces cerevisiae* Pdr12p ATP-binding cassette transporter. *Yeast* **20:** 575–585.

Horrocks, P., K. Dechering, and M. Lanzer. 1998. Control of gene expression in *Plasmodium falciparum*. *Mol Biochem Parasitol* **95:** 171–181.

Horrocks, P., M. Jackson, S. Cheesman, J.H. White, and B.J. Kilbey. 1996. Stage specific expression of proliferating cell nuclear antigen and DNA polymerase delta from *Plasmodium falciparum*. *Mol Biochem Parasitol* **79:** 177–182.

Kidgell, C., S. Volkman, J. Daily, J. Borevitz, D. Plouffe, Y. Zhou, J. Johnson, K. LeRoch, O. Sarr, O. Ndir, S. Mboup, S. Batalov, D. Wirth, E. Winzeler. 2006. A systematic map of genetic variation in *Plasmodium falciparum*. *PLoS Pathog* June 2(6): e57. EPub 2006 June 23.

Lanzer, M., S.P. Wertheimer, D. de Bruin, and J.V. Ravetch. 1993. Plasmodium: control of gene expression in malaria parasites. *Exp Parasitol* **77:** 121–128.

Lasonder, E., Y. Ishihama, J.S. Andersen, A.M. Vermunt, A. Pain, R.W. Sauerwein, W.M. Eling, N. Hall, A.P. Waters, H.G. Stunnenberg, and M. Mann. 2002. Analysis of the *Plasmodium falciparum* proteome by high-accuracy mass spectrometry. *Nature* **419:** 537–542.

LeRoch, K., J. Johnson, Y. Zhou, K. Henson, Y. Yates, H. Vial, and E. Winzeler, 2004. Functional genomics of the antimalarial choline analogue T4 on *Plasmodium falciparum* reveal potential therapeutic targets. Manuscript in preparation.

LeRoch, K., Y. Zhou, S. Batalov, and E.A. Winzeler. 2002. Monitoring the chromosome 2 intraerythrocytic transcriptome of *Plasmodium falciparum* using oligonucleotide arrays. *Am J Trop Med Hyg* **67:** 233–243.

LeRoch, K., Y. Zhou, P.L. Blair, M. Grainger, J.K. Moch, J.D. Haynes, P. De La Vega, A.A. Holder, S. Batalov, D.J. Carucci, and E.A. Winzeler. 2003. Discovery of gene function by expression profiling of the malaria parasite life cycle. *Science* 301, 1503–1508.

Lee, S.G., J.U. Hur, and Y.S. Kim. 2004. A graph-theoretic modeling on GO space for biological interpretation of gene clusters. *Bioinformatics* **20**: 381–388.

Ralph, S.A., G.G. Van Dooren, R.F. Waller, M.J. Crawford, M.J. Fraunholz, B.J. Foth, C.J. Tonkin, D.S. Roos, and G.I. McFadden. 2004. Tropical infectious diseases: metabolic maps and functions of the *Plasmodium falciparum* apicoplast. *Nat Rev Microbiol* **2**: 203–216.

Reinoso-Martin, C., C. Schuller, M. Schuetzer-Muehlbauer, and K. Kuchler. 2003. The yeast protein kinase C cell integrity pathway mediates tolerance to the antifungal drug caspofungin through activation of Slt2p mitogen-activated protein kinase signaling. *Eukaryot Cell* **2**: 1200–1210.

Rosenthal, P.J. 2002. Hydrolysis of erythrocyte proteins by proteases of malaria parasites. *Curr Opin Hematol* **9**: 140–145.

Rosenthal, P.J., P.S. Sijwali, A. Singh, and B.R. Shenai. 2002. Cysteine proteases of malaria parasites: targets for chemotherapy. *Curr Pharm Des* **8**: 1659–1672.

Scherf, A., R. Hernandez-Rivas, P. Buffet, E. Bottius, C. Benatar, B. Pouvelle, J. Gysin, and M. Lanzer. 1998. Antigenic variation in malaria: in situ switching, relaxed and mutually exclusive transcription of var genes during intra-erythrocytic development in *Plasmodium falciparum*. *EMBO J* **17**: 5418–5426.

Schuller, C., Y.M. Mamnun, M. Mollapour, G. Krapf, M. Schuster, B.E. Bauer, P.W. Piper, and K. Kuchler. 2004. Global phenotypic analysis and transcriptional profiling defines the weak acid stress response regulon in *Saccharomyces cerevisiae*. *Mol Biol Cell* **15**: 706–720.

Toronen, P. 2004. Selection of informative clusters from hierarchical cluster tree with gene classes. *BMC Bioinformatics* **5**: 32.

Vial, H.J., P. Eldin, A.G. Tielens, and J.J. van Hellemond. 2003. Phospholipids in parasitic protozoa. *Mol Biochem Parasitol* **126**: 143–154.

Volkman, S.K., D.L. Hartl, D.F. Wirth, K.M. Nielsen, M. Choi, S. Batalov, Y. Zhou, D. Plouffe, K.G. Le Roch, R. Abagyan, and E.A. Winzeler. 2002. Excess polymorphisms in genes for membrane proteins in *Plasmodium falciparum*. *Science* **298**: 216–218.

Wengelnik, K., V. Vidal, M.L. Ancelin, A.M. Cathiard, J.L. Morgat, C.H. Kocken, M. Calas, S. Herrera, A.W. Thomas, and H.J. Vial. 2002. A class of potent antimalarials and their specific accumulation in infected erythrocytes. *Science* **295**: 1311–1314.

Wu, Y., X. Wang, X. Liu, and Y. Wang. 2003. Data-mining approaches reveal hidden families of proteases in the genome of malaria parasite. *Genome Res* **13**: 601–616.

Young, J., Q. Fivelman, P. Blair, P. de la Vega, K. LeRoch, F. Yan, D. Baker, D. Carucci, Y. Zhou, E. Winzeler. 2005. The *Plasmodium falciparum* sexual development transcription. A microarray analysis using ontology-based pattern identification. *Mol Biochem Parisitol* 163(1): 67–79.

Zhou, Y. and R. Abagyan. 2002. Match-only integral distribution (MOID) algorithm for high-density oligonucleotide array analysis. *BMC Bioinformatics* **3**: 3.

Zhou, Y. and R. Abagyan. 2003. Algorithms for high-density oligonucleotide array. *Curr Opin Drug Discov Dev* **6**: 339–345.

Zhou, Y., Young, J., Santrosyan, A., and Winzeler, E. 2004. In silico gene function prediction using ontology-based pattern identification. Submitted.

4 Regional Variations in Intestinal ATP-Binding Cassette Transporter Expression Identified with a Global Error Assessment Model*

David M. Mutch, Anton Petrov, J. Bruce German, Gary Williamson, and Matthew-Alan Roberts

CONTENTS

INTRODUCTION

Emerging comprehensive technologies for examining genes, proteins, and metabolites have led to fundamental changes in our approach to assessing health and identifying disease biomarkers. Indeed, these analytical platforms are now widely

---

* This chapter has been adapted from Mutch et al., 2004 [52], and Mansourian et al., 2004 [56].

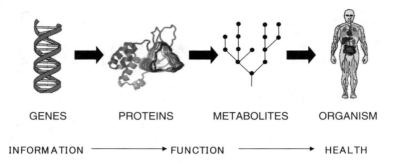

GENES   PROTEINS   METABOLITES   ORGANISM

INFORMATION ──────────▶ FUNCTION ──────────▶ HEALTH

FIGURE 4.1 Integrated metabolism can be defined as the relationships between genes, proteins, and metabolites, which can ultimately define the health status of the organism.

used by both the pharmaceutical and nutritional communities alike. Although the approaches differ, i.e., pharmaceuticals aim to specifically target a dysfunctional gene or protein to treat a disease whereas nutrition aims to predominantly prevent the onset of disease, both fields require a profound understanding of the disease state and its development in order to maintain and/or improve health status. Each analytical platform yields requisite and complementary information implicit in unraveling the mechanisms underlying metabolic disorders. However, despite the obvious relationship between genes, proteins, and metabolites (see Figure 4.1), few examples exist in which these multiple platforms have been integrated at a comprehensive level [1]. Rather, most scientists have concentrated on the field of functional genomics (i.e., those active genes in a biological condition). The sequencing of several genomes, such as human [2,3], mouse [4,5], and rat [6], have provided a complete molecular catalogue that can be used to assay gene function. This knowledge has been exploited with the development of species-specific microarrays composed of all genetic elements present in the aforementioned genomes. Furthermore, the maturing field of gene expression analysis has benefited from the standardization of protocols and data presentation stemming from such initiatives as minimum information about a microarray experiment (MIAME) [7] and Gene Expression Omnibus (GEO) [8,9]. Indeed, the intense perseverance to accurately analyze and interpret the enormous data sets stemming from microarray studies has demonstrated that the field of functional genomics is by far the most characterized and actively utilized global technology on the market.

Despite the tremendous advances in data analysis and supporting bioinformatic software used with the numerous microarray studies performed to date, a critical question remains: Can genomics identify suitable biomarkers to assess both individual and population health status? Encouragingly, the answer appears to be yes in that microarrays, when complemented with additional technologies, can identify genes responsible for a disease state. For example, microarrays have been used to explore the gene expression profile of diseases such as cancer [10], ulcerative colitis [11], Alzheimer's [12,13], celiac [14], etc.; however, these studies have not found that a single dysfunctional gene contributes to the onset of disease. In other words, complex gene networks have been attributed to these disease states. In contrast, such diseases as sitosterolemia and Tangier's disease (both characterized by abnormal lipid transport) have been correlated to mutations in members of the ATP-binding cassette (ABC) family of transporters. Coupling microarray technology, bioinformatics, classical biochemical studies, and genetic mapping revealed that mutations in Abcg5/Abcg8 [15] and Abca1 [16] are implicated in the development of sitosterolemia and Tangier's disease, respectively. However, based on the aforementioned examples, it is difficult to assume *a priori* that a single genetic element underlies a given disease phenotype. Therefore, microarrays provide a complete assessment of all genetic elements in a given species and offers an attractive approach to unravel those molecular mechanisms contributing to a specific disease state.

ABC proteins are the largest known family of transmembrane transporters [17] and are involved in the directional transport of a wide variety of substrates, including sugars, amino acids, glycans, sterols, phospholipids, peptides, proteins, toxins, antibiotics, and xenobiotics across biological membranes [18]. This protein family, comprising approximately 50 transporters in higher mammals, has been further subdivided based on structural similarities and domain sequence homologies into seven subfamilies, ABCA to ABCG [19]. Furthermore, the high degree of conservation between the ABC proteins of different species reinforces their functional importance in the transport of molecular compounds. The structure, molecular organization both within and between species, the known functions of ABC transporters, and how they may contribute to the onset of disease have been thoroughly described in recent reviews [17,20,21].

To fully characterize the biological functions of this large transporter family and, ultimately, their contribution to the onset of disease states, it is critically important to ascertain where in the body these transporters are expressed. Langmann and colleagues described a whole-body gene transcript characterization of all currently known human ABC transporters, using quantitative real-time PCR [22]. The authors analyzed and revealed the expression profiles of these genes in 20 different tissues and concluded that tissues involved in secretory function (adrenal gland), metabolic function (liver), barrier function (small intestine), and development (uterus, testis) had high levels of ABC transporter transcripts. However, the authors did not divide the small intestine into its functionally distinct regions (i.e., duodenum, jejunum, and ileum). It has previously been reported that factors such as disease, pH, motility, bile, and the microbial community vary along the length of the gastrointestinal tract (GIT) and influence drug, and therefore presumably nutrient, bioavailability [23]. Furthermore, it can be assumed that these factors coordinate gut function by mediating gene expression. Indeed, this notion is supported by the findings of a study suggesting that the microbial community coordinates only a subset of GIT gene expression (i.e., 10x more genes differ between GIT regions than within any given region examined with and without a microbiota) [24]. Therefore, this would suggest that each of the aforementioned factors contribute to the regional expression profiles, and eventually the regional functions, in the anatomically distinct organs of the digestive tract, as demonstrated by several laboratories [25–28]. As ABC proteins are highly involved in the transport of nutrients and drug compounds alike, we propose that their regional expression patterns along the intestinal tract are a prerequisite to deciphering their functions in this organ.

In this regard, the present chapter illustrates both the thoroughness and innate sensitivity of microarrays by examining the regional variations of the critically important ABC transporter family along the intestinal tract. This protein family is implicated in the transport of both pharmaceutical and nutritional compounds and certain members have been, as already described briefly, correlated with the onset of metabolic disorders. However, prior to the interpretation of microarray data sets, one must assure that the analysis has been performed with an appropriate biostatistical model. Therefore, this chapter aims to describe the development of a Global Error Assessment (GEA) model for the identification of differentially expressed genes from microarray data sets and its use in the focused analysis of ABC transporter expression along the intestinal tract.

GLOBAL ERROR ASSESSMENT (GEA) MODEL

Many of the first experiments to benefit from the global view of microarrays utilized a simple fold-change (FC) cutoff for the selection of differentially expressed genes. However, evidence now demonstrates that such a selection method makes several assumptions that are out of context with the rest of the experimental and biological data at hand [29]. As a result, the maturing biostatistical community is continually motivated to develop methods permitting the power of this comprehensive platform to be exploited [30–40]. However, the new statistical models, annotation tools, and exploration into the many facets of microarray technology (e.g., number of replicates, pooling, etc.) published on a monthly basis highlight the fact that there is currently no universal method being applied for the analysis of these enormous data sets.

The drive to develop statistical models for the analysis of microarray experiments stems primarily from a single common factor across the majority of studies: a low number of experimental replicates (k). As the use of microarrays is relatively resource intensive for most laboratories, a low k is the experimental norm. A small number of experimental replicates decrease the power of standard statistical tests (e.g., Student's t-test, classical ANOVA, etc.) to differentiate between regulated and non-regulated genes by producing an inaccurate estimate of variance [41,42]. Furthermore, even in the case of achieving reasonable numbers of replicates, there is a continuous desire to derive greater statistical power from the inherent multidimensional, yet simultaneous, measurements characteristic of microarrays. Increasing support for this approach has appeared in the literature [39,43–49]. These publications call for the need to "borrow statistical power" through pooling replicates from different genes together during significance testing. In our present research, we extended the concept of borrowing statistical power for estimating noise variance (as previously described by Mutch et al. [29,50]]) and applied this to an ANOVA-based regulation significance test. This model, termed the *Global Error Assessment* (GEA) model, directly generates a robust estimate of the mean squared error (MSE) or, equivalently, of the standard deviation by estimating a localized error from the measurement information of several hundred neighboring genes with similar expression levels. The robust MSE of this group of neighboring genes is a highly powerful estimate of the denominator of the F statistic used in standard statistical tests. It is this principal difference between GEA and other ANOVA-based tests that enable GEA to more powerfully determine differentially expressed genes.

THE GEA PROTOCOL

The principles underlying GEA methodology resemble those of a typical analysis of variance; however, this method calculates a robust estimation of the within treatment variability. Robustness is achieved by two means: (1) averaging within treatment variability of genes that are expressed at a similar level (i.e., neighboring genes) and (2) using estimates of the average variability instead of classical ones. For this to be accomplished, the following protocol was implemented:

1. Calculate the mean normalized average difference intensity (ADI) from MAS5.0*, the MSA and the MSE for each gene on the microarray platform, where MSE is defined as the standard deviation (SD) of a gene within a single condition and MSA as the SD of a gene between n conditions.
2. Sort genes by ascending mean ADI and group them into bins of 200 consecutive genes (corresponding to approximately 100 bins for an Affymetrix GeneChip). Various bin sizes were examined in order to determine how bin size would affect the GEA model. Bin sizes of 25, 50, 100, and 400 genes were examined, as shown in Figure 4.2. The relationship between variability and expression level remained stable across the range of bin sizes, indicating that small changes in bin size do not have major effects. A bin size of 200 appeared to be optimal because it provides an accurate local estimate of MSE while simultaneously approaching a smoothed trend line.
3. The MSE of the 200 genes in each bin are summarized using a robust estimation:

$$MSE_{Robust} = Median_{i = 1,...,200}(MSE) * df_E/\chi^{-1} (0.5,df_E),$$

where χ^{-1} is the inverse of the one-tailed probability of the chi-squared distribution.
4. For each gene, compute the test statistic:

$$F = MSA/MSE_{Robust},$$

which follows Snedecor's F distribution with degrees of freedom $df_A = n - 1$ and $df_{E,Robust} = 200*(nk - n)$.
5. Select genes for which $MSA > Limit_{Robust, \alpha} = MSE_{Robust}* F^{-1} (1 - \alpha, df_A, df_{E,Robust})$, where α is the significance level.

---

* Note that the GEA methodology does not need to be used exclusively with MAS5.0 software; this model can be applied to the normalized data derived from any analytical platform.

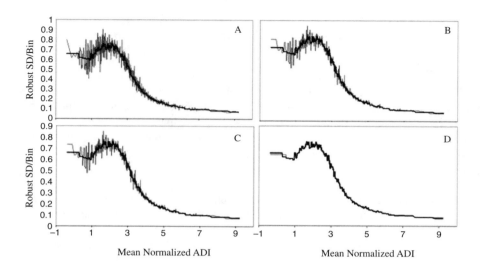

FIGURE 4.2 Bin sizes of 25 (A), 50 (B), 100 (C), and 400 (D) genes are compared with a bin of 200 genes, where the x-axis is the mean normalized ADI per bin and the y-axis the robust standard deviation per bin (i.e., the square root of MSE_{Robust}). (Adapted from Mansourian, R., Mutch, D.M., Antille, N., Aubert, J., Fogel, P., Le Goff, J.M., Moulin, J., Petrov, A., Rytz, A., Voegel, J.J., Roberts, M.A. The global error assessment (GEA) model for the selection of differentially expressed genes in microarray data. *Bioinformatics* 20(16): 2726–2737, 2004.)

COMPARING GEA WITH THE CLASSICAL AND PERMUTATIONAL ANOVA TESTS

Using a well-defined *in vitro* system comprised of nine replicates for each treatment, a comparison between GEA and a classical ANOVA was performed with the goal of demonstrating the enhanced ability of GEA to minimize the number of false positives and identify those genes that are truly differentially regulated. Figure 4.3 illustrates both the characteristic "data cloud" to be expected with a microarray platform and the increased sensitivity of GEA vs. the classical ANOVA to identify true positives around this data cloud. If one simply plots the expression (in log scale) of each gene on the microarray platform for two control samples, one can quickly see that variability is a function of absolute expression, i.e., as indicated by the pear-shaped form of the gray data cloud (Figure 4.3A). When plotting a control sample (x-axis) vs. a treatment sample (y-axis), the great majority of genes still lie within the data cloud; however, genes that are differentially regulated by the treatment diffuse out of the data cloud (open black squares in Figure 4.3B). By choosing a highly confident GEA p-value the question can then be asked, at what confidence level would the classical technique be required to achieve full concordance in the gene selection. In this example, a highly significant, but arbitrary, p-value of $1e^{30}$ was chosen and led to the selection of 531 genes. Black closed squares overlaid on the variability cloud identify these 531 genes and demonstrate that as absolute

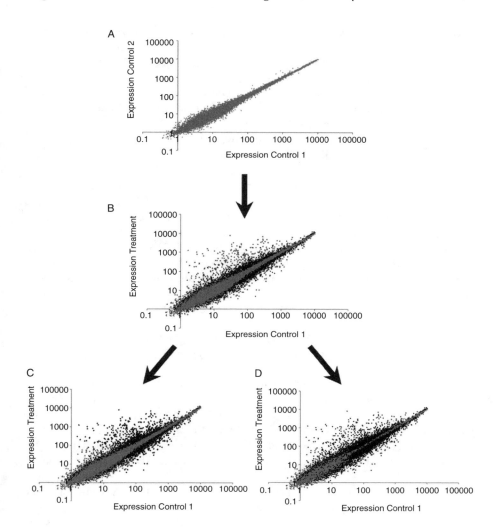

FIGURE 4.3 Comparing GEA and the classical ANOVA. (A) Defining the relationship between absolute gene expression and variability demonstrates the heterogeneous relationship between the two, as shown when plotting control 1 vs. control 2. (B) Plotting control 1 vs. treatment reveals those genes modulated by the treatment (indicated by black dots). (C) GEA attributes significance to 531 genes (identified in black dots, $p < 1e^{30}$) that truly differentiate themselves from the underlying data cloud (in gray). (D) In contrast, the classical ANOVA must be less stringent ($p < .02$) in order to select the same genes as GEA, resulting in many false positives (i.e., those genes clearly lying within the underlying data cloud). (Adapted from Mansourian, R., Mutch, D.M., Antille, N., Aubert, J., Fogel, P., Le Goff, J.M., Moulin, J., Petrov, A., Rytz, A., Voegel, J.J., Roberts, M.A. The global error assessment (GEA) model for the selection of differentially expressed genes in microarray data. *Bioinformatics*. 20(16): 2726–2737, 2004.)

expression increases, GEA is able to confidently select genes closer to the contour of the variability cloud (Figure 4.3C). In order to achieve concordance (523 out of 531 genes), the classical ANOVA must relax to a p-value of .02 or greater. Black closed squares in Figure 4.3D correspond to 3358 genes selected by classical ANOVA with $p > .02$, almost all of which overlap with the underlying variability cloud. It can be concluded from this analysis that GEA does in fact derive increased statistical power from the binned MSE.

A similar comparison was performed between GEA and a permutational analog of ANOVA [51]. The benefit of utilizing this method lies in the attempt to estimate the actual distribution of the test statistic (F) through the use of thousands of computer permutations. Although more robust than the classical ANOVA, it still suffers from a lack of power under conditions of low k. Results from this comparison indicated that the permutational ANOVA performed better than its classical counterpart; however, for concordance to be achieved between the two methods, the permutational ANOVA must relax to a p-value of .003 or greater. This further demonstrates the enhanced statistical power of GEA to discriminate between true and false positives and, ultimately, yield biological information that can be accurately and confidently interpreted.

REGIONAL VARIATIONS IN ABC TRANSPORTER EXPRESSION IN THE INTESTINE

In order to elucidate the expression profiles of ABC transporters along the anterior–posterior (A–P) axis of the intestinal tract, murine RNA samples corresponding to the duodenum, jejunum, ileum, and colon were hybridized to Affymetrix Mu74v2 GeneChips [52]. To date, 43 of 49 murine ABC transporters have been annotated by Affymetrix and are located across the three GeneChips. ABC transporters not yet annotated or present on the GeneChips are Abca8, Abca12, Abca13, Abcb5, Abcb8, Abcc4, and Abcc11.

It is interesting to note that, at the mRNA level, most of the ABC transporters are not differentially expressed along the intestinal tract. Indeed, only eight transporters were identified as differentially expressed in the gut by both the classical ANOVA and GEA ($\alpha < 0.01$ for both statistical tests). Differentially expressed transporters were Abcb2, Abcb3, Abcb9, Abcc3, Abcc6, Abcd1, Abcg5, and Abcg8. When visualizing all ABC transporters on a bivariate plot comparing the small intestine (duodenum, jejunum, and ileum values averaged to obtain a single value) to the colon, where the mean of the natural logarithm of the fold change (M) is plotted against the mean expression value (A), one can immediately observe that the great majority of these transporters (identified by gray dots) lie within the data cloud (Figure 4.4). Additionally, estimation of a lowess regression function predicting the local mean standard deviation demonstrates that the majority of these transporters are not differentially expressed. Furthermore, as the x-axis is representative of absolute expression levels, it is clear that ABC transporters are present in the intestinal tract at various levels, from lowly or not at all expressed up to highly expressed. The aforementioned eight differentially expressed ABC transporters are

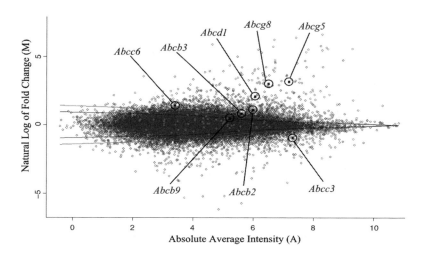

FIGURE 4.4 Plotting absolute average intensity (absolute expression) vs. fold change identifies differentially regulated genes between the small intestine and colon. Lowess curves indicated the two- and three-times standard deviations. Differentially expressed ABC transporters are indicated. (Adapted from Mutch, D.M., Anderle, P., Fiaux, M., Mansourian, R., Vidal, K., Wahli, W., Williamson, G., Roberts, M.A. Regional variations in ABC transporter expression along the mouse intestinal tract. *Physiol Genomics* 17(1): 11–20, 2004.)

indicated and their degree of differential expression placed in context with all genetic elements present on the GeneChips.

VALIDATION 1: GEA ANALYSIS OF MICROARRAY DATA VS. REAL-TIME PCR

Eight ABC transporters were validated using both TaqMan RT-PCR assays on demand and assays by design primer/probe sets (Applied Biosystems, Foster City, CA) [52]. The transporters selected for validation displayed one of the following trends: (1) no change along the intestinal tract (Abca1, Abcc1, Abcc6, Abcd3), (2) an increase along the intestinal tract (Abcb1a), or (3) a decrease along the intestinal tract (Abcd1, Abcg5, Abcg8). Overall, the concordance between the two techniques indicated that trends seen in microarray data could also be seen with RT-PCR; for example, an increase in gene expression along the A–P axis could be seen with both techniques (Table 4.1). Because of the greater dynamic range attainable with RT-PCR, this technique is more often used as a means to confirm trends in microarray data rather than duplicate the fold changes seen with chip experiments [50,53,54]. Therefore, it was not surprising to see that discrepancies in the two data sets arose when examining the statistical significance of fold changes (for both RT-PCR and microarray data) in gene expression levels of the jejunum, ileum, and colon in relation to the duodenum. Microarray data indicated that many of the gene changes observed were not found to be statistically different

TABLE 4.1
Comparative Analysis of Gene Expression Levels

	Microarray			Real-Time PCR		
Gene	Jejunum	Ileum	Colon	Jejunum	Ileum	Colon
Abca1	1.21	1.04	1.51	3.43[a]	3.78[b]	3.22[b]
Abcb1a	1.52	3.29[b]	6.36[b]	1.66	5.68[b]	7.02[b]
Abcc1	1.00	1.42	2.01	1.01	3.56[b]	4.86[b]
Abcc6	4.71	1.08	2.61	3.83[b]	2.58[b]	1.58
Abcd1	2.83[b]	2.37	7.46[b]	2.31[b]	1.48[b]	2.14[b]
Abcd3	1.28	1.16	1.03	1.64	1.93[b]	2.38[b]
Abcg5	1.63[a]	1.17	21.33[b]	1.76	2.44	45.41[b]
Abcg8	2.27[b]	1.08	15.96[b]	2.30[a]	2.19[a]	60.55[b]

Note: Jejunum, ileum, and colon values are compared to the duodenum, which has been arbitrarily set to 1 (and therefore not depicted). A positive value represents an increase in mRNA levels, and a negative value represents a decrease. Significance for microarray results was identified by pairwise GEA analyses. Significance for RT-PCR results was determined using a two-tailed, homoscedastic Student's *t*-test.

[a] $p < 0.01$.

[b] $p < 0.001$.

Source: Data from Mutch, D.M., Anderle, P., Fiaux, M., Mansourian, R., Vidal, K., Wahli, W., Williamson, G., Roberts, M.A. Regional variations in ABC transporter expression along the mouse intestinal tract. *Physiol Genomics* 17(1): 11–20, 2004.

($\alpha < 0.01$), even in circumstances in which an apparent fold change of 4.7 (Abcc6 in the jejunum) is seen. In contrast, RT-PCR was able to identify additional statistically significant differences in situations in which the microarray was not, such as the 3.8 fold change observed for Abcc6 in the jejunum. Findings such as these were expected because of the inherent differences in sensitivities between the two methods [55]. Indeed, identifying 100% of the truly differentially regulated genes in a microarray experiment is still complicated by lowly expressed genes (i.e., corresponding to many transcription factors and receptors), which may be highly variable within or between biological treatments [48]. The development of robust statistical methods, such as GEA, aim to dissociate those lowly expressed genes that are variable within a treatment (i.e., technical variability) from those that are variable between experimental conditions (i.e., biological variability) [48,56]. The present study demonstrated that the GEA methodology has increased sensitivity in comparison to the classical ANOVA for gene selection. Whereas GEA identified Abcb1a and Abcd1 as differentially expressed, the classical ANOVA failed to assign statistical significance to the changes in expression. RT-PCR confirmed that these genes were indeed differentially expressed in the gut, reinforcing the validity of the finding that the GEA method has increased sensitivity for the detection of low-abundance genes.

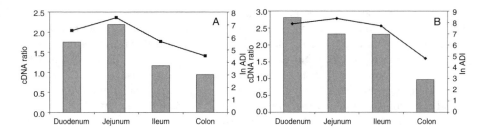

FIGURE 4.5 Comparing expression profiles for Abcd1 (A) and Abcg5 (B) along the intestinal tract, as assayed by cDNA (gray bars) and oligonucleotide microarrays (black line).

VALIDATION 2: GEA ANALYSIS OF MICROARRAY DATA VS. AN *IN SILICO* cDNA DATA SET

An alternative means to validate the Affymetrix data set was through an *in silico* comparison with the publicly available cDNA data set produced by Bates and colleagues [27]. As previously described in the literature, the hybridization specificity for target transcripts and the platform comparisons between spotted cDNA and oligonucleotide microarrays are still being explored [57,58]; however, enough similarities remain to make the comparison useful. Despite the different strain of mouse used by Bates and colleagues, their mice, consisting of only male animals, were sacrificed at a similar age. It should be further noted that the GEM1 cDNA platform (Incyte Genomics) comprised approximately 8000 sequence-verified expression sequence tags (ESTs), corresponding to roughly 25% of the mouse genome [4,5].

The cDNA data set contained expression information for 16 of the 49 ABC transporters. Of the eight ABC transporters identified in our data set as differentially regulated along the gut, only two were present in the cDNA data set: Abcd1 and Abcg5. Figure 4.5 demonstrates that despite the experimental differences, both cDNA and Affymetrix platforms indicate that Abcg5 is expressed at significantly lower levels in the colon and similarly expressed in the duodenum, jejunum, and ileum. This finding is in agreement with RT-PCR data (Table 4.1). Expression-profiling trends revealed that Abcd1 is most highly expressed in the jejunum and most lowly expressed in the colon (Figure 4.5), as confirmed by RT-PCR. The remaining 14 ABC transporters found in the cDNA data set were not differentially expressed along the intestinal tract. This agrees with our high-density oligonucleotide results and further specifies that the members of this family are expressed at various levels in the intestine and that most of these transporters are not differentially expressed.

VALIDATION 3: GEA ANALYSIS OF MICROARRAY DATA VS. PROTEIN EXPRESSION

Inasmuch as agreement between different gene expression platforms reinforces the validity of microarray data, it is only through comparisons with proteins and metabolites that one can achieve the greatest degree of confidence with gene expression data. However, analyzing proteins and metabolites at a comprehensive level is not well established and faces many challenges from the complexity of biological systems. Previous attempts to correlate genes expression with protein profiles have yielded poor results [59,60]. Nevertheless, one of the differentially regulated ABC transporters was qualitatively examined by immunohistochemistry, with the goal of correlating gene and protein expression. Immunohistochemical staining for Abcc3 revealed a similar cellular pattern of expression along the intestinal tract; i.e., protein location is restricted to the villi of the intestine, as no staining was present in the crypts of Lieberkühn (data not shown). In all segments of the intestine, Abcc3 is expressed basolaterally in enterocytes. Abcc3 mRNA is most highly expressed in the colon and to a lesser extent in the various regions of the small intestine (data not shown). Although not quantitative, the high degree of protein staining in the colon visually concords with the mRNA expression profile; however, this qualitative statement is based on the lower exposure time required to capture protein staining in the colon (0.48 sec) compared to the small intestine (between 1.04 and 1.44 sec).

CONCLUSION

Microarray technology enables the simultaneous study of thousands of unique genetic elements and provides insight into the coordinated control of many genes [50]. Furthermore, the comprehensive nature of transcriptomic platforms provides an ideal tool for deciphering the potentially complex gene networks underlying physiological functions. This platform, in conjunction with classical experimental approaches, can yield a powerful tool for both the identification of disease biomarkers and the unraveling of molecular mechanisms underlying biological function. The example described in this chapter outlined the principal findings of a recent study in which the expression profiles of ABC transporters along the murine intestinal tract were explored. Previous work has illustrated that members of this transporter family, Abca1 and Abcg5/g8, are implicated in the onset of Tangier's disease and sitosterolemia, respectively. Although the biological significance of a number of ABC transporters remains to be discovered, several transporters have been characterized and reported to regulate the transport of numerous medicinal and nutritional compounds. In the GIT, these transporters have a potential role in regulating the bioavailability of bioactive compounds. Therefore, creating a catalogue of ABC transporter expression profiles along the intestinal tract will provide the pharmaceutical community information to exploit the mechanisms regulating the absorption of orally consumed compounds.

Few studies have examined the expression profiles of ABC transporters in the intestine in order to elucidate their region-specific functions. Studies by Rost et al., Stephens et al., and Chianale et al. demonstrate the importance of treating the small intestine as functionally distinct regions rather than as a single entity, as differences in mRNA profiles, protein expression, and ATP-transporter function were reported [28,61,62]. Of those ABC transporters identified as differentially expressed in the present study, Abcb9, Abcc6, and Abcd1 have not been previously examined in the intestine and their functions in this organ are currently unknown. Nevertheless, current information concerning their functions permits some speculation regarding their expression profiles in the intestine. For example, Abcb9 (transporter associated with antigen processing-like; TAPL) is closely related to the TAP2 gene and has a similar intestinal expression pattern as the TAP1/TAP2 complex (described in more detail later). This lysosomal protein may play a role in the translocation of peptides from the cytosol into the lysosome for degradation [63,64]. Abcc6 is a confirmed member of the MRP family of drug efflux pumps and has been demonstrated to transport glutathione conjugates, thereby having a potential role in the regulation of xenobiotic bioavailability in the intestine [65]. Finally, Abcd1 is a peroxisomal half transporter that is mutated in adrenoleukodystrophy, which is characterized by a reduced peroxisomal very long chain fatty acid (VLCFA) β-oxidation [17]. As the primary site for dietary fatty acid absorption occurs in the upper intestine, the higher Abcd1 expression levels in the duodenum and jejunum suggest that this peroxisomal transporter is actively involved in the metabolism of dietary VLCFA. Although the definitive functions of intestinal Abcb9, Abcc6, and Abcd1 have yet to be ascribed, two pairs of half transporters, Abcb2/Abcb3 (TAP1/TAP2) and Abcg5/Abcg8, have previously been studied in the intestine and have been implicated in immune responses and sterol transport, respectively.

TAP1 and TAP2 (transporters associated with antigen presentation/processing) have been found to preferentially transport 9 to 12 amino acid peptides into the lumen of the endoplasmic reticulum and load these peptides onto major histocompatibility complex class 1 molecules, which are critical to an immune response [66,67].

Abcg5 and Abcg8, which are associated with sitosterolemia and the selective transport of sterol compounds [15,68], are highly expressed in the small intestine and found at much lower levels in the colon. This suggests that the selection process for the efflux of plant sterols vs. cholesterol from enterocytes back to the intestinal lumen is restricted to the small intestine and would not occur to a significant extent in the large intestine. The relative stability of mRNA expression in the duodenum, jejunum, and ileum would suggest that this active selection process could occur equivalently along the entire length of the small intestine; however, this will need to be examined via functional transport studies.

Abcc3 protein analysis provided additional information that is important in understanding the role of this differentially expressed transporter in the GIT. As indicated above, Rost and colleagues found that Abcc3 (MRP3) was most highly expressed in the colon of the rat intestine [28]. Our findings indicate that MRP3 is also highly expressed in the murine colon at both the mRNA and protein levels. Furthermore, the similar cellular location (basolateral in enterocytes) found between the mouse and rat further supports the notion of a high degree of conservation for ABC transporters among eukaryotes. This suggests that MRP3 may have a similar role in the ATP-dependent transport of 17β-glucuronosyl estradiol, glucuronsosyl bilirubin, monovalent bile salts (taurocholate and glycocholate), and sulfated bile salts (i.e., taurochenodeoxycholate-3-sulfate, taurolithocholate-3-sulfate) from the enterocyte to the blood in all higher mammals.

In conclusion, GEA methodology was able to identify significant differences in the expression levels of ABC transporters along the intestinal tract. As revealed through a complementary analysis by real-time PCR, GEA attributed significance to regional variations in expression where the classical ANOVA did not. Concordance with an alternate expression-profiling platform reinforces the high degree of conservation for ABC transporters among eukaryotes. Furthermore, with regard to Abcc3, concordance between gene and protein profiles suggested that mRNA is a suitable means to assay function. Therefore, in addition to demonstrating the enhanced statistical power of GEA, this study revealed that transporters implicated in regulating the bioavailability of bioactive compounds are differentially regulated along the intestinal tract, thereby providing the scientific community with specific targets to modulate the absorption of medicinal and nutritional compounds alike. Comparing ABC transporter expression and function in various disease states (e.g., inflammatory bowel disease and cholestasis) to the steady state described in this chapter may yield molecular targets that will permit disease-associated complications, such as nutrient malabsorption and inflammation, to be addressed.

REFERENCES

1. Griffin, J.L., Bonney, S.A., Mann, C., Hebbachi, A.M., Gibbons, G.F., Nicholson, J.K., Shoulders, C.C., Scott, J. An integrated reverse functional genomic and metabolic approach to understanding orotic acid-induced fatty liver. *Physiol Genomics* 17(2): 140–149, 2004.
2. McPherson, J.D., Marra, M., Hillier, L. et al. A physical map of the human genome. *Nature* 409(6822): 934–941, 2001.
3. Venter, J.C., Adams, M.D., Myers, E.W. et al. The sequence of the human genome. *Science* 291(5507): 1304–1351, 2001.
4. Okazaki, Y., Furuno, M., Kasukawa, T. et al. Analysis of the mouse transcriptome based on functional annotation of 60,770 full-length cDNAs. *Nature* 420(6915): 563–573, 2002.
5. Waterston, R.H., Lindblad-Toh, K., Birney, E. et al. Initial sequencing and comparative analysis of the mouse genome. *Nature* 420(6915): 520–562, 2002.
6. Gibbs, R.A., Weinstock, G.M., Metzker, M.L. et al. Genome sequence of the Brown Norway rat yields insights into mammalian evolution. *Nature* 428(6982): 493–521, 2004.
7. Brazma, A., Hingamp, P., Quackenbush, J. et al. Minimum information about a microarray experiment (MIAME)-toward standards for microarray data. *Nat Genet* 29(4): 365–371, 2001.
8. Edgar, R., Domrachev, M., Lash, A.E. Gene expression omnibus: NCBI gene expression and hybridization array data repository. *Nucl Acids Res* 30(1): 207–210, 2002.
9. Wheeler, D.L., Church, D.M., Lash, A.E., Leipe, D.D., Madden, T.L., Pontius, J.U., Schuler, G.D., Schriml, L.M., Tatusova, T.A., Wagner, L., Rapp, B.A. Database resources of the National Center for Biotechnology Information. *Nucl Acids Res* 29(1): 11–16, 2001.
10. Rhodes, D.R., Yu, J., Shanker, K., Deshpande, N., Varambally, R., Ghosh, D., Barrette, T., Pandey, A., Chinnaiyan, A.M. Large-scale meta-analysis of cancer microarray data identifies common transcriptional profiles of neoplastic transformation and progression. *Proc Natl Acad Sci U S A 2004*, 101(25): 9309–9314.
11. Lawrance, I., Fiocchi, C., Chakravarti, S. Ulcerative colitis and Crohn's disease: distinctive gene expression profiles and novel susceptibility candidate genes. *Hum Mol Genet* 10(5): 445–456, 2001.
12. Loring, J.F., Wen, X., Lee, J.M., Seilhamer, J., Somogyi, R. A gene expression profile of Alzheimer's disease. *DNA Cell Biol* 20(11): 683–695, 2001.

13. Hata, R., Masumura, M., Akatsu, H., Li, F., Fujita, H., Nagai, Y., Yamamoto, T., Okada, H., Kosaka, K., Sakanaka, M., Sawada, T. Up-regulation of calcineurin Abeta mRNA in the Alzheimer's disease brain: assessment by cDNA microarray. *Biochem Biophys Res Commun* 284(2): 310–316, 2001.

14. Diosdado, B., Wapenaar, M.C., Franke, L., Duran, K.J., Goerres, M.J., Hadithi, M., Crusius, J.B., Meijer, J.W., Duggan, D.J., Mulder, C.J., Holstege, F.C., Wijmenga, C. A microarray screen for novel candidate genes in coeliac disease pathogenesis. *Gut* 53(7): 944–951, 2004.

15. Berge, K.E., Tian, H., Graf, G.A., Yu, L., Grishin, N.V., Schultz, J., Kwiterovich, P., Shan, B., Barnes, R., Hobbs, H.H. Accumulation of dietary cholesterol in sitosterolemia caused by mutations in adjacent ABC transporters. *Science* 290(5497): 1771–1775, 2000.

16. Lawn, R.M., Wade, D.P., Garvin, M.R., Wang, X., Schwartz, K., Porter, J.G., Seilhamer, J.J., Vaughan, A.M., Oram, J.F. The Tangier disease gene product ABC1 controls the cellular apolipoprotein-mediated lipid removal pathway. *J Clin Invest* 104(8): R25–31, 1999.

17. http://www.ncbi.nlm.nih.gov/books/bv.fcgi?call=bv.View..ShowTOC&rid=mono_001. TOC&depth=2. Accessed 2002.

18. Gottesman, M.M., Ambudkar, S.V. Overview: ABC transporters and human disease. *J Bioenerg Biomembr* 33(6): 453–458, 2001.

19. Dean, M., Rzhetsky, A., Allikmets, R. The human ATP-binding cassette (ABC) transporter superfamily. *Genome Res* 11(7): 1156–1166, 2001.

20. Chan, L.M., Lowes, S., Hirst, B.H. The ABCs of drug transport in intestine and liver: efflux proteins limiting drug absorption and bioavailability. *Eur J Pharm Sci* 21(1): 25–51, 2004.

21. Holland, B.I., Kuchler, K., Higgins, C.F., Cole, S.P.C. *ABC proteins from Bacteria to Man.* Academic Press, London, 2002.

22. Langmann, T., Mauerer, R., Zahn, A., Moehle, C., Probst, M., Stremmel, W., Schmitz, G. Real-time reverse transcription-PCR expression profiling of the complete human ATP-binding cassette transporter superfamily in various tissues. *Clin Chem* 49(2): 230–238, 2003.

23. Dressman, J.B., Bass, P., Ritschel, W.A., Friend, D.R., Rubinstein, A., Ziv, E. Gastrointestinal parameters that influence oral medications. *J Pharm Sci* 82(9): 857–872, 1993.

24. Mutch, D.M., Simmering, R., Donnicola, D., Fotopoulos, G., Holzwarth, J.A., Williamson, G., Corthesy-Theulaz, I. Impact of commensal microbiota on murine gastrointestinal tract gene ontologies. *Physiol Genomics* 19(1): 22–31, 2004.

25. Saitoh, H., Aungst, B.J. Possible involvement of multiple P-glycoprotein-mediated efflux systems in the transport of verapamil and other organic cations across rat intestine. *Pharm Res* 12(9): 1304–1310, 1995.

26. Makhey, V.D., Guo, A., Norris, D.A., Hu, P., Yan, J., Sinko, P.J. Characterization of the regional intestinal kinetics of drug efflux in rat and human intestine and in Caco-2 cells. *Pharm Res* 15(8): 1160–1167, 1998.

27. Bates, M.D., Erwin, C.R., Sanford, L.P., Wiginton, D., Bezerra, J.A., Schatzman, L.C., Jegga, A.G., Ley-Ebert, C., Williams, S.S., Steinbrecher, K.A., Warner, B.W., Cohen, M.B., Aronow, B.J. Novel genes and functional relationships in the adult mouse gastrointestinal tract identified by microarray analysis. *Gastroenterology* 122(5): 1467–1482, 2002.

28. Rost, D., Mahner, S., Sugiyama, Y., Stremmel, W. Expression and localization of the multidrug resistance-associated protein 3 in rat small and large intestine. *Am J Physiol Gastrointest Liver Physiol* 282(4): G720–726, 2002.

29. Mutch, D.M., Berger, A., Mansourian, R., Rytz, A., Roberts, M.A. Identifying differentially expressed genes from microarray data with the limit fold change (LFC) model. In Hardiman, G, Ed., *Microarrays Methods and Applications: Nuts and Bolts*, DNA Press, Eagleville, PA, 2003, pp. 193–204.

30. Kapushesky, M., Kemmeren, P., Culhane, A.C., Durinck, S., Ihmels, J., Korner, C., Kull, M., Torrente, A., Sarkans, U., Vilo, J., Brazma, A. Expression Profiler: next generation — an online platform for analysis of microarray data. *Nucl Acids Res* 32: W465–470, 2004.

31. Tadesse, M.G., Ibrahim, J.G. A Bayesian hierarchical model for the analysis of Affymetrix arrays. *Ann N Y Acad Sci* 1020: 41–48, 2004.

32. He, W. A spline function approach for detecting differentially expressed genes in microarray data analysis. *Bioinformatics* 20(17): 2954–2963, 2004.

33. Breitling, R., Amtmann, A., Herzyk, P. Iterative Group Analysis (iGA): a simple tool to enhance sensitivity and facilitate interpretation of microarray experiments. *BMC Bioinformatics* 5(1): 34, 2004.

34. Wu, F.X., Zhang, W.J., Kusalik, A.J. Modeling gene expression from microarray expression data with state-space equations. *Pac Symp Biocomput* 9: 581–592, 2004.

35. Zhang, B., Schmoyer, D., Kirov, S., Snoddy, J. GOTree Machine (GOTM): a Web-based platform for interpreting sets of interesting genes using gene ontology hierarchies. *BMC Bioinformatics* 5(1): 16, 2004.

36. Mansson, R., Tsapogas, P., Akerlund, M., Lagergren, A., Gisler, R., Sigvardsson, M. Pearson correlation analysis of microarray data allows for the identification of genetic targets for early B-cell factor. *J Biol Chem* 279(17): 17905–17913, 2004.

37. Datta, S., Satten, G.A., Benos, D.J., Xia, J., Heslin, M.J. An empirical Bayes adjustment to increase the sensitivity of detecting differentially expressed genes in microarray experiments. *Bioinformatics* 20(2): 235–242, 2004.

38. Wang, S., Ethier, S. A generalized likelihood ratio test to identify differentially expressed genes from microarray data. *Bioinformatics* 20(1): 100–104, 2004.

39. Draghici, S., Kulaeva, O., Hoff, B., Petrov, A., Shams, S., Tainsky, M.A. Noise sampling method: an ANOVA approach allowing robust selection of differentially regulated genes measured by DNA microarrays. *Bioinformatics* 19(11): 1348–1359, 2003.

40. Anderle, P., Duval, M., Draghici, S., Kuklin, A., Littlejohn, T.G., Medrano, J.F., Vilanova, D., Roberts, M.A. Gene expression databases and data mining. *Biotechniques* Mar (Suppl.): 36–44, 2003.

41. Baldi, P., Long, A.D. A Bayesian framework for the analysis of microarray expression data: regularized t-test and statistical inferences of gene changes. *Bioinformatics* 17(6): 509–519, 2001.

42. Cuim, X., Churchill, G.A. Statistical tests for differential expression in cDNA microarray experiments. *Genome Biol* 4(4): 210, 2003.

43. Baggerly, K.A., Coombes, K.R., Hess, K.R., Stivers, D.N., Abruzzo, L.V., Zhang, W. Identifying differentially expressed genes in cDNA microarray experiments. *J Comput Biol* 8(6): 639–659, 2001.

44. Claverie, J.M. Computational methods for the identification of differential and coordinated gene expression. *Hum Mol Genet* 8(10): 1821–1832, 1999.

45. Hess, K.R., Zhang, W., Baggerly, K.A., Stivers, D.N., Coombes, K.R. Microarrays: handling the deluge of data and extracting reliable information. *Trends Biotechnol* 19(11): 463–468, 2001.

46. Jain, N., Thatte, J., Braciale, T., Ley, K., O'Connell, M., Lee, J.K. Local-pooled-error test for identifying differentially expressed genes with a small number of replicated microarrays. *Bioinformatics* 19(15): 1945–1951, 2003.

47. Kamb, A., Ramaswami, M. A simple method for statistical analysis of intensity differences in microarray-derived gene expression data. *BMC Biotechnol* 1(1): 8, 2001.

48. Lin, H., Stoehr, J.P., Nadler, S.T., Schueler, K.M., Yandell, B.S., Attie, A.D. Adaptive gene picking with microarray data: detecting important low abundance signals. In *The Analysis of Gene Expression Data: Methods and Software*. Springer-Verlag, New York, 2003.

49. Nadon, R., Shi, P., Skandalis, A., Woody, E., Hubschle, H., Susko, E., Ramm, P., Rghei, N. Statistical interference methods for gene expression arrays, *Proceedings of SPIE, BIOS 2001*, Microarrays. In *Optical Technologies and Informatics*, 2001, pp. 46–55.

50. Mutch, D.M., Berger, A., Mansourian, R., Rytz, A., Roberts, M.A. The limit fold change model: a practical approach for selecting differentially expressed genes from microarray data. *BMC Bioinformatics* 3(1): 17, 2002.

51. Dudoit, S., Yang, Y.H., Speed, T.P., Callow, M.J. Statistical methods for identifying differentially expressed genes in replicated cDNA microarray experiments. *Statistica Sinica* 12(1):111–139, 2002.

52. Mutch, D.M., Anderle, P., Fiaux, M., Mansourian, R., Vidal, K., Wahli, W., Williamson, G., Roberts, M.A. Regional variations in ABC transporter expression along the mouse intestinal tract. *Physiol Genomics* 17(1): 11–20, 2004.

53. Wurmbach, E., Yuen, T., Ebersole, B.J., Sealfon, S.C. Gonadotropin releasing hormone receptor-coupled gene network organization. *J Biol Chem* 276(50): 47195–47201, 2001.

54. Kendziorski, C.M., Zhang, Y., Lan, H., Attie, A.D. The efficiency of pooling mRNA in microarray experiments. *Biostatistics* 4(3): 465–477, 2003.

55. Yuen, T., Wurmbach, E., Pfeffer, R.L., Ebersole, B.J., Sealfon, S.C. Accuracy and calibration of commercial oligonucleotide and custom cDNA microarrays. *Nucl Acids Res* 30(10): e48, 2002.

56. Mansourian, R., Mutch, D.M., Antille, N., Aubert, J., Fogel, P., Le Goff, J.M., Moulin, J., Petrov, A., Rytz, A., Voegel, J.J., Roberts, M.A. The global error assessment (GEA) model for the selection of differentially expressed genes in microarray data. *Bioinformatics*. 20(16): 2726–2737, 2004.

57. Mecham, B.H., Klus, G.T., Strovel, J., Augustus, M., Byrne, D., Bozso, P., Wetmore, D.Z., Mariani, T.J., Kohane, I.S., Szallasi, Z. Sequence-matched probes produce increased cross-platform consistency and more reproducible biological results in microarray-based gene expression measurements. *Nucl. Acids Res* 32(9): e74, 2004.

58. Mah, N., Thelin, A., Lu, T., Nikolaus, S., Kuhbacher, T., Gurbuz, Y., Eickhoff, H., Kloppel, G., Lehrach, H., Mellgard, B., Costello, C.M., Schreiber, S. A comparison of oligonucleotide and cDNA-based microarray systems. *Physiol Genomics* 16(3): 361–370, 2004.

59. Munoz, E.T., Bogard, L.D., Deem, M.W. Microarray and EST database estimates of mRNA expression levels differ: the protein length versus expression curve for C. elegans. *BMC Genomics* 5(1): 30, 2004.

60. Gygi, S.P., Rochon, Y., Franza, B.R., Aebersold, R. Correlation between protein and mRNA abundance in yeast. *Mol Cell Biol* 19(3): 1720–1730, 1999.

61. Chianale, J., Vollrath, V., Wielandt, A.M., Miranda, S., Gonzalez, R., Fresno, A.M., Quintana, C., Gonzalez, S., Andrade, L., Guzman, S. Differences between nuclear run-off and mRNA levels for multidrug resistance gene expression in the cephalocaudal axis of the mouse intestine. *Biochim Biophys Acta* 1264(3): 369–376, 1995.

62. Stephens, R.H., O'Neill, C.A., Warhurst, A., Carlson, G.L., Rowland, M., Warhurst, G. Kinetic profiling of P-glycoprotein-mediated drug efflux in rat and human intestinal epithelia. *J Pharmacol Exp Ther* 296(2): 584–591, 2001.

63. Kobayashi, A., Hori, S., Suita, N., Maeda, M. Gene organization of human transporter associated with antigen processing-like (TAPL, ABCB9): analysis of alternative splicing variants and promoter activity. *Biochem Biophys Res Commun* 309(4): 815–822, 2003.

64. Zhang, F., Zhang, W., Liu, L., Fisher, C.L., Hui, D., Childs, S., Dorovini-Zis, K., Ling, V. Characterization of ABCB9, an ATP binding cassette protein associated with lysosomes. *J Biol Chem* 275(30): 23287–23294, 2000.

65. Homolya, L., Varadi, A., Sarkadi, B. Multidrug resistance-associated proteins: export pumps for conjugates with glutathione, glucuronate or sulfate. *Biofactors* 17(1-4): 103–114, 2003.

66. Tsukada, C., Miyaji, C., Kawamura, H., Miyakawa, R., Yokoyama, H., Ishimoto, Y., Miyazawa, S., Watanabe, H., Abo, T. Characterization of extrathymic CD8 alpha beta T cells in the liver and intestine in TAP-1 deficient mice. *Immunology* 109(3): 343–350, 2003.

67. Lankat-Buttgereit, B., Tampe, R. The transporter associated with antigen processing: function and implications in human diseases. *Physiol Rev* 82(1): 187–204, 2002.

68. Igel, M., Giesa, U., Lutjohann, D., von Bergmann, K. Comparison of the intestinal uptake of cholesterol, plant sterols, and stanols in mice. *J Lipid Res* 44(3): 533–538, 2003.

5 Toxicogenomics in Drug Safety Evaluation: Bridging Drug Discovery and Development

Alex Y. Nie, Michael K. McMillian,
and Peter G. Lord

CONTENTS

INTRODUCTION

Biological responses to xenobiotics are frequently manifest at the transcriptional level. Hence, differential gene expression studies are highly applicable to both pure and applied toxicology. The sequencing of the human genome and in particular the sequencing of genomes of laboratory species along with the development of DNA microarrays has allowed for more comprehensive and rapid investigations of gene expression relevant to toxicology. Increasing knowledge of gene expression responses, gained using DNA microarray technology, also brings about the ability to screen chemicals and drugs for toxic potential, because it has been recognized that such responses can be predictive of toxicity by revealing early biological responses to xenobiotics [1,2]. Gene expression also indicates at the molecular-level changes to the biology of a cell or organ that may lead to or constitute the likely basis of a toxicity [3,4].

The term *toxicogenomics* has been applied to the use of genomics and transcriptional profiling in toxicology [3]. In the academic sector, toxicogenomics approaches are being used to understand toxic mechanisms of environmental chemicals and drugs [5–8]. In the chemical industry, these approaches are being used in screening for toxicity and for hazard identification [9,10]. The pharmaceutical sector has invested in toxicogenomics for the purposes of drug candidate selection (on the basis of predictive screening for toxicity), hazard identification, and risk assessment [11–14].

A variety of gene array platforms from commercial and in-house sources have been used, including nylon cDNA macroarrays [15–17], glass-based cDNA microarrays [18–21], and oligonucleotide chips [22,23]. Whole genome microarrays have found good use for gene hunting and hypothesis generation. More focused microarrays, particularly spotted arrays, are by nature better for screening purposes because of lower cost and lower data content.

Focused microarrays have been developed in several ways. In the early days of toxicogenomics, arrays were assembled from genes chosen from the literature as being relevant to specific toxic mechanisms (Clontechniques Vol XVI, January 2001, 24,25). For example, perturbations of biological processes such as proliferation and apoptosis are considered to be critical in chemical carcinogenesis [26–28]. Hence, genes implicated in these processes can be used to predict or examine carcinogenic responses. Stress response genes indicate exposure to xenobiotics and to some extent their expression can predict the outcome [29–33]. Changes in expression of genes for several enzyme families can be interpreted as predictive or indicative of toxicity. Induction of genes for drug-metabolizing enzymes and drug transporters gives a good indication of potential ADME characteristics [34]. Genes have additionally been selected empirically from experiments using well-characterized toxins. These have generally used gene-hunting techniques such as differential display [35,36] or suppression subtraction hybridization [37,38] but have also used whole genome arrays [34].

In the pharmaceutical industry, the interest in microarrays for toxicology has been driven mainly by the desire to better screen candidate drugs for their toxicity [14]. There are huge benefits to be made from resolving toxicity issues early in the drug development process both from the perspective of making cost savings and from accelerating the development program. Typically, microarrays are used in short-term *in vitro* or *in vivo* studies of drugs. These studies provide information for making decisions on progressing a drug through the development process. Drugs from the same compound series may be ranked for toxic potency if they manifest the same toxicity. Alternatively candidates may be selected or deselected on the basis of prediction of a potential toxicity.

Applying microarrays in toxicology has not been simple. Changes in gene transcription represent a part of a dynamic response to xenobiotics; however, it is not practicable to monitor a complete time-course of the response, particularly with *in vivo* studies. In spite of this, the microarray approach has proven to be effective in identifying changes in gene expression associated with toxicity [19,21–23,39]. Following more in-depth study, defined links between specific transcriptional responses and subsequent toxicity will be determined, but for the time being it must be recognized that much of these data are not immediately interpretable.

The implementation of more screens for predicting toxicity in early phases of drug development has led to much debate about their relevance to conventional (regulatory) drug safety assessment. [11,40,41]. Genomic data have predictive value but, in general, do not constitute definitive endpoints for toxicity. One way of viewing such data is that they generate hypotheses that are then tested by conventional toxicity assays. Ideally, gene expression information can predict an outcome before it is pathologically manifest.

The challenges of using microarrays in toxicology are varied but are being addressed and overcome. Many of the technical issues are not specific to toxicological applications. Issues of data normalization, statistical processing, and data visualization are generic to microarray experiments [42,43]. The available data format standards lend themselves to toxicogenomics experiments with little need for adaptation. It is more in the capture of nongenomic experimental information where attention needs to be paid so that the genomic data can be properly evaluated in context with the biological data. There are a number of initiatives in establishing data standards for toxicogenomics (comprehensively reviewed by Mattes and colleagues; see Reference 44). Collectively, these initiatives should resolve the issues of data formatting raised by FDA in their recent draft guidance on submission of pharmacogenomic data to the regulatory agencies (http://www.fda.gov/cder/guidance/5900dft.doc). The ILSI/HESI Committee on the use of genomics in risk assessment has been in the unique position of providing shared learning, consensus building, and a debating forum to enable the toxicogenomic community to address the challenges of using gene microarrays [45]. A public database of toxicogenomic data was a high priority of the collaboration. With the help of this consortium of companies, academics, and regulatory agencies, the EBI has proposed the MIAME/TOX guidance (http://www.mged.org/Workgroups/MIAME/miame_1.1.html) for a toxicogenomics experiment database. The committee has also provided a focus for industry, academics, and regulators alike to gain an in-depth understanding of the complexities of each microarray platform and approaches to data analysis.

A continuing challenge is at the philosophical level. Making the distinction between predictive data vs. definitive data is proving to be problematic in the regulatory setting of drug risk assessment. Much of the concern about using gene expression data in a risk assessment revolves around the difficulty in accepting that such data are mostly nondefinitive. The data provide mechanistic information that is used to develop hypotheses that are testable by further experimentation. It is useful to compare transcriptional analysis with traditional toxicological analysis. Microarray experiments generate much more data than the more conventional toxicology experiments. For example, clinical chemistry analysis may comprise around 20 different measurements per animal per time-point in an *in vivo* study [46,47]. A microarray experiment may comprise 30,000 different measurements per RNA sample, of which several hundred may show statistically significant differences between control and treated tissue. Additionally, clinical chemistry measurements have a long history of use in drug risk assessment, whereas microarray data interpretation in a biological context is at a very early stage of understanding. Clinical chemistry effects are truly endpoints in that they reflect events such as cell damage, e.g., ALT [48]. Genomic measurements often reflect the beginnings of a toxic response and therefore represent starting points rather than endpoints.

Engagement of industry with regulators to understand both the power and limitations of multiple gene expression analysis will help to ensure that microarray technology is used to its greatest advantage in drug development.

TOXICOGENOMICS EXPERIMENTAL DESIGN

Although much has been written about experimental design for microarray experiments [49–54], questions of biological interest and costs generally determine the approaches taken. A common biological question with candidate drugs is how they may induce liver toxicities. Most types of liver toxicities are detected only after repeated administration of drug candidate compound; some types are relatively benign and reversible, others are rodent-specific, and still others are likely to cause serious problems in man. These latter compounds must be detected and their potential toxicity understood to cease their continued development into pharmaceuticals, whereas development can continue if the toxicity of compounds is rodent-specific or if the therapeutic benefits of mildly toxic compounds outweigh their potential safety risk. At present, decisions on the continued development of drug candidates that cause liver toxicities in rodents or dogs are largely made based on histopathology and clinical chemistry results. Toxicogenomics offers a number of advantages when used in conjunction with conventional toxicological methods. Transcriptional responses of hepatic parenchymal cells in response to treatment with drug compound are largely adaptive (dying or dead cells stop making mRNA), and frequently give clues to developing pathology long before histological changes occur. Thus, changes in expression of important genes can be used as "predictor biomarkers" rather than the "indicator biomarkers" of conventional toxicology where damage is assessed. A feature of prediction is that few if any gene signatures are 100% accurate at predicting damage. Incidentally, mRNA changes usually reflect stress responses rather than damage responses, and often toxicity is prevented by adaptive changes, so toxicogenomics data are best used to point out likely problems that can subsequently be examined with conventional methods.

Another major advantage of toxicogenomics is that just knowing the genes and pathways affected by a hepatotoxic compound frequently increases understanding of relevant toxic mechanisms. PPARα agonists, which are peroxisome proliferators and nongenotoxic carcinogens in rodent liver, provide a good example of a class of toxicants amenable to investigation through transcriptional profiling. These compounds exert their pharmacological effects by activating a transcription factor, and most of their toxicities appear to reflect exaggerated pharmacology; for example, the more efficacious in activating PPARα, the more likely a compound is to induce tumors [55]. Many enzyme inductions by peroxisome proliferators, particularly those enzymes involved in β-oxidation of fatty acids, were characterized long before PPARα was identified as the transcription factor responsible. Data obtained using microarrays are largely confirmatory of such historic data, but additional relationships with other genes and pathways become apparent, particularly when comparing a large number of gene responses to a large number of diverse hepatotoxicants. For example, many hepatic genes have been found to be oppositely regulated by PPARα agonists and macrophage activators [21]. These differential transcriptional effects probably contribute to anti-inflammatory effects observed with many PPARα agonists [56].

A single gene change can sometimes be an adequate predictor for a type of hepatotoxicity, but more frequently a combination of gene changes is necessary to establish a signature for a particular type of hepatotoxicity. A great deal of time, effort, and money have been spent characterizing gene signatures for a number of common hepatotoxicities—peroxisome proliferation, macrophage activator-induced necrosis and fibrosis, oxidative stress and reactive-metabolite-induced necrosis, phospholipidosis, microvesicular and macrovesicular steatosis, cholestasis, bile duct damage, venoocclusion, genotoxic and nongenotoxic carcinogenesis, and hepatomegaly—with varying degrees of success. Hepatotoxicants are seldom "pure," and a single compound often produces several types of toxicity. For example, if enough structurally distinct PPARα agonists are compared, distinctions can be made between class-specific and compound-specific gene responses. For instance, several strong PPARα agonists repress CYP8B1 (sterol 12a-hydroxylase), a rate-limiting enzyme in bile acid synthesis and one of several gene changes associated with intrahepatic cholestasis [57,58], but this mRNA effect and cholestasis are not observed with all PPARα agonists. Similarly, few hepatotoxicities are "pure." Severe bile duct damage induces extrahepatic cholestasis, and many necrosis inducers induce steatosis at low dose or early time points. To be most useful, selection of gene signature sets (and eventually the best marker genes for routine screening) must be reasonably specific for the outcome (particularly as to reversible or nonreversible); genes that elucidate rodent-specific or independent outcomes can be particularly useful (as in comparing mouse and human PPARa transcription factor effects [55]).

Our approach to identifying gene signatures has been to take as many well-characterized hepatotoxicants as possible, administer to three male rats per group as a single maximal tolerated dose, snap-freeze liver at necropsy at 24 h, prepare RNA, label RNA as probe, hybridize to cDNA microarrays (with four replicate spots per liver sample), normalize microarray data, and then analyze data for signature genes. The end result is effectively cherry-picking the best genes for routine screening by PCR. Supervised clustering of a large number of different paradigm compounds in a given toxicity class lessens the chance of spurious gene correlations frequently observed using just one or two paradigm compounds as a training set. A group of three animals represents the minimum biological replicates for interpreting the effect of a paradigm hepatotoxicant (five is better, but almost doubles the cost). A single high dose of a paradigm hepatotoxicant generally provides useful gene changes (multiple doses are again preferable but expensive). In most cases, gene responses to low doses are difficult to distinguish from controls. The obvious exceptions are compounds where the toxicities reflect exaggerated pharmacology, for example, PPARα agonists, steroids, and other transcriptional effectors; low doses could separate toxic from pharmacological gene responses. The 24-h time period allows establishment of a broad, survey type of database where most hepatotoxicities can be predicted before they occur. Preparation of high-quality RNA is critical; enough RNA is prepared for repeated uses of the database samples, and the archive is only as good as the RNA stored. Microarray preparation and RNA labeling and hybridization have improved markedly in the last few years. Four replicate spots generally provide sufficient data per gene per sample. The normalization and analysis involved in establishing gene signatures for hepatotoxicities is a cumulative, creative process that is improving and incorporating weaker (often low abundance mRNA) but important gene changes. Although results derived from unbiased data analysis might seem preferable, supervised clustering of paradigm compounds (and some genes) makes use of a solid literature on hepatotoxicants, and this knowledge-based approach allows subtle but important distinctions to be made that might be missed (or overfit) by a purely statistical approach.

It is important to note that the above experimental design applies only to the 100 or so well-characterized paradigm hepatotoxicants that populate our database. Toxicogenomic experiments on in-house drug candidates are generally run at multiple doses (but again for 24 h in 3 male rats per group). Parallel studies provide exposure data for these compounds (absence of toxicity often reflects poor bioavailability) and 5 to 14 d later rats are necropsied, and histopathology and clinical chemistry results are obtained (which provide a reality check for 24-h gene signature predictions). Frequently, a well-investigated compound of the same class is compared at the same time. Thus from a toxicogenomics perspective, we can rank order in-house drug candidates with a known therapeutic.

CASE STUDY: PPARα AGONIST EXAMPLE

To illustrate the data analysis process and demonstrate how toxicogenomics helps with drug safety assessment but to obviate the need for complicated figures and tables, we have selected a small portion of the database as a sample data set that includes 7 PPARα agonists and 13 reference compounds representing 8 other toxicity classes (Table 5.1). One proprietary compound (Compound X, designed as a novel PPARα activator) has been used for testing.

Due to the large number of samples in the study, animal handling and dosing, RNA preparation and labeling, and array manufacturing and hybridization had to be split into batches. These extraneous factors may and have been observed in earlier experiments to cause variations. Therefore, the raw microarray data need to be preprocessed to correct for such effects and biases. The preprocessing starts with a base two logarithmic transformation, a Spline normalization among four technical replicates. An averaging of them into one chip, a sorted Spline normalization among chips, run on the same day, followed by a linear normalization and background flooring steps, and finally log ratios are generated by subtracting the geometric mean of the vehicle controls on the same day from each chip [21]. This normalization procedure places all of the chips in the study to the same scale to compare (Figure 5.1). It is important to note that after normalizing and summarizing, the 444 raw data chips are condensed into only 111 samples.

Data analysis is facilitated with good visualization tools such as clustering algorithms and heat maps [59] and multidimensional projections of principle component analysis (PCA). Figure 5.2 shows that PPARα agonists are not separable from vehicle controls or the reference compounds using all genes on the chip. Gene selection is thus required and is important in removing genes that do not carry relevant information while keeping ones that respond specifically and uniformly to a class of treatments (PPARα agonists in the present example).

In a typical toxicogenomic study, a group of treatments from a certain toxicological class are compared to ones that are not in that class, and a list of genes and their associated

TABLE 5.1
PPARα Agonists and the Reference Compounds

Compound	DOSE (mg/kg)	PPARα activation	Carcinogenicity	Cholestasis	Fibrosis	Necrosis	Necrosis–Macrophage Activation	Necrosis–Oxidative Stress	NSAID	Phospholipidosis	Steatosis	Venoocclusion	Training/ Testing
Benzafibrate	500	X											Training
Benzbromarone	200	X											Training
Bromobenzene	900									X			Testing
Clofibrate	600	X		X		X		X			X		Training
DiEH phthalate	1000	X											Training
Diflunisal	750			X		X			X		X		Testing
ErythroMC Estolate	1500			X						X	X		Training
Ethinyl estradiol	500		X	X		X		X			X		Training
Fenbufen	250					X			X				Testing
Flurbiprofen	40					X			X				Training
LPS	3		X		X	X	X						Training
Perfluoro decanoate	50	X		X									Testing
Perfluoro octanoate	150	X		X									Training
Perhexilene	2010									X			Training
Phalloidin	1			X									Testing
Phenobarbital	225		X										Training
Rifampin	600		X	X								X	Training
Sulindac	400			X					X				Testing
Acrine	50					X						X	Training
WY14643	100	X		X									Testing
Vehicle control	0												Training/ Testing

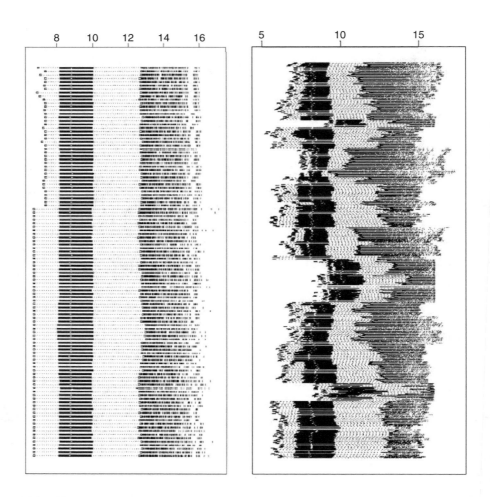

FIGURE 5.1 Boxplots of raw and normalized data sets. Raw data (on the right-hand side) show marked batch-to-batch differences. The normalization procedure removes most of these batch effects and keeps samples comparable across the database. There is a noticeable difference between the first third of the samples (on the left-hand side) and the rest, reflecting a microarray platform change. To take advantage of the large number of samples in each batch, the normalization is done with all of the samples in the toxicogenomics database instead of the selected PPARα agonist sample data set.

mathematical rules can be determined to discriminate them. Then the same gene list and discriminant rules can be used to classify new treatments. If a new treatment is classified as a close resembler to the treatments in the toxicological class, it is predicted as a toxic treatment in that class, otherwise it is not. It is generally believed that before the gene list can be used to test new treatment, it needs to be tested with blinded samples that are not previously used to generate the gene list and rules [60]. The needs for such an independent testing will be discussed in the following paragraphs.

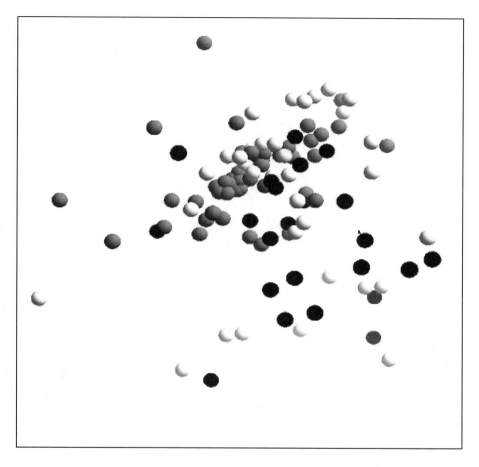

FIGURE 5.2 PCA plot with all of the genes on the chip. PPARα agonists (black) are not visually separable from vehicle controls (grey) or other reference compounds (white).

There are a large number of discriminant methods available [61], and basically all of them can be applied in toxicogenomics. Just to name a few: k-nearest neighbor, linear and quadratic discriminant analysis, and linear discriminant analysis following PCA. In this chapter, we are using PCA with the largest two principal components followed by linear discriminant analysis. However, instead of drawing a line in the two-dimensional PCA space to separate treatment groups, we will just plot them on a figure and visualize the separabilities. If samples are visually separable on the PCA plot, they are separable by discriminant analysis with the principal components that are used to generate the plot, but even if they are not visually separable, there still is a chance that the samples can be separated using more principal components.

In the following section, six typical gene selection approaches are used on this data set and compared for their performances. Prior to gene selection, the PPARα agonist sample data set is partitioned into training and testing subsets. All the compounds are sorted alphabetically, and every third compound (except WY14643,

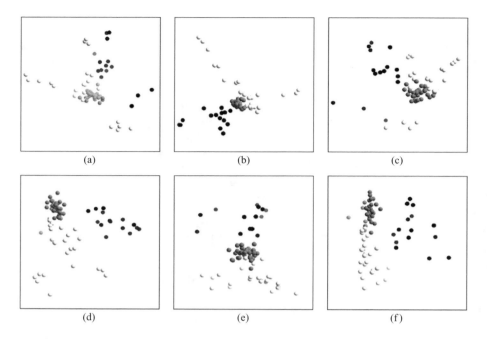

FIGURE 5.3 PCA plots with selected genes. Only the training samples are plotted. The color scheme is the same as in Figure 5.2.

which is the second in group) is selected for testing. The vehicle controls are also split so that one third of them are for testing too. This partitioning results in a 73-sample training and a 38-sample testing data sets. Gene selection is solely done on the 73 training samples, and a PCA plot based on the selected genes for each approach is generated in Figure 5.3.

Unsupervised gene selection procedures take no treatment classification into consideration while selecting genes and therefore are least likely to be biased by treatment classification. The most intuitive way of selecting genes using an unsupervised procedure is to use a cutoff fold change value, i.e., any gene over that decreased or increased level is chosen. Using a fourfold cutoff value yields 205 genes from the 73 training samples. The two-dimensional PCA plot in Figure 5.3a shows that with this simple approach, PPARα agonists can be well separated from vehicle controls and most of the reference samples.

Ranking all of the genes by their ability to separate different treatments and keep the same treatments together is the second unsupervised gene selection approach. This approach selects genes with small variance within treatments and great variance between treatments; ANOVA is normally used for this purpose. The top 50 genes from the ANOVA test are used to generate the PCA plot in Figure 5.3b.

Unlike the unsupervised approaches, supervised gene selection approaches are largely dependent on treatment classification, and genes that are coregulated by a specific group of treatments are selected. The strength of supervised gene selection is specificity to a certain toxicological end point, such as peroxisome proliferation in the present example, and this approach is therefore more likely to select genes that best separate PPARα-agonist-treated samples from other samples. A major limitation of such supervised approaches, especially when complicated discriminant functions are used on such large sets of gene expression data, is that the "best" genes will always be selected, but these genes may only be predictive for the samples from which they are selected. Cross-validation is normally used to avoid such false findings [60]. The 38 testing samples in this sample data set will be used to cross-validate the genes selected using the 73 training samples.

Using a cutoff value, typically 1.5-, 2-, or 4-fold, to filter out genes that do not have big changes in the target treatment group (PPARα agonists here), has been widely used in literature. We use 4-fold as the cutoff value in the present example, and 93 genes pass the criteria (Figure 5.3c). However, genes that are biologically important but do not change by a large magnitude are missed by this approach. Ranking all genes by comparing the difference in between and within treatment class variance takes advantage of statistical power and is preferable to a simple fold change cutoff. Student's t-test is a powerful but simple method and is our fourth gene selection method in this study. Fifty genes that rank the highest in comparing all PPARα agonists with vehicle controls in the training set are selected by Student's t-test (Figure 5.3d). Similarly, 50 genes are selected by ANOVA that compares PPARα agonists to both vehicle controls and all the other reference compounds (Figure 5.3e); this yields good specificity to the peroxisome proliferator class, and the selected genes tend to show robust, coregulated responses. For PPARα agonists, enzymes involved in β-oxidation of fatty acids are coinduced and share common biochemical pathways [62,63].

The univariate gene selection approaches described so far evaluate each gene independently, and gene lists are compiled from the best individual genes. In contrast, a multivariate approach evaluates a group of genes at a time, until a best or close-to-best group of genes is found for separating the treatment class of interest (PPARα agonists) from other samples [64]. A multivariate gene selection approach takes advantage of the combinatory power of all genes in the list. In multivariate gene selection approaches, a large number of combinations of genes are searched, and the search can be exhaustive, i.e., every possible combination is evaluated and compared (which is practically unfeasible) or heuristically improved by algorithms such as forward selection, backward elimination, or genetic algorithm. At each step during the search, the performance of the gene set is computed as the posterior error from linear discriminant analysis. In forward selection, performance improves while more genes are added into the gene list at each step, and the search can be stopped either when perfect separation is achieved or the number of genes reaches a preset number. Forward selection was used to select 25 genes (Figure 5.3f). This number of genes was decided experimentally to balance the search time and performance of the gene lists. A characteristic of multivariate approaches to gene selection is selection against redundant (coregulated and covariant) gene changes; whereas these approaches select against genes overrepresented in a single biochemical pathway, multivariate gene selection is more likely to pick up other important genes that behave differently across the treatment groups.

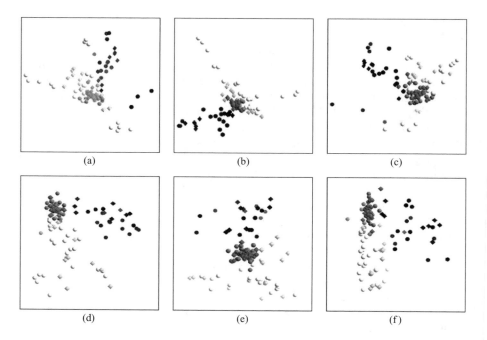

FIGURE 5.4 PCA plots with selected genes. Both training (round) and testing (diamond) samples are plotted. The color scheme is the same as in Figure 5.2.

The 38 testing samples left out from the gene selection approaches are now brought back to test the selected gene lists (Figure 5.4). The figure shows that the performance of the selected gene lists on the testing samples is comparable to that on the training samples, with a few testing non-PPARα agonists, frequently "misclassified" as PPARα agonists. These "misclassified" samples fenbufen and diflunisal are NSAIDs (nonsteroidal anti-inflammatory drugs); many if not most NSAIDs have been demonstrated to bind to and activate PPAR receptors, and several reportedly act as peroxisome proliferators at high doses [65–84]. Therefore, instead of invalidating the predictions, it is reasonable to regard both fenbufen and diflunisal as PPARα agonists (these two NSAIDs also clustered with PPARα agonists using the full database). In addition to all gene selection approaches predicting PPARα agonists and non-PPARα agonists quite accurately, it is noticed that supervised gene selection methods marginally outperformed unsupervised ones (Figure 5.4).

The selected gene lists are shown in Table 5.2. Genes are sorted by the number of their appearances, using the six different gene selection methods. Many of these genes, especially those on the top such as acyl-CoA oxidase and CYP4b1, are well known for their roles in the peroxisome proliferation process and have been annotated by us in a previous publication [21]. A few genes that repeatedly show up in a number of lists such as the rat androgen binding protein may provide fresh insight into the mechanisms of action of PPARα agonists, although their roles in peroxisome proliferation have not been well documented previously.

TABLE 5.2
Genes Selected by the Six Different Methods

Gene Accession	Gene Name	Gene Selection Method					
		1	2	3	4	5	6
NM_017340	Acyl-CoA oxidase	X	X	X	X	X	
NM_013214	Brain acyl-CoA hydrolase	X	X	X	X	X	
NM_016999	Cytochrome P450, subfamily 4B, polypeptide 1	X	X	X	X	X	
NM_031315	Cytosolic acyl-CoA thioesterase 1	X	X	X	X	X	
Gene 07	Proprietary gene	X	X	X		X	X
NM_031561	cd36 Antigen	X		X	X	X	
BG671569	EST	X		X	X		X
Gene 05	Proprietary gene	X	X	X	X		
Gene 11	Proprietary gene	X	X			X	X
M38759	Rat androgen binding protein (ABP) mRNA, complete cds	X	X	X		X	
NM_022407	Aldehyde dehydrogenase family 1, member A1	X	X	X			
NM_022298	Alpha-tubulin	X		X	X		
NM_031703	Aquaporin 3	X		X		X	
M33936	Cytochrome P450 4A3	X		X		X	
NM_017156	Cytochrome P450, 2b19	X	X	X			
NM_022936	Cytosolic epoxide hydrolase	X		X			X
NM_017306	Dodecenoyl-coenzyme A delta isomerase	X		X		X	
AA817759	EST	X	X	X			
BE110688	EST	X		X		X	
NM_031589	Glucose-6-phosphatase, transport protein 1	X		X	X		
NM_017080	Hydroxysteroid 11-beta dehydrogenase 1	X		X		X	
NM_013122	Insulin-like growth factor binding protein 2	X		X	X		
M11794	Metallothionein	X		X		X	
J00696	Orosomucoid 1	X	X	X			
Gene 10	Proprietary gene				X	X	X
Gene 16	Proprietary gene				X	X	X
Gene 29	Proprietary gene				X	X	X
NM_031531	Serine protease inhibitor	X	X	X			
AF157026	Solute carrier family 34 (sodium phosphate), member 2	X		X		X	
NM_006082	Tubulin, alpha, ubiquitous	X	X	X			
NM_006082	Tubulin, alpha, ubiquitous	X	X	X			
NM_006082	Tubulin, alpha, ubiquitous	X	X	X			
NM_006082	Tubulin, alpha, ubiquitous	X	X	X			
M89902	3-Hydroxybutyrate dehydrogenase (heart, mitochondrial)	X		X			
NM_017075	Acetyl-Coenzyme A acetyltransferase 1				X	X	
X05341	Acetyl-Coenzyme A acyltransferase 2 (mitochondrial 3-oxoacyl-Coenzyme A thiolase)				X	X	

(continued)

TABLE 5.2 (Continued)
Genes Selected by the Six Different Methods

Gene Accession	Gene Name	1	2	3	4	5	6
NM_016990	Adducin 1, alpha	X			X		
NM_013218	Adenylate kinase 3	X		X			
NM_013215	Aflatoxin B1 aldehyde reductase	X	X				
NM_012488	Alpha-2-macroglobulin	X	X				
NM_022298	Alpha-tubulin	X					X
NM_012824	Apolipoprotein C-I	X		X			
NM_021577	Argininosuccinate lyase	X		X			
NM_013113	ATPase Na+/K+ transporting beta 1 polypeptide	X		X			
NM_012504	ATPase, Na+K+ transporting, alpha 1	X		X			
NM_030850	Betaine-homocysteine methyltransferase	X		X			
AB010632	Carboxylesterase 2 (intestine, liver)	X		X			
NM_012532	Ceruloplasmin	X			X		
NM_017177	Choline/ethanolamine kinase					X	X
NM_016994	Complement component 3	X		X			
NM_016995	Complement component 4 binding protein, beta	X		X			
NM_022536	Cyclophilin B	X		X			
U46118	Cytochrome P450 3A9	X		X			
M35266	Cytosolic cysteine dioxygenase 1	X		X			
NM_031853	Diazepam binding inhibitor				X	X	
NM_022594	Enoyl coenzyme A hydratase 1	X				X	
NM_012844	Epoxide hydrolase 1	X	X				
AA848338	EST					X	X
AA858661	EST				X	X	
AA945076	EST	X		X			
AA945149	EST	X		X			
AI013902	EST	X		X			
AI059602	EST	X		X			
AW915938	EST	X		X			
AW918678	EST	X		X			
BG667982	EST	X		X			
BI285007	EST	X	X				
M20629	Esterase 2	X		X			
NM_001402	Eukaryotic translation elongation factor 1 alpha 1	X		X			
X86561	Fibrinogen, alpha polypeptide	X		X			
NM_012792	Flavin containing monooxygenase 1	X		X			
NM_017251	Gap junction membrane channel protein beta 1	X		X			
M96674	Glucagon receptor				X	X	
NM_031580	Glucose regulated protein, 58 kDa				X		X
NM_013098	Glucose-6-phosphatase, catalytic	X		X			
NM_013096	Hemoglobin, alpha 1	X		X			
NM_012964	Hyaluronan mediated motility receptor	X		X			

TABLE 5.2 (Continued)
Genes Selected by the Six Different Methods

Gene Accession	Gene Name	1	2	3	4	5	6
NM_012964	Hyaluronan mediated motility receptor	X		X			
NM_032082	Hydroxyacid oxidase (glycolate oxidase) 3	X		X			
NM_053329	Insulin-like growth factor binding protein, acid labile subunit	X		X			
NM_019242	Interferon-related developmental regulator 1	X		X			
NM_017321	Iron-responsive element-binding protein				X	X	
NM_012608	Membrane metallo endopeptidase	X		X			
X60822	Methionine adenosyltransferase I, alpha	X		X			
NM_017028	Myxovirus (influenza virus) resistance 2	X		X			
AF014503	Nuclear protein 1	X		X			
NM_022381	Proliferating cell nuclear antigen	X	X				
NM_012998	Prolyl 4-hydroxylase, beta polypeptide				X	X	
Gene 01	Proprietary gene	X					X
Gene 03	Proprietary gene	X		X			
Gene 04	Proprietary gene	X		X			
Gene 06	Proprietary gene	X		X			
Gene 08	Proprietary gene	X		X			
Gene 15	Proprietary gene	X	X				
Gene 17	Proprietary gene				X	X	
Gene 18	Proprietary gene	X		X			
Gene 20	Proprietary gene	X				X	
Gene 22	Proprietary gene					X	X
Gene 31	Proprietary gene	X		X			
Gene 33	Proprietary gene	X	X				
Gene 35	Proprietary gene	X		X			
NM_013065	Protein phosphatase 1, catalytic subunit, beta isoform	X		X			
NM_017039	Protein phosphatase 2a, catalytic subunit, alpha isoform	X		X			
NM_019140	Protein tyrosine phosphatase, receptor type, D	X		X			
AF306457	RAN, member RAS oncogene family	X			X		
NM_012695	Rat senescence marker protein 2A gene, exons 1 and 2	X		X			
NM_012695	Rat senescence marker protein 2A gene, exons 1 and 2	X		X			
M91235	Rat VL30 element mRNA	X		X			
NM_022514	Ribosomal protein L27	X			X		
X82669	RT1 class Ib gene(Aw2)	X		X			
NM_012656	Secreted acidic cysteine rich glycoprotein				X		X
U58857	Serine (or cysteine) proteinase inhibitor, clade B, member 5	X		X			

(*continued*)

TABLE 5.2 (Continued)
Genes Selected by the Six Different Methods

Gene Accession	Gene Name	Gene Selection Method					
		1	2	3	4	5	6
M83143	Sialyltransferase 1				X	X	
NM_019269	Solute carrier family 22, member 5				X	X	
AF157026	Solute carrier family 34 (sodium phosphate), member 2					X	X
AF249673	Solute carrier family 38, member 2	X		X			
U19485	Spp-24 precursor				X	X	
J02585	Stearoyl-Coenzyme A desaturase 1	X		X			
D50559	Sterol-C4-methyl oxidase-like	X				X	
NM_031834	Sulfotransferase family 1A, phenol-preferring, member 1	X	X				
NM_031834	Sulfotransferase family 1A, phenol-preferring, member 1	X		X			
L31883	Tissue inhibitor of metalloproteinase 1	X	X				
NM_012681	Transthyretin	X		X			
NM_006082	Tubulin, alpha, ubiquitous	X		X			
NM_006082	Tubulin, alpha, ubiquitous	X	X				
NM_006082	Tubulin, alpha, ubiquitous	X		X			
NM_006082	Tubulin, alpha, ubiquitous	X		X			
NM_006082	Tubulin, alpha, ubiquitous	X		X			
NM_006082	Tubulin, alpha, ubiquitous	X		X			
NM_006082	Tubulin, alpha, ubiquitous	X		X			
NM_006082	Tubulin, alpha, ubiquitous	X	X				
J02589	UDP glycosyltransferase 2 family, polypeptide B	X		X			
Y07744	UDP-N-acetylglucosamine-2-epimerase/ N-acetylmannosamine kinase	X		X			
M96548	Zinc finger protein 354A	X		X			
D00569	2,4-Dienoyl CoA reductase 1, mitochondrial					X	
NM_017268	3-Hydroxy-3-methylglutaryl-Coenzyme A synthase 1	X					
NM_017268	3-Hydroxy-3-methylglutaryl-Coenzyme A synthase 1	X					
M33648	3-Hydroxy-3-methylglutaryl-Coenzyme A synthase 2	X					
NM_016986	Acetyl-coenzyme A dehydrogenase, medium chain					X	
NM_012891	Acyl-Coenzyme A dehydrogenase, very long chain	X					
M64780	Agrin	X					
NM_031731	Aldehyde dehydrogenase family 3, subfamily A2					X	
NM_012899	Aminolevulinate, delta-, dehydratase	X					
NM_024484	Aminolevulinic acid synthase 1	X					
NM_024148	Apurinic/apyrimidinic endonuclease 1	X					

TABLE 5.2 (Continued)
Genes Selected by the Six Different Methods

Gene Accession	Gene Name	Gene Selection Method					
		1	2	3	4	5	6
NM_031839	Arachidonic acid epoxygenase	X					
NM_012911	Arrestin, beta 2	X					
J03753	ATPase, Ca++ transporting, plasma membrane 1	X					
AJ277881	ATP-binding cassette, sub-family C (CFTR/MRP), member 1	X					
NM_012828	Calcium channel, voltage-dependent, beta 3 subunit	X					
NM_013146	Caldesmon 1	X					
NM_022399	Calreticulin				X		
NM_031565	Carboxylesterase 1	X					
NM_031559	Carnitine palmitoyltransferase 1, liver	X					
NM_012930	Carnitine palmitoyltransferase 2					X	
U05341	Cell cycle protein p55CDC		X				
U52948	Complement component 9	X					
NM_032061	Contactin associated protein 1				X		
NM_031690	Crystallin, beta B3	X					
NM_017148	Cysteine and glycine-rich protein 1				X		
NM_031572	Cytochrome P450 15-beta gene	X					
NM_012541	Cytochrome P450, 1a2	X					
NM_031543	Cytochrome P450, subfamily 2E, polypeptide 1	X					
U64030	Deoxyuridinetriphosphatase (dUTPase)	X					
AY026512	Dynein-associated protein RKM23	X					
U31668	E2F transcription factor 5		X				
NM_012551	Early growth response 1	X					
Y07783	ER transmembrane protein Dri 42				X		
AA817749	EST	X					
AA817964	EST	X					
AA818163	EST		X				
AA818342	EST	X					
AA851329	EST		X				
AA858661	EST				X		
AA892234	EST	X					
AA899344	EST	X					
AA899344	EST	X					
AA900340	EST					X	
AA944161	EST	X					
AA945615	EST	X					
AA946508	EST	X					
AI228159	EST				X		
AI230381	EST	X					
AI233916	EST		X				
AI317842	EST		X				

(continued)

TABLE 5.2 (Continued)
Genes Selected by the Six Different Methods

Gene Accession	Gene Name	1	2	3	4	5	6
AI406939	EST						X
AW143388	EST	X					
BE095878	EST		X				
BE108882	EST	X					
BE109520	EST	X					
BF281192	EST					X	
BF284879	EST	X					
BF524965	EST		X				
BF555189	EST				X		
BG378729	EST						X
BG378729	EST	X					
BG663025	EST						X
BG672085	EST	X					
BI278268	EST				X		
BI278598	EST			X			
BI278612	EST			X			
BI278780	EST	X					
BI282736	EST	X					
BI284279	EST	X					
BI285402	EST						X
BI296125	EST				X		
BI303631	EST						X
NM_012947	Eukaryotic elongation factor-2 kinase	X					
NM_019356	Eukaryotic translation initiation factor 2, subunit 1 (alpha)	X					
NM_031840	Farensyl diphosphate synthase	X					
D90109	Fatty acid Coenzyme A ligase, long chain 2	X					
NM_017332	Fatty acid synthase					X	
U05675	Fibrinogen, beta polypeptide				X		
NM_019143	Fibronectin 1		X				
AF281018	Flap structure-specific endonuclease 1		X				
NM_022928	G protein-coupled receptor kinase 2, groucho gene related (Drosophila)					X	
NM_017006	Glucose-6-phosphate dehydrogenase	X					
NM_017305	Glutamate cysteine ligase, modifier subunit	X					
NM_017014	Glutathione S-transferase, mu 1	X					
X02904	Glutathione S-transferase, pi 2	X					
NM_017013	Glutathione-S-transferase, alpha type2	X					
M17412	Growth and transformation-dependent protein	X					
NM_012966	Heat shock 10 kDa protein 1	X					
NM_031970	Heat shock 27kDa protein 1	X					
NM_012580	Heme oxygenase 1	X					

TABLE 5.2 (Continued)
Genes Selected by the Six Different Methods

Gene Accession	Gene Name	1	2	3	4	5	6
BG668317	Hsp90 alpha	X					
D16478	Hydroxyacyl-Coenzyme A dehydrogenase/ 3-ketoacyl-Coenzyme A hiolase/ enoyl-Coenzyme A hydratase (trifunctional protein), alpha subunit	X					
D16479	Hydroxyacyl-Coenzyme A dehydrogenase/ 3-ketoacyl-Coenzyme A thiolase/ enoyl-Coenzyme A hydratase (trifunctional protein), beta subunit	X					
NM_031512	Interleukin 1 beta		X				
AF003835	Isopentenyl-diphosphate delta isomerase					X	
NM_012741	K-kininogen, differential splicing leads to HMW Kngk				X		
NM_012811	Milk fat globule-EGF factor 8 protein				X		
NM_017083	Myosin 5B				X		
NM_017000	NAD(P)H dehydrogenase, quinone 1	X					
NM_031130	Nuclear receptor subfamily 2, group F, member 1	X					
NM_031553	Nuclear transcription factor-Y beta						X
NM_022521	Ornithine aminotransferase	X					
NM_022694	P105 coactivator				X		
NM_031975	Parathymosin	X					
D88666	Phosphatidylserine-specific phospholipase A1		X				
M76591	Phosphoglycerate mutase 1	X					
NM_019237	Procollagen C-proteinase enhancer protein		X				
AF121217	Procollagen, type I, alpha 2					X	
NM_012929	Procollagen, type II, alpha 1	X					
NM_021766	Progesterone receptor membrane component 1	X					
Gene 02	Proprietary gene	X					
Gene 09	Proprietary gene	X					
Gene 12	Proprietary gene	X					
Gene 13	Proprietary gene					X	
Gene 14	Proprietary gene						X
Gene 19	Proprietary gene	X					
Gene 21	Proprietary gene						X
Gene 23	Proprietary gene	X					
Gene 24	Proprietary gene		X				
Gene 25	Proprietary gene		X				
Gene 26	Proprietary gene		X				
Gene 27	Proprietary gene		X				
Gene 28	Proprietary gene				X		
Gene 30	Proprietary gene	X					
Gene 32	Proprietary gene	X					

(*continued*)

TABLE 5.2 (Continued)
Genes Selected by the Six Different Methods

Gene Accession	Gene Name	Gene Selection Method					
		1	2	3	4	5	6
Gene 34	Proprietary gene	X					
Gene 36	Proprietary gene		X				
NM_033236	Proteasome (prosome, macropain) 26S subunit, ATPase 2						X
D32249	Protein carrying the RING-H2 sequence motif				X		
NM_017347	Protein kinase, mitogen activated 3 (extracellular-signal-regulated kinase 1, ERK1)	X					
NM_019249	Protein tyrosine phosphatase, receptor type, F				X		
X15800	Pyruvate kinase, muscle	X					
AA858574	R. norvegicus mRNA for lamin A	X					
L18948	S100 calcium-binding protein A9 (calgranulin B)	X					
NM_012656	Secreted acidic cysteine rich glycoprotein				X		
NM_012657	Serine protease inhibitor	X					
NM_017170	Serum amyloid P-component						X
NM_012751	Solute carrier family 2, member 4	X					
NM_017166	Stathmin 1	X					
NM_017051	Superoxide dismutase 2	X					
NM_019350	Synaptotagmin 5	X					
NM_013026	Syndecan 1				X		
NM_012887	Thymopoietin		X				
NM_006082	Tubulin, alpha, ubiquitous	X					
NM_006082	Tubulin, alpha, ubiquitous	X					
NM_006082	Tubulin, alpha, ubiquitous	X					
NM_006082	Tubulin, alpha, ubiquitous		X				
NM_006082	Tubulin, alpha, ubiquitous	X					
NM_006082	Tubulin, alpha, ubiquitous		X				
AB011679	Tubulin, beta 5	X					
NM_019376	Tyrosine 3-monooxygenase/tryptophan 5-monooxygenase activation protein, gamma polypeptide	X					
NM_031603	Tyrosine 3-monooxygenase/tryptophan 5-monooxygenase activation protein, epsilon polypeptide	X					
J02589	UDP glycosyltransferase 2 family, polypeptide B				X		
NM_031325	UDP-glucose dehydrogenase	X					
NM_031836	Vascular endothelial growth factor				X		
NM_012603	v-myc Avian myelocytomatosis viral oncogene homolog		X				

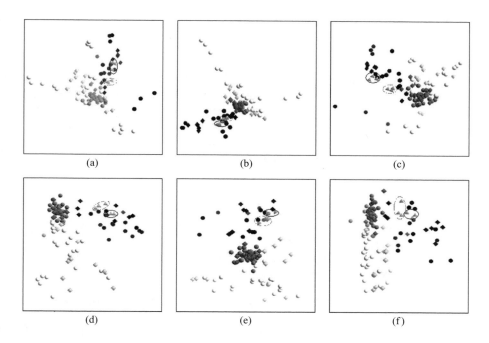

FIGURE 5.5 PCA plots with selected genes. Two difference doses, 25 mg/kg (circled with dotted line) and 250 mg/kg (circled with solid line), of Compound X (triangle) are plotted relative to the training (round) and testing (diamond) samples. Both doses of Compound X are colocated with the PPARα agonists. The color scheme is the same as in Figure 5.2.

Because the selected gene lists have been tested and validated by the testing samples, it naturally follows that these gene lists can be used to classify the real testing compound, Compound X. As shown in Figure 5.5, Compound X is predicted as a PPARα agonist with all six of the selected gene lists. Examining the clinical, chemical, and pathological reports that were made available after this prediction was done, it was clearly demonstrated that Compound X is truly a PPARα ligand and causes peroxisome proliferation.

The six gene selection methods differ in many ways. Unsupervised methods are least prone to classification bias but are less sensitive to weak toxicological responses. Statistical methods emphasize both the magnitude and the consistency of fold changes. Although univariate methods tend to select all genes responding to the toxicological treatments and thus can be useful in understanding the underlying mechanisms of such changes, multivariate methods ignore redundant genes and therefore are more specific in prediction. All of these methods, however, worked well on the PPARα sample data set and arrived at the same conclusion when testing Compound X. The important thing in analyzing toxicogenomic data is not to pre-decide what method to use, but to test and cross-validate them on real data to see which ones work best. After the methods are cross-validated, they tend to give the correct prediction, using new treatment samples.

SUMMARY AND CONCLUSIONS

The field of toxicogenomics has matured rapidly, and interpretation of toxicogenomic data is becoming much easier as available data analysis methods generally lead to the same conclusions when conducted properly. In addition, it is recognized that experimental design is very important in yielding results that can be interpreted in the best biological and toxicological context. It is becoming apparent that a relatively small proportion of genes is discriminatory for toxic responses. Differential gene expression experiments are emerging as valuable additions to the battery of conventional assays routinely used in drug risk assessment.

REFERENCES

1. Farr, S., Dunn, R.T., II. Concise review: gene expression applied to toxicology. *Toxicol Sci* 50: 1–9, 1999.
2. MacGregor, J.T., Farr, S., Tucker, J.D., Heddle, J.A., Tice, R.R., Turteltaub, K.W. New molecular endpoints and methods for routine toxicity testing. *Fundam Appl Toxicol* 26: 156–173, 1995.
3. Nuwaysir, E.F., Bittner, M., Trent, J., Barrett, J.C., Afshari, C.A. Microarrays and toxicology: the advent of toxicogenomics. *Mol Carcinog* 24: 153–159, 1999.
4. Rockett, J.C., Dix, D.J. Application of DNA arrays to toxicology. *Environ Health Perspect* 107: 681–685, 1999.
5. Adachi, T., Ono, Y., Koh, K.B., Takashima, K., Tainaka, H., Matsuno, Y., Nakagawa, S., Todaka, E., Sakurai, K., Fukata, H., Iguchi, T., Komiyama, M., Mori, C. Long-term alteration of gene expression without morphological change in testis after neonatal exposure to genistein in mice: toxicogenomic analysis using cDNA microarray. *Food Chem Toxicol* 42: 445–452, 2004.
6. Chhabra, R.S., Bucher, J.R., Wolfe, M., Portier, C. Toxicity characterization of environmental chemicals by the U.S. National Toxicology Program: an overview. *Int J Hyg Environ Health* 206: 437–445, 2003.
7. Mori, C., Komiyama, M., Adachi, T., Sakurai, K., Nishimura, D., Takashima, K., Todaka, E. Application of toxicogenomic analysis to risk assessment of delayed long-term effects of multiple chemicals, including endocrine disruptors in human fetuses. *EHP Toxicogenomics* 111: 7–13, 2003.
8. Walsh, P.J., Bookman, R.J., Zaias, J., Mayer, G.D., Abraham, W., Bourdelais, A.J., Baden, D.G. Toxicogenomic effects of marine brevetoxins in liver and brain of mouse. *Comp Biochem Physiol B Biochem Mol Biol* 136: 173–182, 2003.
9. Pennie, W.D., Kimber, I. Toxicogenomics; transcript profiling and potential application to chemical allergy. *Toxicol In Vitro* 16: 319–326, 2002.
10. Snape, J.R., Maund, S.J., Pickford, D.B., Hutchinson, T.H. Ecotoxicogenomics: the challenge of integrating genomics into aquatic and terrestrial ecotoxicology. *Aquat Toxicol* 67: 143–154, 2004.
11. Lord, P.G. Progress in applying genomics in drug development. *Toxicol Lett* 149: 371–375, 2004.
12. Peterson, R.L., Casciotti, L., Block, L., Goad, M.E., Tong, Z., Meehan, J.T., Jordan, R.A., Vinlove, M.P., Markiewicz, V.R., Weed, C.A., Dorner, A.J. Mechanistic toxicogenomic analysis of WAY-144122 administration in Sprague-Dawley rats. *Toxicol Appl Pharmacol* 196: 80–94, 2004.

13. Rodi, C.P., Bunch, R.T., Curtiss, S.W., Kier, L.D., Cabonce, M.A., Davila, J.C., Mitchell, M.D., Alden, C.L., Morris, D.L. Revolution through genomics in investigative and discovery toxicology. *Toxicol Pathol* 27: 107–110, 1999.

14. Ulrich, R., Friend, S.H. Toxicogenomics and drug discovery: will new technologies help us produce better drugs? *Nat Rev Drug Discov* 1: 84–88, 2002.

15. Baker, V.A., Harries, H.M., Waring, J.F., Duggan, C.M., Ni, H.A., Jolly, R.A., Yoon, L.W., De Souza, A.T., Schmid, J.E., Brown, R.H., Ulrich, R.G., Rockett, J.C. Clofibrate-induced gene expression changes in rat liver: a cross-laboratory analysis using membrane cDNA arrays. *Environ Health Perspect* 112: 428–438, 2004.

16. Crosby, L.M., Hyder, K.S., DeAngelo, A.B., Kepler, T.B., Gaskill, B., Benavides, G.R., Yoon, L., Morgan, K.T. Morphologic analysis correlates with gene expression changes in cultured F344 rat mesothelial cells. *Toxicol Appl Pharmacol* 169: 205–221, 2000.

17. Harries, H.M., Fletcher, S.T., Duggan, C.M., Baker, V.A. The use of genomics technology to investigate gene expression changes in cultured human liver cells. *Toxicol In Vitro* 15: 399–405, 2001.

18. Afshari, C.A., Nuwaysir, E.F., Barrett, J.C. Application of complementary DNA microarray technology to carcinogen identification, toxicology, and drug safety evaluation. *Cancer Res* 59: 4759–4760, 1999.

19. Burczynski, M.E., McMillian, M., Ciervo, J., Li, L., Parker, J.B., Dunn, R.T., II, Hicken, S., Farr, S., Johnson, M.D. Toxicogenomics-based discrimination of toxic mechanism in HepG2 human hepatoma cells. *Toxicol Sci* 58: 399–415, 2000.

20. Lord, P.G., Barne, K.A., Kramer, K., Bacon, E.J., Mooney, J., O'Brien, S., Bugelski, P.J. cDNA microarrays in investigative toxicology: a study of differential gene expression in compound induced cardiac hypertrophy. *Comments Toxicol* 7: 381–392, 2001.

21. McMillian, M., Nie, A.Y., Parker, J.B., Leone, A., Kemmerer, M., Bryant, S., Herlich, J., Yieh, L., Bittner, A., Liu, X., Wan, J., Johnson, M.D. Inverse gene expression patterns for macrophage activating hepatotoxicants and peroxisome proliferators in rat liver. *Biochem Pharmacol* 67: 2141–2165, 2004.

22. Waring, J.F., Ciurlionis, R., Jolly, R.A., Heindel, M., Ulrich, R.G. Microarray analysis of hepatotoxins in vitro reveals a correlation between gene expression profiles and mechanisms of toxicity. *Toxicol Lett* 120: 359–368, 2001.

23. Waring, J.F., Jolly, R.A., Ciurlionis, R., Lum, P.Y., Praestgaard, J.T., Morfitt, D.C., Buratto, B., Roberts, C., Schadt, E., Ulrich, R.G. Clustering of hepatotoxins based on mechanism of toxicity using gene expression profiles. *Toxicol Appl Pharmacol* 175: 28–42, 2001.

24. Nadadur, S.S., Schladweiler, M.C., Kodavanti, U.P. A pulmonary rat gene array for screening altered expression profiles in air pollutant-induced lung injury. *Inhal Toxicol* 12: 1239–1254, 2000.

25. Pennie, W.D., Woodyatt, N.J., Aldridge, T.C., Orphanides, G. Application of genomics to the definition of the molecular basis for toxicity. *Toxicol Lett* 120: 353–358, 2001.

26. Bisgaard, H.C., Nagy, P., Santoni-Rugiu, E., Thorgeirsson, S.S. Proliferation, apoptosis, and induction of hepatic transcription factors are characteristics of the early response of biliary epithelial (oval) cells to chemical carcinogens. *Hepatology* 23: 62–70, 1996.

27. Columbano, A., Endoh, T., Denda, A., Noguchi, O., Nakae, D., Hasegawa, K., Ledda-Columbano, G.M., Zedda, A.I., Konishi, Y. Effects of cell proliferation and cell death (apoptosis and necrosis) on the early stages of rat hepatocarcinogenesis. *Carcinogenesis* 17: 395–400, 1996.

28. Schulte-Hermann, R., Bursch, W., Marian, B., Grasl-Kraupp, B. Active cell death (apoptosis) and cellular proliferation as indicators of exposure to carcinogens. *IARC Sci Publ.* 273–285, 1999.

29. Rau, M.A., Whitaker, J., Freedman, J.H., Di Giulio, R.T. Differential susceptibility of fish and rat liver cells to oxidative stress and cytotoxicity upon exposure to prooxidants. *Comp Biochem Physiol C Toxicol Pharmacol* 137: 335–342, 2004.

30. Itoh, K., Yamamoto, M. Molecular mechanism of adaptive response to electrophiles. *Seikagaku* 76: 339–348, 2004.

31. Alam, J., Cook, J.L. Transcriptional regulation of the heme oxygenase-1 gene via the stress response element pathway. *Curr Pharm Des* 9: 2499–2511, 2003.

32. Sharp, F.R., Sagar, S.M. Alterations in gene expression as an index of neuronal injury: heat shock and the immediate early gene response. *Neurotoxicology* 15: 51–59, 1994.

33. Janssen, Y.M., Van Houten, B., Borm, P.J., Mossman, B.T. Cell and tissue responses to oxidative damage. *Lab Invest* 69: 261–274, 1993.

34. Gerhold, D., Lu, M., Xu, J., Austin, C., Caskey, C.T., Rushmore, T. Monitoring expression of genes involved in drug metabolism and toxicology using DNA microarrays. *Physiol Genomics* 5: 161–170, 2001.

35. Garcia-Allan, C., Lord, P.G., Loughlin, J.M., Orton, T.C., Sidaway, J.E.. Identification of phenobarbitone-modulated genes in mouse liver by differential display. *J Biochem Mol Toxicol* 14: 65–72, 2000.

36. Thai, S.F., Allen, J.W., DeAngelo, A.B., George, M.H., Fuscoe, J.C. Detection of early gene expression changes by differential display in the livers of mice exposed to dichloroacetic acid. *Carcinogenesis* 22: 1317–1322, 2001.

37. Crunkhorn, S.E., Plant, K.E., Gibson, G.G., Kramer, K., Lyon, J., Lord, P.G., Plant, N.J. Gene expression changes in rat liver following exposure to liver growth agents: role of Kupffer cells in xenobiotic-mediated liver growth. *Biochem Pharmacol* 67: 107–118, 2004.

38. Rockett, J.C., Swales, K.E., Esdaile, D.J., Gibson, G.G. Use of suppression-PCR subtractive hybridisation to identify genes that demonstrate altered expression in male rat and guinea pig livers following exposure to Wy-14,643, a peroxisome proliferator and non-genotoxic hepatocarcinogen. *Toxicology* 144: 13–29, 2000.

39. Waring, J.F., Gum, R., Morfitt, D., Jolly, R.A., Ciurlionis, R., Heindel, M., Gallenberg, L., Buratto, B, Ulrich, R.G. Identifying toxic mechanisms using DNA microarrays: evidence that an experimental inhibitor of cell adhesion molecule expression signals through the aryl hydrocarbon nuclear receptor. *Toxicology* 181–182: 537–550, 2002.

40. Lesko, L.J., Salerno, R.A., Spear, B.B., Anderson, D.C., Anderson, T., Brazell, C., Collins, J., Dorner, A., Essayan, D., Gomez-Mancilla, B., Hackett, J., Huang, S.M., Ide, S., Killinger, J., Leighton, J., Mansfield, E., Meyer, R., Ryan, S.G., Schmith, V., Shaw, P., Sistare, F., Watson, M., Worobec, A. Pharmacogenetics and pharmacogenomics in drug development and regulatory decision making: report of the first FDA-PWG-PhRMA-DruSafe Workshop. *J Clin Pharmacol* 43: 342–358, 2003.

41. Petricoin, E.F., III, Hackett, J.L., Lesko, L.J., Puri, R.K., Gutman, S.I., Chumakov, K., Woodcock, J., Feigal, D.W., Jr., Zoon, K.C., Sistare, F.D. Medical applications of microarray technologies: a regulatory science perspective. *Nat Genet* 32(Suppl.): 474–479, 2002.

42. Amaratunga, D.J.C. Preprocessing microarray data. *Exploration and Analysis of DNA Microarray and Protein Array Data*. Hoboken, NJ: John Wiley & Sons, 2004, pp. 60–81.

43. Tong, W., Harris, S., Cao, X., Fang, H., Shi, L., Sun, H., Fuscoe, J., Harris, A., Hong, H., Xie, Q., Perkins, R., Casciano, D. Development of public toxicogenomics software for microarray data management and analysis. *Mutat Res* 549: 241–253, 2004.

44. Mattes, W.B., Pettit, S.D., Sansone, S.A., Bushel, P.R., Waters, M.D. Database development in toxicogenomics: issues and efforts. *Environ Health Perspect* 112: 495–505, 2004.

45. Pennie, W., Pettit, S.D., Lord, P.G. Toxicogenomics in risk assessment: an overview of an HESI collaborative research program. *Environ Health Perspect* 112: 417–419, 2004.

46. Thompson, M.B. Clinical pathology in the National Toxicology Program. *Toxicol Pathol* 20: 484–488, 1992; discussion 488–489.

47. Weingand, K., Brown, G., Hall, R., Davies, D., Gossett, K., Neptun, D., Waner, T., Matsuzawa, T., Salemink, P., Froelke, W., Provost, J.P., Dal Negro, G., Batchelor, J., Nomura, M., Groetsch, H., Boink, A., Kimball, J., Woodman, D., York, M., Fabianson-Johnson, E., Lupart, M., Melloni, E. Harmonization of animal clinical pathology testing in toxicity and safety studies. The Joint Scientific Committee for International Harmonization of Clinical Pathology Testing. *Fundam Appl Toxicol* 29: 198–201, 1996.

48. Carakostas, M.C., Banerjee, A.K. Interpreting rodent clinical laboratory data in safety assessment studies: biological and analytical components of variation. *Fundam Appl Toxicol* 15: 744–753, 1990.

49. Yoo, C., Cooper, G.F. An evaluation of a system that recommends microarray experiments to perform to discover gene-regulation pathways. *Artif Intell Med* 31: 169–182, 2004.

50. Townsend, J.P. Resolution of large and small differences in gene expression using models for the Bayesian analysis of gene expression levels and spotted DNA microarrays. *BMC Bioinformatics* 5: 54, 2004.

51. Holzman, T., Kolker, E. Statistical analysis of global gene expression data: some practical considerations. *Curr Opin Biotechnol* 15: 52–57, 2004.

52. Kerr, M.K., Churchill, G.A. Experimental design for gene expression microarrays. *Biostatistics* 2: 183–201, 2001.

53. Leung, Y.F., Cavalieri, D. Fundamentals of cDNA microarray data analysis. *Trends Genet* 19: 649–659, 2003.

54. Kerr, M.K. Design considerations for efficient and effective microarray studies. *Biometrics* 59: 822–828, 2003.

55. Cheung, C., Akiyama, T.E., Ward, J.M., Nicol, C.J., Feigenbaum, L., Vinson, C., Gonzalez, F.J. Diminished hepatocellular proliferation in mice humanized for the nuclear receptor peroxisome proliferator-activated receptor alpha. *Cancer Res* 64: 3849–3854, 2004.

56. Okaya, T., Lentsch, A.B. Peroxisome proliferator-activated receptor-alpha regulates postischemic liver injury. *Am J Physiol Gastrointest Liver Physiol* 286: G606–612, 2004.

57. Koopen, N.R., Post, S.M., Wolters, H., Havinga, R., Stellaard, F., Boverhof, R., Kuipers, F., Princen, H.M. Differential effects of 17alpha-ethinylestradiol on the neutral and acidic pathways of bile salt synthesis in the rat. *J Lipid Res* 40: 100–108, 1999.

58. Wang, L., Han, Y., Kim, C.S., Lee, Y.K., Moore, D.D. Resistance of SHP-null mice to bile acid-induced liver damage. *J Biol Chem* 278: 44475–44481, 2003.

59. Quackenbush, J. Computational analysis of microarray data. *Nat Rev Genet* 2: 418–427, 2001.

60. Ambroise, C., McLachlan, G.J. Selection bias in gene extraction on the basis of microarray gene-expression data. *Proc Natl Acad Sci U S A* 99: 6562–6566, 2002.

61. Ramaswamy, S., Tamayo, P., Rifkin, R., Mukherjee, S., Yeang, C.H., Angelo, M., Ladd, C., Reich, M., Latulippe, E., Mesirov, J.P., Poggio, T., Gerald, W., Loda, M., Lander, E.S., Golub, T.R. Multiclass cancer diagnosis using tumor gene expression signatures. *Proc Natl Acad Sci U S A* 98: 15149–15154, 2001.

62. Cornwell, P.D., De Souza, A.T., Ulrich, R.G. Profiling of hepatic gene expression in rats treated with fibric acid analogs. *Mutat Res* 549: 131–145, 2004.

63. Chu, R., Lim, H., Brumfield, L., Liu, H., Herring, C., Ulintz, P., Reddy, J.K., Davison, M. Protein profiling of mouse livers with peroxisome proliferator-activated receptor alpha activation. *Mol Cell Biol* 24: 6288–6297, 2004.

64. Weiss, S.M., Kulikowski, C. Feature selection. *Computer Systems That Learn: Classification and Prediction Methods from Statistics, Neural Nets, Machine Learning, and Expert Systems*, 1991.

65. Lehmann, J.M., Lenhard, J.M., Oliver, B.B., Ringold, G.M., Kliewer, S.A. Peroxisome proliferator-activated receptors alpha and gamma are activated by indomethacin and other non-steroidal anti-inflammatory drugs. *J Biol Chem* 272: 3406–3410, 1997.

66. Watanabe, T., Suga, T. Effects of some anti-inflammatory drugs on biochemical values and on hepatic peroxisomal enzymes of rat. *J Pharmacobiodyn* 8: 1060–1067, 1985.

67. Kojo, H., Fukagawa, M., Tajima, K., Suzuki, A., Fujimura, T., Aramori, I., Hayashi, K, Nishimura, S. Evaluation of human peroxisome proliferator-activated receptor (PPAR) subtype selectivity of a variety of anti-inflammatory drugs based on a novel assay for PPAR delta(beta). *J Pharmacol Sci* 93: 347–355, 2003.

68. Sastre, M., Dewachter, I., Landreth, G.E., Willson, T.M., Klockgether, T., van Leuven, F., Heneka, M.T. Nonsteroidal anti-inflammatory drugs and peroxisome proliferator-activated receptor-gamma agonists modulate immunostimulated processing of amyloid precursor protein through regulation of beta-secretase. *J Neurosci* 23: 9796–9804, 2003.

69. Paik, J.H., Ju, J.H., Lee, J.Y., Boudreau, M.D., Hwang, D.H. Two opposing effects of non-steroidal anti-inflammatory drugs on the expression of the inducible cyclooxygenase: mediation through different signaling pathways. *J Biol Chem* 275: 28173–28179, 2000.

70. Jaradat, M.S., Wongsud, B., Phornchirasilp, S., Rangwala, S.M., Shams, G., Sutton, M., Romstedt, K.J., Noonan, D.J., Feller, D.R. Activation of peroxisome proliferator-activated receptor isoforms and inhibition of prostaglandin H(2) synthases by ibuprofen, naproxen, and indomethacin. *Biochem Pharmacol* 62: 1587–1595, 2001.

71. Rekka, E., Ayalogu, E.O., Lewis, D.F., Gibson, G.G., Ioannides, C. Induction of hepatic microsomal CYP4A activity and of peroxisomal beta-oxidation by two non-steroidal anti-inflammatory drugs. *Arch Toxicol* 68: 73–78, 1994.

72. Lee, M.J., Gee, P., Beard, S.E. Detection of peroxisome proliferators using a reporter construct derived from the rat acyl-CoA oxidase promoter in the rat liver cell line H-4-II-E. *Cancer Res* 57: 1575–1579, 1997.

73. Rivero, A., Monreal, J.I., Gil, M.J. Peroxisome enzyme modification and oxidative stress in rat by hypolipidemic and antiinflammatory drugs. *Rev Esp Fisiol* 50: 259–268, 1994.

74. Shimada, T., Koitabashi, A., Fujii, Y., Hashimoto, T., Hosaka, K., Tabei, K., Namatame, T., Yoneda, M., Hiraishi, H., Terano, A. PPARgamma mediates NSAIDs-induced upregulation of TFF2 expression in gastric epithelial cells. *FEBS Lett* 558: 33–38, 2004.

75. Gelman, L., Fruchart, J.C., Auwerx, J. An update on the mechanisms of action of the peroxisome proliferator-activated receptors (PPARs) and their roles in inflammation and cancer. *Cell Mol Life Sci* 55:932–943.

76. Pang, L., Nie, M., Corbett, L., Knox, A.J. Cyclooxygenase-2 expression by nonsteroidal anti-inflammatory drugs in human airway smooth muscle cells: role of peroxisome proliferator-activated receptors. *J Immunol* 170: 1043–1051, 2003.

77. Chen, M., Pych, E., Corpron, C., Harmon, C.M. Regulation of CD36 expression in human melanoma cells. *Adv Exp Med Biol* 507: 337–342, 2002.

78. Rossi Paccani, S., Boncristiano, M., Baldari, C.T. Molecular mechanisms underlying suppression of lymphocyte responses by nonsteroidal anti-inflammatory drugs. *Cell Mol Life Sci* 60: 1071–1083, 2003.

79. Ayrton, A.D., Ioannides, C., Parke, D.V. Induction of the cytochrome P450 I and IV families and peroxisomal proliferation in the liver of rats treated with benoxaprofen: possible implications in its hepatotoxicity. *Biochem Pharmacol* 42: 109–115, 1991.

80. Foxworthy, P.S., Perry, D.N., Eacho, P.I. Induction of peroxisomal beta-oxidation by nonsteroidal anti-inflammatory drugs. *Toxicol Appl Pharmacol* 118: 271–274, 1993.

81. Jiang, C., Ting, A.T., Seed, B. PPAR-gamma agonists inhibit production of monocyte inflammatory cytokines. *Nature* 391: 82–86, 1998.

82. He, T.C., Chan, T.A., Vogelstein, B., Kinzler, K.W. PPARdelta is an APC-regulated target of nonsteroidal anti-inflammatory drugs. *Cell* 99: 335–345, 1999.

83. Adamson, D.J., Frew, D., Tatoud, R., Wolf, C.R., Palmer, C.N. Diclofenac antagonizes peroxisome proliferator-activated receptor-gamma signaling. *Mol Pharmacol* 61: 7–12, 2002.

84. Makita, T., Hakoi, K. Proliferation and alteration of hepatic peroxisomes and reduction of ATPase activity on their limiting membrane after oral administration of acetylsaliatic acid (aspirin) for four weeks to male rats. *Ann N Y Acad Sci* 748: 640–644, 1995.

6 The Next Generation of Automated Microarray Platforms for a Multiplexed CYP2D6 Assay

Phillip Kim, Yung-Kang Ken Fu, Vijay Mahant, Fareed Kureshy, Gary Hardiman, and Jacques Corbeil

CONTENTS

INTRODUCTION

One recent advance and fundamental shift in medicine has been the advent of personalized medicine. Improvements in DNA microarray technology have generated data on a scale that, for the first time, permits detailed scrutiny of the human genome, thereby providing the infrastructure to understand the genetic basis of diseases. These advances have the potential to enhance healthcare management by improving disease diagnosis and implementing treatments adapted to each patient. The traditional approach of "one drug fits all" is no longer acceptable. The advent of custom optimization of drugs and dosages results in better treatment efficacy and enhanced safety [1,2]. The emergence of pharmacogenomics has provided a basis with which to evaluate individual variations in response to treatments. Pharmacogenetic testing aims at determining the underlying genotypic and phenotypic differences in the pharmacodynamics and pharmacokinetics of drug metabolism and, as a result, will become an integral part of the therapeutic monitoring and health management of patients.

Drug metabolism occurs predominantly in the liver and involves cytochrome P450 enzymes. The most common drug metabolism variations among individuals are due to polymorphisms of cytochrome P450 genes [3,4]. CYP2D6 is one of the best-studied cytochrome P450 enzymes and is responsible for the metabolism of the majority of pharmaceuticals presently on the market, including a wide range of agents such as beta-blockers, antidepressants, antipsychotics, and opioids (Table 6.1). Currently, more than 70 variant alleles of the CYP2D6 gene have been identified, and additional alleles will likely be expanded [5]. Some of these influence drug metabolism directly.

The CYP2D6 locus is localized on Chromosome 22 in humans and consists of the active CYP2D6 gene and two pseudogenes, CYP2D7P and CYP2D8P [6]. Specific mutations in the CYP2D6 gene result in ultra rapid metabolizer (UM), extensive metabolizer (EM), or poor metabolizer (PM) phenotypes. For instance, deficient hydroxylation (i.e., PM) of debrisoquinone affects 5 to 10% of Caucasian, 1 to 2% of Asian, and 5% of African American populations [7,8,9]. A PM phenotype may result in an adverse reaction upon administration of drugs in standard doses or undesirable drug–drug interactions using multiple-drug therapeutics [10]. Genotyping will become an integral part of evaluating the drug metabolic status of an individual prior to drug administration. Microarray technology is now being applied to clinical diagnostics and genotyping. However, the emergence and success of microarray utilization in the clinic will depend on the ability of this technology to meet the rigorous requirements applied to human diagnostics, and cost-effective, automated, high-throughput microarray platforms are clearly needed to meet the demands of monitoring multiple genotypes as well as the stringent requirements for clinical assays.

In this chapter, we discuss the characteristics of the INFINITI™ system designed for the clinical laboratory and present the CYP2D6 assay we have developed. A challenge with many CYP2D6 genotyping methods is to analyze in a "multiplex" manner using a large number of allelic variants in a single tube and to integrate all the complex steps employed in molecular testing such as allele-specific primer extension (ASPE), hybridization, washing, and signal detection. With the INFINITI™ system, the discrete processes of sample handling, reagent management, hybridization, and detection have been integrated into a totally self-contained automated platform.

The CYP2D6 assay is designed to detect the most prevalent and informative CYP2D6 allele variants. The target regions of the CYP2D6 gene are amplified via a multiplex PCR reaction with specific primers and reaction conditions that can

TABLE 6.1
Examples of Drugs Metabolized by CYP2D6

Antidepressants	Antiarrhythmics	Anticonvulsants	Antihistamines	Analgesics	Antihypertensive	Stimulants
Imipramine	Quinidine	Carbazepins	Loratadine	Codeine	Metoprolol	Metamphetamine
Nortriptyline	Procinamide	Promethazine	—	Tramedol	Propanolol	—
Amitryptyline	Mexiletine	Haloperidol	—	Hydrocodone	Timolol	—
Doxepin	Encainide	—	—	Propoxyphene	—	—
Bupropion	—	—	—	—	—	—
Trazodone	—	—	—	—	—	—
Quinidine	—	—	—	—	—	—
Mexiletine	—	—	—	—	—	—
Flecainide	—	—	—	—	—	—
Fluoxetine	—	—	—	—	—	—
Paroxetine	—	—	—	—	—	—

discriminate CYP2D6 from its pseudogenes. The PCR multiplex reaction is followed by the incorporation of fluorescently labeled nucleotides by primer extension and by the hybridization of the labeled products to an array of immobilized oligonucleotides on the biochip.

THE INFINITI™ PLATFORM

The platform comprises four major elements, which are discussed in the following subsections.

THE INFINITI ANALYZER

The INFINITI™ Analyzer is an automated, continuous-flow, microarray platform (Figure 6.1). The analyzer contains two lasers (red and green), a thermal stringency station, and a thermal cycling unit for performing assays requiring varying temperature conditions or cycles for ASPE and additional applications. The system is designed to operate in a continuous random access mode. To avoid cross contamination of samples or reagents, disposable pipette tips are used for each step in the assay. The analyzer has no tubing or pump and does not require any priming of reagents, therefore eliminating leakage and minimizing microbial contamination growth in the analyzer.

To perform a test such as an ASPE, an operator loads the microplate with the PCR products on the temperature cycling unit, the appropriate microarray magazines, and the Intellipac™ (described below), which contains all the necessary reagents.

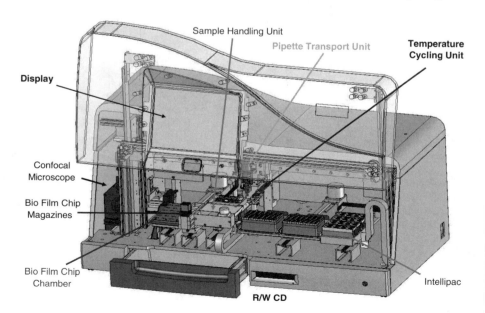

FIGURE 6.1 The INFINITI™ Analyzer showing location of its components and reagents.

The assay begins with an inventory check to ensure that all the appropriate consumables such as pipette tips, BioFilmChips™, and wash buffers are loaded onto the platform. Upon passing the reagent requirements test, the analyzer generates a work list and processes 24 BioFilmChips simultaneously.

The system's scheduler enables seamless processing without any manual intervention. The turnaround time of an assay depends on the type of assay being performed. The time to first results using primer extension and hybridization is 1 to 2 h, depending on the assay. The analyzer can store data, perform data analysis, and generate a report. The results can be formatted and personalized to meet any requirements.

BioFilmChip™

Glass-slide-based microarrays have been widely employed in research. However, glass slides are not practical for routine clinical laboratory use because they are difficult to handle by automation. Their open configuration is not suitable to hold and maintain reaction mixtures without evaporation. To solve these issues, a novel film-based microarray with a reaction chamber was developed. The key component of the novel chip is the BioFilm™, which consists of multiple layers of porous hydrogel matrices 8~10-μm thick on a polyester solid base. This provides an aqueous microenvironment that is highly compatible with biological materials. The second layer incorporates a proprietary composition that reduces intrinsic fluorescence, thereby improving assay sensitivity and eliminating potential artifacts due to "hot spots." The linker layer is used for immobilization of biological molecules such as oligonucleotides, antibodies, or antigens. These molecules can be coupled covalently using glutaraldehyde, imidoester, and epoxides, or noncovalently using the highly specific streptavidin–biotin interaction [11]. The BioFilm™ is then "sandwiched" between a thermal conducting base and a plastic chamber.

The BioFilmChip microarray is configured with 15 $\times$ 16 arrays (240 spots) per chip, which is practical for most current diagnostic applications. Multiple chips can be utilized for gene expression and protein expression analyses that require thousands of spots. The BioFilmChip microarray can be printed using contact (quill pin) or noncontact (piezoelectric) methods [12]. The BioFilmChips are washed to remove uncoupled materials after printing, and then dried and stored desiccated at 2 to 8°C. The BioFilmChips are housed in a magazine.

Qmatic™

The Qmatic Software Manager is the brain of the system; it provides unprecedented flexibility and simplicity for performing many different types of assays simultaneously. This proprietary software manages the complex tasks performed by the INFINITI™ Analyzer. This is accomplished by the integration and synchronization of assay protocols, including regulation of temperature cycling, delivery of reagents by robotics (aspiration and dispensing), detection, data analysis, and handling and reporting results. It identifies samples, queries the Laboratory Information System (LIS) for the assays to be performed, and prompts the operator to load the assay components, such as the BioFilmChip Microarray Magazine and Intellipac. In addition, it also calculates and monitors usage of reagents required for each assay.

Intellipac

Intellipac is the reagent management module and the communication link between INFINITI™ and the onboard reagent management system. Intellipac has eight reservoirs, each containing specific assay reagents, and it has a built-in chip with 128 KB of memory. This chip provides an electronic storage of assay protocols and reagents lot numbers. Upon entering the specific test, the assay information is downloaded into the analyzer. Pertinent data, such as the expiration date of reagents, the volume of reagents previously used, the history of prior usage, and the operator information, are stored and updated with each utilization. Intellipac is equipped with a sliding lid for enclosing the reservoirs to minimize evaporation.

CYP2D6 GENOTYPING

Single-Tube Multiplexed PCR Amplification

DNA samples were extracted using the QIAamp DNA Blood Mini Kit (QIAGEN, Valencia, CA) from 300 μl of whole blood. Three target regions of the CYP2D6 gene were amplified using three sets of primers designed to amplify specific target regions using Platinum Taq DNA polymerase (Invitrogen, Carlsbad, CA) without amplifying pseudogenes. This can be achieved using 10 to 50 ng of isolated genomic DNA. PCR amplicons were sized using 1% agarose gel electrophoresis. The specificity of PCR reaction was confirmed by the absence of any nonspecific PCR products due to pseudogenes. The 5'-end fragment of 1649 bp (for the detection of *41), the middle 1545 bp fragment (for the detection of *10, *12, and *17 alleles), and the 3'-end 1505 bp fragment (for the detection of *29, *6, *8, *4, *3, *9, *2, and *7) were coamplified (Figure 6.2). The reaction buffer conditions used eliminated the separation of unused primers and unincorporated dNTPs post-PCR, and allowed continuous ASPE processing in a single tube. The thermal cycling conditions were 95°C for 3 min, 30 cycles of 95°C for 30 sec, 62°C for 30 sec, 72°C for 3 min, and a final holding at 4°C.

BioFilm™ Chip Printing Format

A set of capture oligonucleotides (18 to 24 mers) was synthesized with 3' biotin (Integrated DNA Technology, Coralville, IA), and these were printed on the streptavidin-coated BioFilmChip· using a custom contact arrayer in triplicates. Each spot

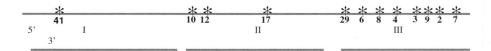

FIGURE 6.2 Three different regions of CYP2D6 gene were amplified using a single tube multiplexed PCR reaction. The corresponding allele variants contained in these specific regions are shown.

has a diameter of 150 to 180 μm and contains approximately 10^8 capture molecules. The distance between the centers of each of the spots was 400 μm. The capture probe for mutant alleles *2 (2850C > T), *3 (2549Adel), *4 (1846G > A), *6 (1707Tdel), *7 (2935A > C), *8 (1785G > T), *9 (2613–2615delAGA), *10 (100C > T), *12 (124G > A), *17 (1023C > T), *29 (1659G > A), and *41 (1584C > G), and their corresponding wild-type alleles were printed in triplicates to form allelic blocks (see Figure 6.3B for detail).

CYP2D6 ASSAY PROCEDURE AND ASSAY VALIDATION

Upon completion of the PCR amplification, the microplate was loaded onto the thermal cycling unit on board INFINITI™ Analyzer. The system automatically scanned Intellipac and the magazine holding the CYP2D6 BioFilmChips. This inventory process ensured that the material required for each run was present and monitored utilization of each component throughout the entire process. Upon engaging the initial inventory process, the operator generated a "work list." Qmatic Software determined the critical assay parameters and scheduled the process sequences. It initiated the ASPE process by applying ASPE reagent mix from Intellipac. After the ASPE reaction of 40 cycles of 95°C for 15 sec, and 53°C for 15 sec with intermediate ramping rates of 1.5°C/sec heating and 1°C/sec cooling, the robotic dispensing unit dispensed the hybridization buffer into each reaction and subsequently applied the reaction mixture onto the BioFilmChips in the hybridization chamber. Each allele-specific primer was designed with unique but very-well-characterized oligo-zip molecules linked to their 5' end. These oligo-zip fragments of allele-specific probes participated in the subsequent hybridization step with universal hybridization conditions, and the genotype-independent oligo-zip and zip-capture interaction ensured the elimination of false signal from potential cross-hybridization often seen in other nucleic acid detection systems. Regardless of the polymerase-mediated extension status, all allele-specific probes bind to their corresponding zip-capture molecules on the BioFilmChips surface. This universal hybridization process was performed for 30 min at 39°C (±0.5°C), and the BioFilmChips were subsequently washed and transported automatically to the optical detectors, where the signal from each oligo-zip spot was analyzed automatically and a report was generated. The assay was validated via a blind study using 46 samples, after which the key was revealed by comparison against a reference method.

RESULTS AND DISCUSSION

The time to the first reportable result was approximately 90 min, and subsequent results were generated every 5 min. The positive hybridization control spots (based on cy3, not shown), ensured the reagent dispensing and assay performance. The analyzer examined both cy5 (for assay signals) and cy3 (for control signals) to qualify each chip. A typical scanned image is shown in Figure 6.3A. However, genotyping of clinical samples was performed in a randomly scattered pattern as shown in Figure 6.3B. The local and global background subtractions were performed

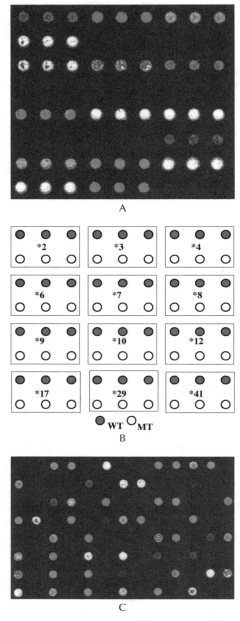

FIGURE 6.3 (A) Cy-5 laser scanned image showing homozygous mutations for *2 (2850C > T) with three mutant spots, and a heterozygous genotype for *17 (1023C > T), and *29 (1659G > A). (B) Allelic cell organization in triplicate for both the wild type and corresponding SNP. (C) Randomly scattered CYP2D6 chip in triplicate configuration. Both local and global background subtraction processes are integrated into the data analysis software.

by the Qmatic software. The results (Figure 6.3C) were analyzed based on triplicates, and displayed the analyte, ratio, % CV, and the patient's genotype (homozygote, heterozygote, or wild-type).

A patient sample displayed was determined to be homozygous mutant for *2 (2850C > T) (three mutant spots), and heterozygous for *17 (1023C > T) and *29 (1659G > A) (six positive spots in their allelic cells), with the following possible genotypes: *2, *17, *29 or *2, *29, and *17. The patient was homozygous wild-type for other alleles (*3, *4, *6, *7, *8, *9, *10, and *12), with only three positive signals.

We performed 2D6 genotyping validation studies using 46 ethnically diverse DNA samples. We had 100% correlation (Table 6.2) on all the SNP genotype calls.

TABLE 6.2
Summary of Assay Validation Studies[*]

Sample ID	*2	*3	*4	*6	*7	*8	*9	*10	*12	*17	*29	*41	Genotype
AG001	X											X	*41
AG002						X							*10
AG003	X									X			*2,*29
AG004	XX								X				*2,*17
AG005			XX					XX					*4,*4
AG0013													*1,*1
AG0014								X					*10
AG0015	X		X					X				X	*4,*41
AG0016								X					*10
AG0017								X					*10
AG0025	XX											XX	*41,*41
AG0026								XX					*5,*10
AG0027	X												*2
AG0028	X												*2
AG0029	XX									XX			*17,*17
AG0037								XX					*10,*10
AG0038	X							X					*2,*10
AG0039													*1,*1
AG0040												X	*41
AG0041	X									X			*17
AG0049													*1,*1
AG0050								X					*10
AG0051	X											X	*41
AG0052	XX											XX	*41,*41
AG0053	XX											XX	*41,*41
AG0061	X					X				X			*9,*17
AG0062								XX					*10,*10
AG0063			X					X					*4
AG0064			X					X					*4
AG0066													*1,*1
AG0074													*1,*1
AG0075								XX					*10,*10
AG0076								X					*10

(continued)

TABLE 6.2 (Continued)
Summary of Assay Validation Studies*

Sample ID	*2	*3	*4	*6	*7	*8	*9	*10	*12	*17	*29	*41	Genotype
AG0077	X											X	*41
AG0078													*1,*1
AG0086			XX					XX					*5,*4
AG0087	XX									X			*2,*17
AG0088													*1,*5
AG0089			X					X					*4
AG0090													*1,*1
AG0032L	X	X						X		X			*4,*17
AG0081L	X											X	*41
AG0021L	XX									XX			*17,*17
AG0010L	X	X						X		X			*4,*17
AG0011L	X	X						X		X			*4,*17
AG0012L	X												*2

* All 46 samples had 100 concordant genotype calls. *5 (deletion) for samples AG0026, AG0086, and AG0088, and *2XN (duplication) for a sample AG0040 were determined by accompanying deletion and duplication analysis based on PCR & ASPE process. Gene deletion and duplication determination process are being integrated into SNP-multiplex reaction along with all the other SNPs in a single tube reaction format.

Furthermore, microarray-based deletion and duplication detection also showed 100% concordance with the reference method. Additional deletion (for *5) and duplication (*2XN) analysis showed three *5 samples and a *2XN sample. Currently, SNP and deletion or duplication detection processes are performed separately, and the goal is to multiplex them into a single tube (see also Figure 6.4).

CONCLUSION

Modern therapeutic approaches will require knowledge of a patient's pharmacogenetic profile to assist the physician in prescribing the appropriate medication at the most effective dosage. This approach should therefore enhance the effectiveness and safety of treatments for many diseases. Pharmacogenetic testing and profiling may soon become routine as the demand for cost effectiveness, high throughput, and rapid turnaround are realized. Because patients' CYP2D6 phenotype can have a profound impact on their response to drug treatments, it is critical to have rapid and economical molecular diagnostic tools to determine individuals' CYP2D6 genotypes.

Currently, the methods most widely used for genetic testing, such as the CFTR test, typically require simultaneous analysis of multiple mutations, a process which is laborious and highly complex. Unquestionably, microarray technology offers a more practical approach and is gaining wider acceptance in the clinical environment.

The CYP2D6 assay is a high-throughput molecular diagnostic test, optimized for use on the completely automated INFINITI™ microarray platform. The most

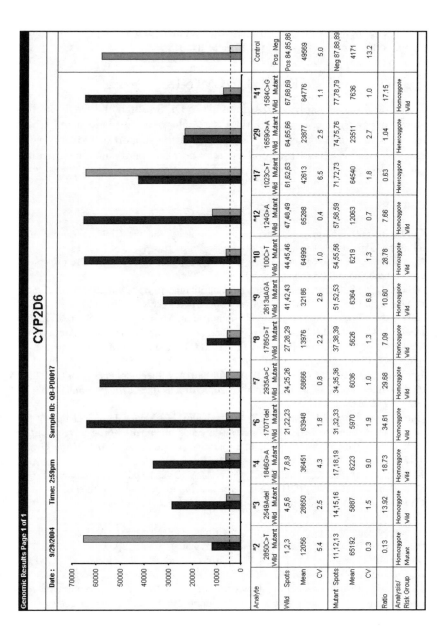

FIGURE 6.4 (*A color version follows page 204*) Result analysis and presentation for the CYP2D6 assay.

common point mutations and small deletions in CYP2D6 can be readily detected using this system. As the platform is completely automated, a large number of samples can be processed simultaneously, with minimal human intervention. The CYP2D6 assay should provide great benefits to clinical laboratories performing patient pharmacogenetic profiling for therapy management and to the pharmaceutical industry conducting drug discovery and clinical trials.

ACKNOWLEDGMENTS

The authors would like to express their gratitude to Dr. S. Singh and Dr. J. Streifel for their technical assistance. Jacques Corbeil is a holder of a Canada Research Chair in Medical Genomics and expresses his gratitude for the support of the University of California Center for AIDS Research Genomics core laboratory and, in particular, Dr. S. Rought.

REFERENCES

1. Goldstein, D.B. Pharmacogenetics in the laboratory and the clinic. *N Engl J Med* 348: 553–556, 2003.
2. Mancinelli, L., Croni, M., and Sadee, W. Pharmacogenomics: the promise of personalized medicine. *AAPS PharmSci* 2(1): 1–13, 2000.
3. Nebert, D.W. Polymorphisms in drug-metabolizing enzymes: what is their clinical relevance and why do they exist? *Am J Hum Genet* 60: 265–271, 1997.
4. Daly, A.K., Brockmoller, J., Broly, F., Eichelbaum, M., Evans, W.E., Gonzalez, F.J., Huang, J.D., Idle, J.R., Ingelman-Sundberg, M., Ishizaki, T., Jacqz-Aigrain, E., Meyer, U.A., Nebert, D.W., Steen, V.M., Wolf, C.R., Zanger, U.M. Nomenclature for human CYP2D6 alleles. *Pharmacogenetics* 6: 193–201, 1996.
5. Bertilsson, L., Dahl, M.L., Dalen, P., Al-Shurbaji, A. Molecular genetics of CYP2D6: clinical relevance with focus on psychotropic drugs. *Br J Clin Pharmacol.* 53(2): 111–122, 2002.
6. Gaedigk, A., Blum, M., Gaedigk, R., Eichelbaum, M., Meyer, U.A. Deletion of the entire cytochrome P450 CYP2D6 gene as a cause of impaired drug metabolism in poor metabolizers of the debrisoquine/sparteine polymorphism. *Am J Hum Genet.* 48(5): 943–950, 1991.
7. Kimura, S., Umeno, M., Skoda, R.C., Meyer, U.A., Gonzalez, F.J. The human debrisoquine 4-hydroxylase (CYP2D) locus: sequence and identification of the polymorphic CYP2D6 gene, a related gene, and a pseudogene. *Am J Hum Genet.* 45(6): 889–904, 1989.
8. Kitada, M. Genetic polymorphism of cytochrome P450 enzymes in Asian populations: focus on CYP2D6. *Int J Clin Pharmacol Res.* 23(1): 31–35, 2003.
9. Mancinelli, L., Cronin, M., Sadee, W. Pharmacogenomics: the promise of personalized medicine. *AAPS PharmSci.* 2(1): E4, 2000.
10. Cascorbi, I. Pharmacogenetics of cytochrome p4502D6: genetic background and clinical implication. *Eur J Clin Invest.* 33(Suppl. 2): 17–22, November 2003.
11. Bayer, E.A. and Wilchek, M. The use of the avidin-biotin complex ASA tool in molecular biology. *Methods Biochem Anal.* 26: 1–45, 1980.
12. Schena, M., Heller, R.A., Theriault, T.P., Konrad, K., Lachenmeier, E., and Davis, R.W. Microarrays: biotechnology's discovery platform for functional genomics. *Trends Biotechnol.* 16: 301–306, 1998.

7 Evaluation of Skeletal Muscle and Adipose Tissue Biopsy Procedures for Microarray-Based Gene-Expression Profiling

*M.B.A. van Doorn, M.J.B. Kemme, M. Ouwens,
E.J. van Hoogdalem, X. Liu, Q.S. Li, C.R. Jones,
M.L. de Kam, J. Burggraaf, and A.F. Cohen*

CONTENTS

INTRODUCTION

Since the publication of the first microarray experiment in 1995 [1], this technology has rapidly evolved and has now become one of the major tools for global gene-expression profiling. The results of recent microarray studies have greatly added to our understanding of e.g., disease pathways [2–5], mechanisms of drug action [6], and tumor classification [7,8] underlining the increasingly important role of gene-expression profiling in biomedicine.

Microarray analysis (using cDNA spotted arrays) typically requires 5 to 10 μg of good-quality RNA per sample to obtain accurate and reproducible results. Animal and human studies using cell cultures or therapeutically removed tissues (e.g., tumor material) have little or no constraints in obtaining sufficient RNA quantity as the tissue supply is usually large. In contrast, clinical (drug intervention) trials are restricted in the amount and type of healthy tissue that can be obtained. It is therefore of great importance to make use of tissue biopsy procedures that yield sufficient high-quality RNA while having a minimal burden on the study subjects, which would allow these procedures to be performed repeatedly.

There are currently no published studies that have systematically evaluated skeletal muscle and adipose tissue biopsy procedures in terms of tolerability as well as RNA yield and quality. To this end, we selected minimally invasive muscle (modified Bergström needle procedure [9]) and adipose tissue (hollow needle aspiration) procedures as most promising candidates to be evaluated for potential future application in clinical trials studying gene expression in humans. We wanted to balance the tolerability and RNA yield of these candidate methodologies against the RNA yield of white blood cells as obtained by venapuncture. The evaluation of all sampling methodologies was extended to array hybridization, data normalization, as well as outlier detection and removal in order to cover the complete sample processing flow.

Therefore, the objectives of this study were:

1. To evaluate the tolerability of the selected minimally invasive human skeletal muscle and adipose tissue biopsy procedures
2. To evaluate the selected RNA extraction methods for muscle, adipose tissue, and white blood cells in terms of RNA yield and quality
3. To evaluate the selected cDNA microarray hybridization and normalization procedures as well as data reproducibility and outlier detection or removal methodologies

SUBJECTS AND METHODS

The study protocol was approved by the Medical Ethical Committee of the Leiden University Medical Center and performed according to the principles of ICH-GCP, the Helsinki Declaration, and Dutch law.

Six healthy male and six healthy female subjects with an abdominal skin fold thickness of ≥1 cm were included (Table 7.1). Subjects were excluded if they (had) used any medication other than oral contraceptives or occasional paracetamol within 2 weeks of the expected study start.

TABLE 7.1
Demographics

Subject	Gender	Age (Years)	BMI (kg/m²)	W/H Ratio	Skin-Fold Thickness (cm)
1	F	22	25.4	0.72	2.8
2	F	40	27.1	0.78	3.4
3	M	36	34.1	1.05	4.0
4	F	22	22.8	0.71	2.1
5	M	22	24.7	0.94	2.8
6	M	22	29.2	0.91	3.9
7	F	22	22.5	0.73	1.5
8	M	23	21.5	0.87	1.0
9	M	18	24.7	0.84	3.5
10	F	20	19.0	0.67	1.7
11	F	41	33.6	0.96	4.0
12	M	24	28.7	0.86	3.0

Note: BMI = body mass index; W/H ratio = ratio of waist/hip circumference.

The study period consisted of a medical screening followed by a single study day and two follow-up visits after 3 and 7 d, respectively. On the study day, a skeletal muscle biopsy, a subcutaneous adipose tissue biopsy, and a 10-ml blood sample were collected. On the two follow-up visits, the biopsy sites were inspected and the biopsy procedures were evaluated with a tolerability questionnaire.

SKELETAL MUSCLE BIOPSY PROCEDURE

The subject was in a comfortable (semi) recumbent position with one knee supported in 20° flexion allowing the quadriceps muscle to relax. A skin area 10 cm proximal from the patella on the ventrolateral side of the upper leg was disinfected using a chlorhexidine solution. The skin area and muscle tissue were anesthetized by local infiltration of 2- to 5-ml lidocaine HCl 1% and disinfected again. Subsequently, sterile gloves were put on and a sterile surgical aperture cloth was spread over the disinfected area (the area was covered using a disposable surgical drape). The trocard was retracted from the biopsy needle (modified Bergström muscle biopsy needle [9], Department of Movement Sciences, Maastricht University, the Netherlands), and a tube (Connecta Plus 3, Becton & Dickinson) was attached to the needle connecting it to a 20-ml syringe. A small incision (approximately 5-mm width) was made in the skin, subcutaneous tissue, and muscle fascia of the vastus lateralis muscle. The biopsy needle was introduced through the skin and fascia into the muscle. Vacuum was applied to the needle by pulling the syringe plunger and holding it in place to a negative pressure at the 15-ml mark. The cutter was moved 2 cm upwards allowing the muscle tissue to be drawn into the needle opening. Subsequently, the cutter was moved down swiftly and was rotated several times to assure the muscle sample had been cut free completely. Three cuts were made in several directions by rotating the needle 45° after each cut. Subsequently, the vacuum was released by gently letting down the plunger. The needle was retracted, and moderate pressure was applied to the puncture site. The incision in the skin was closed with suture tape, and a pressure bandage (Cohefix, 4m × 10cm) was applied for 24 h.

After retracting the needle, the muscle sample was pushed out of the cutter on a sterile surface and transferred into a preweighed sterile cryo-vial. The sample was weighed on a calibrated electronic balance, immediately snap-frozen in liquid nitrogen, and stored at 80°C.

ADIPOSE TISSUE BIOPSY

The subject was in a relaxed (semi) recumbent position. A skin area 10 cm lateral from the umbilicus was disinfected using a chlorhexidine solution. The skin and subcutaneous tissue of the sample site were anesthetized by local infiltration of approximately 2 ml lidocaine HCl 1%. The skin-fold was lifted and a 14 G needle, connected to a 50-ml syringe, was introduced into the subcutaneous adipose tissue. Vacuum was applied to the syringe, and the needle was passed several times in a horizontal plane through the subcutaneous adipose tissue, drawing it into the needle. The needle was retracted while maintaining the vacuum, and moderate pressure was applied to the biopsy site. Approximately 20 ml of saline at 37°C was drawn into the syringe to wash out excess blood from the adipose tissue. The needle was replaced by a sterile combi-cap (Luer lock) and the plunger was removed from the syringe with the cap facing downwards. A sterile funnel with a nylon filter (Millipore nylon net filter NY8H, Ø 47 mm, 180-μm pore size) was placed in an Erlenmeyer flask and the contents of the syringe were poured out in the filter leaving the adipose tissue sample as washed residue. The adipose tissue was removed from the filter with two sterile spatulas and transferred to a polypropylene cryo-vial. The sample was weighed on a calibrated electronic balance, snap-frozen immediately in liquid nitrogen, and stored at −80°C.

BLOOD SAMPLING AND LEUKOCYTE ISOLATION

A 10-ml blood sample was collected by venapuncture in a sodium heparin coated tube. Samples were immediately stored on ice, and leukocytes were isolated by lysing the red blood cells in 40 ml EL-buffer (Qiagen, Westburg, the Netherlands) within 30 min. Subsequently, white blood cells were washed twice with 20 ml of EL-buffer, lysed in RLT-buffer (Qiagen), and stored at 80°C until extraction of total RNA.

TOLERABILITY ASSESSMENT

Tolerability was assessed using the following questionnaires.

Directly after each biopsy:
- How would you rank the discomfort experienced from the procedure? Not painful, slightly painful (comparable to a venipuncture), moderately painful, or very painful?

At follow-up (1 week after the biopsy procedures):
- How long did you experience discomfort after the biopsy procedure, and if you did, how would you rank the discomfort? Slight, moderate (interfering with normal daily activities), or severe (immobilization)?
- Would you object to (fictively) having the biopsies two more times, now knowing how it feels?

RNA EXTRACTION

For RNA isolation from skeletal muscle and adipose tissue, biopsies were homogenized in RNABee (Tel-Test, Inc., Friendwood, TX), (1 ml RNABee/50 mg biopsy weight) using an Ultraturrax T25 mixer (Janke & Kunkel, IKA Labortechnik, Staufen, Germany). RNA was extracted by the addition of 0.2 ml chloroform to 1 ml homogenate, and precipitated by the addition of 1 volume isopropanol in the presence of 20 μg glycogen as carrier. Following washing with 75% ethanol, RNA was dissolved in diethylpyrocarbonate-treated water. DNA was removed from all RNA samples by DNAse treatment followed by cleanup using an RNeasy minikit according to the manufacturer's instructions (Qiagen, Venlo, The Netherlands). For leukocytes, total RNA was extracted using an RNAeasy midikit (Qiagen). Purity and quantity of the RNA was determined by spectrophotometric analysis at 260 and 280 nm. Integrity of the RNA was verified by agarose gel electrophoresis.

MICROARRAY HYBRIDIZATION

Human cDNAs were spotted onto Corning GAPS slides using an Amersham Biosciences Generation III spotter. Each clone was spotted in duplicate or triplicate on the array, with the exception of control clones that were spotted 4 to 12 times. The in-house microarray contained 7426 clones that represented 5762 unique human genes. The Cy3-labeled cDNA probe preparation, hybridization, and subsequent washes of the arrays were performed as previously described [10]. All arrays were scanned in a ScanArray 4000 (Perkin Elmer Life Sciences, Boston, MA). Quantification was performed using Imagene (Biodiscovery, Marina del Rey, CA).

MICROARRAY STATISTICAL ANALYSIS

CORRELATIONS

Spearman's correlations were calculated between skin-fold thickness and duration of discomfort (adipose tissue only), and between sample weight and duration of discomfort (skeletal muscle and adipose tissue). In addition, Spearman's correlations were calculated between skin-fold thickness and RNA yield (skeletal muscle and adipose tissue).

All calculations were performed using SAS for Windows V8.2 (SAS Institute, Inc., Cary, NC).

Normalization and Elimination of Outlier Data Points

Depending on the RNA yield, each RNA sample was hybridized to two to three microarray slides run in parallel, yielding duplicate or triplicate data points for each sample. For analysis, the intensities from each chip were scaled to the 75th percentile (the 75th percentile value of each chip was set to 100), followed by $\log_2$ transformation. A nonlinear normalization procedure based on smooth spline [11] was applied on the duplicate or triplicate data sets for each sample. A data point was flagged as an outlier if the fold change between the data point and the median of the duplicate or triplicate data points was greater than 1.5-fold and was removed for further analysis. The remaining data points ($\log_2$ transformed intensities) for each clone were averaged in each sample and the resulting mean was designated as the $\log_2$ transformed expression intensity in the sample.

Microarray Data Normalization among Samples within Each Tissue

A sorted nonlinear smooth spline normalization procedure [11] was applied on all samples from the same tissue. After normalization, the ratio between the 95th percentile and the 5th percentile of the expression intensities in each sample was calculated as the dynamic range in that sample. Concordance correlation coefficients were calculated among all samples from the same tissue as described previously [12].

Clustering Analysis

Normalized $\log_2$-transformed expression intensities of samples from all three tissues were subjected to clustering analysis for visualization purposes. The average of the $\log_2$ transformed intensities in all samples, defined as the $\log_2$ transformed geomean, was set to zero for each clone. Clones with at least twofold change between one sample and the geomean of all samples were selected for clustering analysis using GeneCluster™ software and visually represented using Treeview™ version 1.6 (Stanford University, Stanford, CA).

RESULTS

Tolerability Biopsy Procedures

The results of the tolerability questionnaires are summarized in Table 7.2. It should be noted that four subjects experienced discomfort for more than 72 h after the muscle biopsy procedure; three subjects reported low-intensity discomfort (slightly bruised feeling; no immobilization) at the muscle biopsy site up to 4 d after the procedure. In one other subject, although suture tape had been applied, the small cut had sprung open, which resulted in delayed wound healing and prolonged discomfort up to 2 weeks. In addition, eight subjects reported a hematoma at the adipose tissue biopsy site that remained visible for approximately 1 week but this did not result in any discomfort.

There was a significant inverse correlation between skin-fold thickness and duration of discomfort after the adipose tissue biopsy procedure ($r = -.68$; $p = .02$).

TABLE 7.2
Summary Table Tolerability Questionnaires

Discomfort during Biopsy Procedure

	None	Mild	Moderate	Severe
Adipose tissue	6 (50%)	6 (50%)	0 (0%)	0 (0%)
Muscle tissue	1 (8.3%)	7 (58.3%)	4 (33.3%)	0 (0%)

Duration (h) of Discomfort after Biopsy Procedure

	None	<48 h	<72 h	>72 h
Adipose tissue	5 (41.7%)	3 (25%)	3 (25%)	1 (8.3%)[a]
Muscle tissue	0 (0%)	6 (50%)	2 (16.7%)	4 (33.3%)[a]

[a] Low-intensity discomfort.

There was no significant correlation between sample weight and duration of discomfort after the biopsy procedure for muscle or adipose tissue ($r = -.036$; $p = .25$ and $r = .01$; $p = .97$), respectively.

None of the subjects objected to a (fictitious) repeat of the adipose tissue biopsy procedure two more times, while three subjects objected to repeating the muscle biopsy procedure two more times. Of these three subjects, one subject thought the muscle biopsy procedure had been too unpleasant, and the two other subjects objected because they had felt too restricted in their normal daily activities in the 2 d following the procedure.

BIOPSY WEIGHT AND RNA YIELD

For all three tissues, the average biopsy weight (range) and RNA yield (range) are summarized in Table 7.3. It should be noted that only half of each muscle sample was analyzed for RNA quantity. This enabled us to repeat the RNA extraction procedure using an alternative RNA extraction method in case the quality of the

TABLE 7.3
Summary Table Biopsy Weight and RNA Yield

Tissue	Average Biopsy Weight (mg) Mean (Range)	Average RNA Content (μg) Mean (Range)	Average RNA (μg)/Weight (mg) Ratio Mean (Range)
Muscle	121.2 (79–202.0)	16.9 (4.1–41.8)	0.139 (0.051–0.246)
Adipose tissue	197.3 (61–430.0)	3.9 (1.3–8.5)	0.021 (0.013–0.026)
Leucocytes	N.a.[a]	13.0 (2.3–28.8)	N.a.[a]

[a] Not available; 10-ml blood sample.

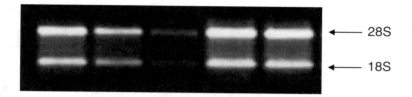

FIGURE 7.1 Example of agarose gel electrophoresis of RNA extracted from one blood sample. The intensity of 28S ribosomal RNA is approximately twice that of the 18S RNA band, indicating integrity of the RNA.

RNA was unsatisfactory. The ratios of the spectrophotometric readings at 260 and 280 nm were 2.04 ± 0.12, 1.97 ± 0.09, and 2.01 ± 0.08 for RNA extracted from muscle, adipose tissue, and leukocytes, respectively, indicating purity of the RNA. No visible RNA degradation or genomic DNA contamination was observed on agarose gel electrophoresis (Figure 7.1).

For adipose tissue, there was a significant correlation between biopsy weight and RNA yield ($r = .95$; $p < .0001$), but there was no significant correlation between skin-fold thickness and RNA yield ($r = .10$; $p = .75$) or skin-fold thickness and biopsy weight ($r = .25$; $p = .43$). For muscle tissue, the correlation between total biopsy weight and RNA yield showed a positive trend but was not statistically significant ($r = .46$; $p = .13$).

EVALUATION OF GENE EXPRESSION PROFILES IN BIOPSY TISSUES AND WHITE BLOOD CELLS

The gene expression profiles in muscle, adipose tissue, and white blood cells were evaluated by hybridization to an in-house custom made cDNA microarray that covers 5762 unique human genes. We observed a higher expression dynamic range in leucocytes from white blood cells, as compared to that in adipose and muscle tissue; the average dynamic ranges among nonoutlier samples (defined below) were 21.8, 14.0, and 4.9 in blood, adipose tissue, and muscle tissue, respectively. Two blood samples and one adipose sample exhibited an atypical, narrow dynamic range as compared to their peers (Figure 7.2). These three samples were therefore labeled as outlier samples.

Pairwise scatter plots and concordance correlation coefficients were generated among all biopsy samples within each tissue type (Figure 7.3 and Figure 7.4). The averages (and standard deviations) of the concordance correlation coefficients observed between nonoutlier samples were 0.971 (0.010), 0.952 (0.021), and 0.949 (0.017) in blood, adipose tissue, and muscle tissue, respectively, indicating high reproducibility of the expression profiles among nonoutlier samples within each tissue type. In contrast, the averages of concordance correlation coefficients were only 0.36 and 0.12 in two blood outliers, and 0.87 in one adipose outlier, after comparing to nonoutlier samples from the same tissue type. The low concordance correlation coefficients observed between outlier samples and nonoutlier samples

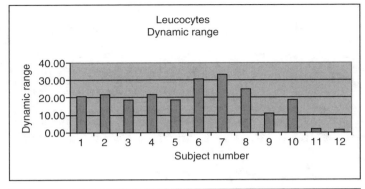

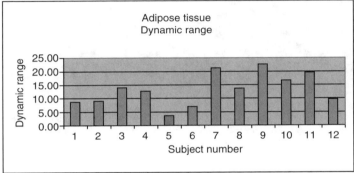

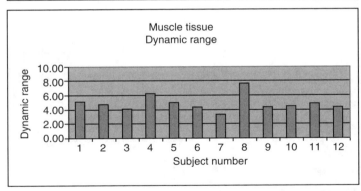

FIGURE 7.2 Dynamic range of gene expression intensities in leukocytes and tissue samples, including three samples (in red) identified as putative outliers.

indicate the discrepancy of the underlying expression profiles and further confirm the expressional abnormality in these outlier samples.

Clustering analysis of all 36 samples based on 4303 differentially expressed clones revealed three major clusters that correspond to the three tissue types, with the exception of two blood samples and two adipose samples that were clustered with muscle samples (Figure 7.5). Three of the four incorrectly clustered samples

Biochips as Pathways to Drug Discovery

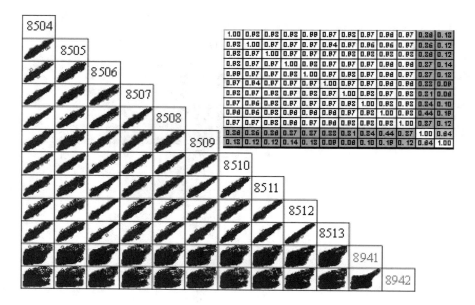

FIGURE 7.3 Scatter plots and concordance correlations scheme of gene expression intensities among white blood cell samples. Samples 8941 and 8942 (in red) are identified as putative outliers.

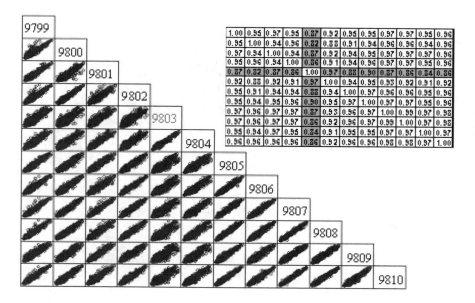

FIGURE 7.4 Scatter plots and concordance correlations scheme of gene expression intensities among adipose tissue samples. Sample 9803 (in red) is identified as putative outlier.

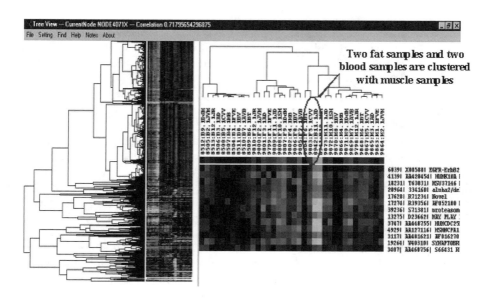

FIGURE 7.5 (*A color version follows page 204*) Clustering on 36 samples reveals 4 spurious samples. Gene number (at least twofold change in one sample and geo-mean): 4303.

were outlier samples as determined by the analyses of expression dynamic range and concordance correlation coefficients, whereas the fourth incorrectly clustered sample contained adipose tissue that also exhibited low dynamic range (6.8 vs. the average of 14.0) and low concordance correlation coefficients (0.92 vs. the average of 0.96), as compared to other nonoutlier adipose tissues. Three tissue-specific clusters were confirmed in clustering analysis after removal of these four incorrectly clustered samples (Figure 7.6).

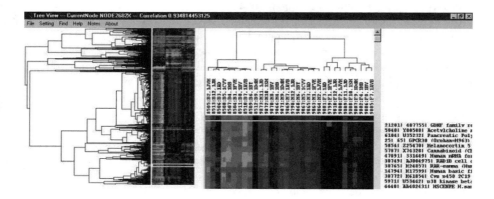

FIGURE 7.6 (*A color version follows page 204*) Clustering on the remaining 32 samples after excluding 4 spurious samples. Gene number (at least twofold change in one sample and geo-mean): 3709.

DISCUSSION

This study evaluated the tolerability of minimally invasive skeletal muscle and adipose tissue biopsy procedures, and the resulting RNA yields and quality for subsequent cDNA microarray analysis. In addition, we evaluated the RNA yield and quality of white blood cells and our methods for microarray hybridization and detection and removal of outliers in the microarray data. In general, the tissue biopsy procedures were well tolerated (Table 7.2).

The sampling of adipose tissue was virtually painless, and discomfort reported during the procedure was primarily related to the introduction of the local anesthetic injection needle. None of the subjects objected to a fictitious repeat of the procedures two more times. The procedure caused a superficial hematoma (in 67% of the subjects) that disappeared after 1 week but did not result in any discomfort. There was a significant inverse correlation between abdominal skin-fold thickness and duration of discomfort after the adipose tissue biopsy procedure. This probably relates to the increased difficulty and duration of the biopsy procedure in leaner subjects with a borderline (approximately 1 cm) abdominal skin-fold thickness. Therefore, subjects with a relatively large abdominal skin-fold thickness are preferred, in order to further minimize the duration of discomfort after the adipose tissue biopsy. The muscle biopsy caused more discomfort during and after the procedure. This was probably related to the more invasive nature of the procedure. The discomfort experienced during the biopsy session was, as for adipose tissue, primarily related to the local anesthetization procedure. Because of adequate anesthesia, performing the biopsy procedure itself was virtually painless. Furthermore, in one subject the surgical tape had been unable to support the small cut sufficiently, which resulted in wound dehiscence and therefore delayed healing. After this observation it was decided to close the biopsy site with a single suture (4×0 ethilon) for all remaining subjects.

Three subjects objected to participating when the procedure would have been repeated two more times. One subject had found the procedure unpleasant, which probably related to some difficulty in relaxing the quadriceps muscle during the procedures. In addition, two subjects complained of pain and discomfort when straining the muscle on the day after the procedure, which restricted them in their daily (sports) activities. Hence, it is of great importance that the subjects are adequately relaxed during the muscle biopsy procedure and should be advised to avoid strenuous exercise (e.g., climbing stairs, sports) the day after the procedure as they may be hindered by the small lesion in the quadriceps muscle. Hence, our data allow future investigators to provide accurate informed consent for potential subjects. In addition, for future studies we suggest closing the biopsy site with a single suture to avoid wound dehiscence. Taking these points into account, it is expected that performing the muscle biopsies more than once in a single study will be tolerable for all subjects. In addition, we now also have data from a considerable number of healthy volunteers and diabetic patients who have undergone the biopsy procedures two or three times per study, and tolerability was in line with the expectations raised in this study.

In the preparatory phase of this study, it was established that cDNA microarray analysis with triplicate hybridizations requires at least 5 μg total RNA per sample but more (up to 10 μg) is preferable to compensate for potential RNA degradation (data not shown). The muscle samples yielded sufficient RNA (average 17 μg, minimum was 4.1 μg), considering only half of each sample was analyzed for RNA content (Table 7.3). In contrast, the adipose tissue samples yielded insufficient RNA quantities (average 3.9 μg), with two samples yielding only 1.4 μg RNA. This was probably related to the comparatively low sample weight (84 and 86 mg), because RNA yield proved to correlate with biopsy weight in adipose tissue. As the RNA yield was relatively low (average 0.021 μg/mg), care should be taken that sufficient adipose tissue is sampled. The projected minimal sample weight for adipose tissue, calculated from the lowest RNA and weight ratio observed in this study (0.013 μg/mg), is around 400 mg yielding approximately 5 μg total RNA. This appears feasible from a technical and tolerability point of view because there was no correlation between biopsy size and duration of discomfort within the 61 to 430 mg range (Table 7.3). Furthermore, it is expected that sampling beyond 400 mg (up to 1000 mg) to obtain even higher RNA yields will not result in any additional discomfort.

The 10-ml blood samples harboring white blood cells yielded sufficient RNA in most cases. One sample had a very low RNA yield (2.3 μg) compared to its peers where the average RNA yield was 13.0 μg (Table 7.3). This sample was designated as an outlier by dynamic range, concordance, and cluster analysis but not by gel electrophoresis analysis, which showed no clear evidence of RNA degradation. The low RNA yield in this case may have resulted from inefficient RNA extraction, which did not allow triplicate or even duplicate microarray hybridizations. Another blood sample was considered a putative outlier although the RNA yield of this sample appeared sufficient (13.8 μg). This was possibly related to RNA degradation during cy3 labeling or just prior to array hybridization. Another possibility could be an inferior-quality cDNA microarray slide. However, quality control showed no chip abnormalities. More recently collected, extensive experience with all described methodologies in two clinical pharmacology studies led to the conclusion that failure rate of the techniques is reduced importantly when methods are applied in a semi routine setting. Therefore, although the chances of accumulating missing data points because of occasional RNA degradation or inefficient RNA extraction can be reduced by collecting samples in duplicate. However, there seems to be little justification to do so in an experienced setting.

In summary, the muscle and adipose tissue biopsy procedures evaluated in this study were well tolerated. The muscle biopsy procedure yielded sufficient amounts of good quality RNA, evidenced by gel electrophoresis as well as subsequent dynamic range, concordance correlation, and cluster analyses. The adipose tissue biopsy procedure however, yielded insufficient RNA quantities to allow duplicate or triplicate hybridizations for all samples, which resulted in two putative outlier samples. The white blood samples yielded sufficient RNA in most cases, and may be taken *in duplo* to prevent unnecessary missing data points in an inexperienced setting. Therefore, we conclude that the procedures evaluated in this study are suitable for use in future clinical microarray studies, provided that adipose tissue biopsy weight is minimally 400 mg and the skeletal muscle biopsy weight is minimally 100 mg. Our methodology for microarray data normalization, and elimination of outlier data points and subsequent clustering analysis proved to be an appropriate approach that is expected to be fully applicable to compare gene expression profiles of healthy volunteers and patients at baseline, and to study drug effects in a clinical pharmacology setting.

REFERENCES

1. Schena, M., Shalon, D., Davis, R.W., and Brown, P.O. Quantitative monitoring of gene-expression patterns with a complementary-DNA microarray. *Science* **270**, 467–470, 1995.
2. Brown, P.O. and Botstein, D. Exploring the new world of the genome with DNA microarrays. *Nat Genet* **21**, 33–37, 1999.
3. Crow, M.K. and Wohlgemuth, J. Microarray analysis of gene expression in lupus. *Arthritis Res Ther* **5**, 279–287, 2003.
4. Kaab, S. et al. Global gene expression in human myocardium-oligonucleotide microarray analysis of regional diversity and transcriptional regulation in heart failure. *J Mol Med* **82**, 308–316, 2004.
5. van der Pouw Kraan, T.C. et al. Rheumatoid arthritis is a heterogeneous disease: evidence for differences in the activation of the STAT-1 pathway between rheumatoid tissues. *Arthritis Rheum.* **48**, 2132–2145, 2003.
6. Waddell, S.J. et al. The use of microarray analysis to determine the gene expression profiles of Mycobacterium tuberculosis in response to anti-bacterial compounds. *Tuberculosis. (Edinb.)* **84**, 263–274, 2004.
7. Dyrskjot, L. et al. Identifying distinct classes of bladder carcinoma using microarrays. *Nat. Genet.* **33**, 90–96, 2003.
8. Neo, S.Y. et al. Identification of discriminators of hepatoma by gene expression profiling using a minimal dataset approach. *Hepatology* **39**, 944–953, 2004.
9. Bergstrom, J. Percutaneous needle biopsy of skeletal muscle in physiological and clinical research. *Scand J Clin Lab Invest* **35**, 609–616, 1975.
10. Luo, L. et al. Gene expression profiles of laser-captured adjacent neuronal subtypes. *Nat Med* **5**, 117–122, 1999.
11. Shaw, K.J. et al. Comparison of the changes in global gene expression of *Escherichia coli* induced by four bactericidal agents. *J Mol Microbiol. Biotechnol.* **5**, 105–122, 2003.
12. Lin, L.I. A concordance correlation coefficient to evaluate reproducibility. *Biometrics* **45**, 255–268, 1989.

8 ChIP-on-Chip: Analysis of Genomewide Protein Binding and Posttranslational Modifications

Daniel Robyr

CONTENTS

ABSTRACT

The mapping of an increasing number of genomes and the recent development of DNA microarrays have allowed researchers to greatly enhance their pace of discovery and raised several questions genomewide; these were not feasible a decade ago. DNA microarrays are mainly used to study global changes in gene expression patterns within the cell under various experimental conditions. Gene expression is a complex network of regulatory cascades and interconnectivity. The analysis of protein binding genomewide, rather than its effect on gene expession, provides a useful tool to circumvent interpretation issues related to these indirect effects. This approach is achieved by a combination of chromatin immunoprecipitation and hybridization to DNA microarrays

123

(ChIP-on-chip). The purpose of this review is to discuss the strengths of ChIP-on-chip experiments, alone and in combination with standard expression arrays.

INTRODUCTION

The completion of the genetic map of the first free-living organism in 1995 (*Haemophilia influenzae*) [1] marked an important milestone in the genomic era. It was soon followed by a flurry of publications of complete genome sequences of other organisms, from the budding yeast *Saccharomyces cerevisiae* [2] to higher eukaryotes, including the fruit fly *Drosophila melanogaster* [3]; these culminated in the publication of the human genome in 2001 [4,5]. The challenge for current and future generations in the postgenomic era consists in making sense of these primary DNA sequences containing coding, regulatory, and "junk" DNA regions. Scientists have traditionally and successfully used a gene-by-gene approach that is both time-consuming and does not take advantage of the available data. For fulfillment of their promises, various genome projects required new technological advances that could not only provide a global and unbiased view of the biological processes but also greatly increase the pace of discoveries. Pioneer work by Fodor (cofounder of Affymetrix) and later by Patrick O. Brown's group led the way to the development and generalization of high-density DNA microarrays [6,7] adapted to genomewide studies. In the past, DNA microarrays have become an invaluable tool for rapid genomewide surveys of gene expression patterns for both biological researchers and clinicians. Expression microarrays are ideal for diagnostic (cancer profiling, disease classification, etc.) studies or for the study of the global effect of drugs or gene mutations on downstream gene regulation. However, the inherent cascade effect of the expression of one gene on others will inevitably hinder a coherent mechanistic interpretation of the data. Indirect effects are best illustrated by the common observation that disruption of a gene encoding a transcriptional repressor can lead to both repression and derepression of gene activity genomewide, whereas only the latter is expected [8,9]. In this case, the issue was alleviated by the development of new tools to determine the transcriptional repressor recruitment sites on the chromosomes [10]. The protein-interacting loci are identified with DNA-binding microarrays that are a combination of chromatin immunoprecipitation (ChIP) and DNA microarrays, also known as *ChIP-on-chip*. In addition, the activity of some proteins, such as histone-modifying enzymes, can readily be surveyed, using the same ChIP-on-chip approach [11–14]. This chapter will focus on applications and various issues concerning ChIP-on-chip experiments with a special emphasis on yeast, because the technique has been applied mostly to this organism so far.

CHIP-ON-CHIP: BRIEF HISTORY AND OVERVIEW OF THE METHOD

A Brief History

In 1985, Solomon and Varshavsky observed that histones could be efficiently and reversibly cross-linked to chromosomal DNA with formaldehyde *in vivo* [15]. They envisioned that this property could be exploited to map protein–DNA interactions *in vivo*; a few years later, they used antibodies to immunoprecipitate cross-linked histones and RNA polymerase II at the *Drosophila hsp70* locus, laying the foundation for chromatin immunoprecipitation (ChIP) [16]. The authors state in the last sentence of their article, "The formaldehyde-based *in vivo* mapping techniques of this work are generally applicable, and can be used both to probe protein–DNA interactions within specific genes and to determine the genomic location of specific chromosomal proteins" [16]. A little more than a decade later, chromatin immunoprecipitation had become the tool of choice for those desiring to study chromatin and protein–DNA interactions *in vivo*. The emergence of DNA microarrays allowed two laboratories to combine both technologies and publish simultaneously genomewide binding maps for several budding yeast transcription factors [17,18].

THE METHOD

Formaldehyde reacts with amino and imino groups found on the side chains of residues such as arginine, histidine, and lysine, creating an intermediate Schiff base, which in turn reacts with another amine group, leading to the final covalent cross-linking of two amino groups. In addition, cross-linking occurs through the amino group of an adenine, cytosine, and guanine. The short range (2 Å) of the reaction, however, prevents efficient cross-linking of proteins that are not close enough to the DNA. Bifunctional imidoester cross-linkers with long spacer arm lengths allow the formation of covalent bonds between more distant primary amine groups on proteins. These include dimethyl adipimidate (DMA), dimethyl pimelimidate (DMP), and dimethyl 3,3'-dithiobispropionimidate (DTBP) with respective spacer lengths of 8.6 Å, 9.2 Å, and 11.9 Å. However, these compounds do not react with DNA and have to be used in conjunction with formaldehyde. This double cross-linking approach assumes that the protein of interest interacts with a formaldehyde-cross-linked DNA-binding protein either directly or indirectly within a larger protein complex. These imidoester compounds were successfully used to cross-link and immunoprecipitate proteins that were resistant to formaldehyde alone [10,19]. Formaldehyde can also be used to cross-link successfully RNA-binding proteins that can subsequently be immunoprecipitated [20].

Several similar ChIP-on-chip methods have been described in detail by different laboratories [21–25]. Here, the basic technical aspects of the approach will only be briefly summarized (Figure 8.1). First, cells are cross-linked with formaldehyde, lysed, and their chromatin content is sheared down to an average size of 500 base pair (bp), either by sonication or by enzymatic digestion. The resolution of the ChIP will depend greatly on the size of the sheared chromatin. Then, highly specific antibodies are used to immunoprecipitate cross-linked chromatin fragments enriched for the targeted protein and "associated" DNA. The antibody specificity is certainly one of the most critical elements of the whole procedure. If possible, these antibodies have to be tested both by ELISA and by chromatin immunoprecipitation against a strain lacking the epitope [26]. In situations in which there are no antibodies raised against the protein of interest, epitope-tagged proteins can be immunoprecipitated with the corresponding antibody. Alternatively, proteins can be affinity-purified using a tag that bypasses the need for an antibody altogether, such as GST (glutathione

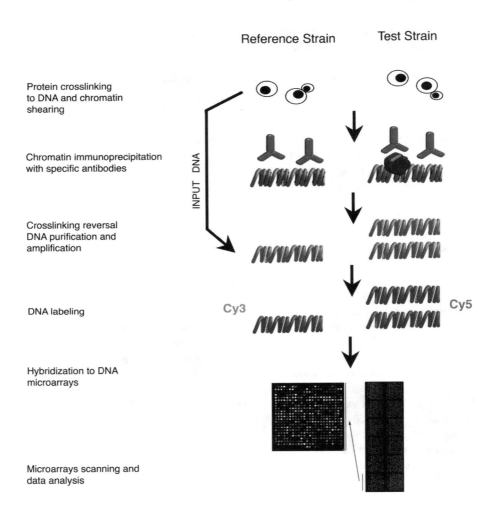

Protein crosslinking
to DNA and chromatin
shearing

Chromatin immunoprecipitation
with specific antibodies

Crosslinking reversal
DNA purification and
amplification

DNA labeling

Hybridization to DNA
microarrays

Microarrays scanning and
data analysis

FIGURE 8.1 *(A color version follows page 204)* General ChIP-on-chip method overview. The "test" and "reference" experiments are run in parallel. The test strain contains the protein of interest (or its epitope-tagged version), whereas the reference strain has neither the protein nor the tag. Chromatin from formaldehyde-treated cells is sheared by sonication to an average size of 500 bp and immunoprecipitated using highly specific antibodies raised against the analyzed protein. After immunoprecipitation and cross-linking reversal, DNA is purified and amplified by PCR. INPUT DNA from the reference strain (cross-linked and sonicated) is used as an alternative in case immunoprecipitation does not yield enough DNA. INPUT DNA is amplified by PCR directly, bypassing the immunoprecipitation step. This approach, however, does not control for any potential weak antibody specificity. Enriched DNA from the test and reference strains is then labeled with the fluorophores Cy3 and Cy5, respectively. The fluorescent DNA probes are subsequently combined and hybridized to a DNA microarray. For any given DNA fragment on the microarray, the ratio of the normalized fluorescent intensities between the two probes reflects the protein enrichment in the test relative to the reference experiment.

S-transferase) or TAP (tandem affinity purification). The tagging of a protein, however, may partially or even fully disrupt its ability to interact with chromosomes. Finally, after cross-linking reversal, DNA is purified, amplified by PCR, and labeled with a fluorophore. One fluorescent dye (e.g., Cy5) is used to label DNA recovered from the "test" strain containing the protein of interest, whereas a second dye (e.g., Cy3) is used for the isogenic control or reference strain lacking that same protein. Background DNA levels recovered from the "reference" strain may be extremely low and, as such, may not be labeled efficiently. Thus, as an alternative control, it is possible to use nonimmunoprecipitated INPUT genomic DNA from the test chromatin after cross-linking and sonication. The problem with this approach, however, is that "INPUT DNA" control does not guarantee against false positives due to weak antibody specificity. Finally, the test- and reference-labeled DNA probes are combined and hybridized onto DNA microarray glass slides containing either intergenic regions (IGRs), open reading frames (ORFs), or both. After scanning for both fluorescent dyes, the normalized ratios of intensities between the two probes reflect the relative enrichment of the protein at different genomic loci. Data normalization and data mining will not be discussed here because it has been reviewed recently, specifically in regard to ChIP-on-chip [27].

YEAST CHIP-ON-CHIP AND BEYOND

The budding yeast *Saccharomyces cerevisiae* was among the first organisms to have its entire genome sequenced [2], primarily because of its relatively small size (12 million bases). As such, it was a prime candidate for early ChIP-on-chip experiments. A full yeast genome can be spotted on a glass microarray with about 12,000 DNA fragments of 1 kb. A similar human microarray, with a genome size roughly 250-fold larger, would require a staggering 3 million spotted elements of 1 kb length, currently not technically achievable on a single chip. Moreover, the low level of repetitive sequences of the yeast genome reduces interpretation uncertainties due to potential cross-hybridization. Unsurprisingly, most published data on ChIP-on-chip are from work on yeast. The proof of the principle came from the binding maps of individual transcription factors such as Gal4, Mbp1, Swi4, Ste12, and Rap1 [17,18,28]. Subsequently, the genomewide binding of over 100 epitope-tagged transcriptional regulators was published [29] in an unprecedented effort to identify interconnection between networks of transcriptional regulators.

CHROMATIN MODIFICATIONS: A CASE STUDY FOR CHIP-ON-CHIP EXPERIMENTS

ChIP-on-chip experiments have been successfully applied to the understanding of histone modifications during gene regulation. These experiments will be discussed in this section in order to best illustrate the complementarity between expression microarrays, binding microarrays, and acetylation microarrays.

In eukaryotes, the packaging of DNA into chromosomes is a crucial component in the regulation of most nuclear processes, including gene regulation, DNA replication, repair, and chromosome recombination. In chromatin, DNA is associated with histone proteins to form nucleosomes, which may hinder or enhance these nuclear events depending on their structure. This may occur through nucleosome positioning, chromatin remodeling, and higher-order chromosome folding, all of which influence DNA accessibility. In addition, histones are the targets of a number of posttranslational modifications such as acetylation, phosphorylation, and methylation. Over the years, considerable amount of effort has been dedicated to the understanding of the purpose and mechanism underlying the regulation of these nuclear events. Histone acetylation is probably the best understood among these molecular switches.

The addition and removal of acetyl groups on lysine residues on histone tails is mediated by histone acetyltransferases and deacetylases, respectively. The numerous members of this class of enzymes have different histone and lysine specificities, and also act at different genomic loci. In yeast, the histone deacetylase Rpd3 is recruited to some promoters by the repressor Ume6 [30], leading to the local deacetylation of all four core histones [26], whereas Hda1 deacetylates histone H3 and H2B at Tup1-repressed loci [31]. In addition, histone acetyltransferases and deacetylases function globally over portions of the genome in an unknown mechanism that does not appear to require targeted recruitment by a transcriptional regulator [32]. However, the histone and gene specificity *in vivo* for some of these enzymes (i.e., Hos1, Hos2, and Hos3) remained elusive, mainly because of the inability to identify their genomic targets using a "one gene at a time" approach. Genomewide studies by expression cDNA microarrays proved unable to unequivocally identify the genes regulated directly by histone deacetylases. For instance, disruption of the gene encoding the histone deacetylase Rpd3 led to more genes' being repressed than derepressed in an expression cDNA microarray [8,9]. This is a common problem with expression cDNA microarrays, in which measurement of gene activity at steady state cannot take into account indirect effects. One way to partially solve this problem would be to take different snapshots over time in an attempt to sort out immediate early-response genes from late-response genes, assuming that the first responders are direct targets. In addition, different activation or repression pathways may converge toward the regulation of a particular gene. Indeed, disruption of a repressor gene may not necessarily lead to a measurable and significant change in gene regulation in the absence of proper activation conditions or because of redundant pathways. Thus, the localization of a gene regulator at a specific genomic locus is a better indicator of its primary function than its potential effect on gene activity. ChIP-on-chip experiments provide an ideal tool. Although Rpd3 is a repressor, its genomewide binding map revealed that it is associated with some genes harboring high transcriptional activity [10]. For instance, Rpd3 is found at the promoter of about half of the highly active ribosomal protein genes in logarithmically growing cells, yet its disruption affects neither their acetylation [12] nor their expression [8,9]. Rpd3 was later shown to play a role in the repression of ribosomal protein genes after inhibition of the nutrient-sensitive Tor pathway with rapamycin [33]. Thus, Rpd3 appears to be poised at ribosomal protein genes, ready to resume its repressive activity in response to poor nutrient availability. This case illustrates the strength of binding microarrays as it revealed functions for Rpd3 that went unnoticed using expression microarrays alone.

On the other hand, failing to detect a protein at a particular genomic locus may inaccurately suggest that the protein does not function at this site. Weak or no cross-linking, epitope masking or even transient interaction of the protein with the chromosome can all contribute to what at first may appear to be a negative result. Unfortunately, there is no easy way to solve this problem other than by altering experiment conditions.

Histones are cross-linked efficiently [15] and the acetylation, methylation, or phosphorylation status of their residues can be monitored by chromatin immuno-precipitation, providing an additional approach to identify histone-modifying enzymes' genomic targets. The availability of very specific antibodies raised against individual acetylated lysine residues on histone tails [26] permits the adaptation of ChIP-on-chip to histone posttranslational modifications [11–13].

Acetylation microarrays have uncovered a "division of labor" for most yeast histone deacetylases [12]. They also allowed the identification of large chromatin domains affected by individual deacetylases such as the subtelomeric domains deacetylated by Hda1 (HAST domains). Importantly, these domains are continuous, encompassing both ORFs and large IGRs. For this reason, their identification would not have been possible using expression arrays. Acetylation microarrays also gave clues for the function of another histone deacetylase, Hos2, whose function was unclear until then. Although Hos2 disruption displayed only a minor effect on acetylation genomewide, the regions affected most appeared to be ORFs, not IGRs [12]. Hos2 function at ORFs was investigated further and, surprisingly, its disruption decreased the activation kinetics of its target genes [34], contradicting the previously accepted notion that all histone deacetylases are transcriptional repressors. Moreover, Hos2 binding on ORFs genomewide correlates positively with gene activity [34].

It is believed that acetylation is globally associated with gene activity and deacetylation with gene repression. For instance, silenced heterochromatin is hypoacetylated as opposed to the hyperacetylated state of transcriptionally competent euchromatin. However, the status and contribution of individual lysine residues in euchromatin during gene transcription was unclear. Acetylation patterns were carefully analyzed at individual genes both under repressive and induced conditions. Because gene regulation comes in a variety of flavors, it may be erroneous to extrapolate data from a single gene to the entire genome. The use of acetylation microarrays allows for the opportunity to ask whether different histone modification patterns are associated with specific functions.

This question was investigated recently in an effort to systematically map genomewide acetylation sites in wild-type cells [13]. This study showed that hyperacetylation of histone H3 lysine 9 (H3K9), H3K18, and H3K27, but hypoacetylation of H4K16, both at promoters and over ORFs, correlates positively with transcription. Most interestingly, specific acetylation patterns turned out to be associated with biologically related gene clusters that respond to particular environment conditions such as heat and cold shock or nitrogen starvation. Specific acetylation patterns may create specialized interaction surfaces for given chromatin regulatory proteins. For instance, chromatin interaction with the bromodomain protein Bdf1 requires a positive charge (deacetylated lysine) at H4K16. Similarly, artificial H4K16 acetylation at the telomere abolishes recruitment of the silencing regulatory protein SIR3 [35,36]. These experiments provide mechanistic insights into gene regulation that were not revealed by expression cDNA microarrays.

CHIP-ON-CHIP AND BIOMEDICAL APPLICATIONS

Profiling of gene expression using cDNA microarrays has been used to classify diseases or tumors with the hope of its being able to refine and customize both diagnosis and treatment. Such an approach, however, would require the ability to accurately compare gene expression between patients and healthy individuals, which is not a trivial matter. Different genetic backgrounds among individuals may alone account for substantial variations in gene expression. RNA sampling and pooling from a large population will level out natural differences but will defeat one main purpose of gene profiling, namely, customized treatment for individuals. Cell homogeneity or lack thereof is another crucial parameter in gene profiling, because crude biopsies often contain many different cell types. Nevertheless, cDNA microarrays have been applied to the profiling of a variety of pathologic conditions, including among others cardiovascular disease, diabetes, breast cancer, or immune diseases (for a review see Weeraratna et al. [37]). Although gene profiling is unequivocally a useful tool for drug and clinical research, its "field" application to both diagnostics and treatment of individual patients will require appropriately devised experimental protocols and data analysis algorithms to separate "noise" from genuine disease-induced expression patterns. ChIP-on-chip will face the same challenges described in this paragraph. A combination of the two arrays approaches (expression and ChIP-on-chip) may be part of the answer.

As mentioned earlier in this chapter, indirect effects are less of a concern for gene profiling in clinical applications, because all that really matters is the change of global expression patterns and not the underlying molecular mechanisms. In that respect, ChIP-on-chip is not necessarily complementary to expression microarrays. ChIP-on-chip, however, may have its own specialized role in areas that cannot be adequately investigated with expression microarrays, such as histone modifications. Covalent histone modifications are linked to tumerogenesis and metastasis mostly through aberrant gene expression [38]. The importance of histone deacetylation has been highlighted in human leukemia [39,40]. Two forms of acute promyelocytic leukemia (APML) are caused by chromosomal translocations, which create oncogenic fusion proteins between the retinoic acid receptor (RAR) and either PML (promyelocytic leukemia) or PFLZ (promyelocytic leukemia zinc finger); these ectopically recruit histone deacetylases, leading to aberrant deacetylation. Naturally, efforts are currently under way to analyze the potential therapeutic action of histone deacetylase inhibitors. Suberoylanilide hydroxamic acid (SAHA) is such an inhibitor whose antitumoral activity appears to commit target cells to apoptosis [41]. In that context, one could envision the use of "acetylation" microarrays to investigate the existence of a histone modification signature, specific to a particular type of cancer; this could be acetylation, phosphorylation, methylation, ubiquitination, sumolation, or ADP-ribosylation. The unraveling of potential tumor-specific signatures at precise genes could not only provide formidable insights into tumorigenesis but also aid the design of highly specific inhibitors, thereby minimizing the risk of side effects.

CONCLUDING REMARKS

The incomplete coverage of current DNA microarrays for higher eukaryotic genomes limits their full extension to ChIP-on-chip in theses organisms. Nevertheless, it is possible to prepare custom microarrays that contain a portion of the genome of interest. Using this approach, several groups have applied ChIP-on-chip to *Drosophila melanogaster* [14] and humans [42,43]. ChIP-on-chip resolution can be greatly improved by using tiled microarrays. The preparation of such microarrays usually leads to a decrease in the coverage of the genome of interest since those would require a staggering amount of synthesized probes that would not fit into a single chip. Thus, researchers who are currently using tiled microarray for ChIP-on-chip experiments have limited themselves to small areas of various genomes. In the near future, much needed higher-density microarrays will certainly be available.

In the postgenomic era, ChIP-on-chip experiments will be a crucial tool for deciphering the regulatory networks within the nucleus and their interaction with the genome.

ACKNOWLEDGMENTS

I am grateful to Adam Sperling for suggestions and critical reading of the manuscript.

REFERENCES

1. Smith, H.O., Tomb, J.F., Dougherty, B.A., Fleischmann, R.D., Venter, J.C. Frequency and distribution of DNA uptake signal sequences in the *Haemophilus influenzae* Rd genome. *Science* 269(5223): 538–540, 1995.
2. Goffeau, A., Barrell, B.G., Bussey, H., Davis, R.W., Dujon, B., Feldmann, H., Galibert, F., Hoheisel, J.D., Jacq, C., Johnston, M., Louis, E.J., Mewes, H.W., Murakami, Y., Philippsen, P., Tettelin, H., Oliver, S.G. Life with 6000 genes. *Science* 274(5287): 546, 563–567, 1996.
3. Adams, M.D., Celniker, S.E., Holt, R.A., Evans, C.A., Gocayne, J.D., Amanatides, P.G., Scherer, S.E., Li, P.W., Hoskins, R.A., Galle, R.F., George, R.A., Lewis, S.E., Richards, S., Ashburner, M., Henderson, S.N., Sutton, G.G., Wortman, J.R., Yandell, M.D., et al. The genome sequence of *Drosophila melanogaster*. *Science* 287(5461): 2185–2195, 2000.
4. Lander, E.S., Linton, L.M., Birren, B., Nusbaum, C., Zody, M.C., Baldwin, J., Devon, K., Dewar, K., Doyle, M., FitzHugh, W., Funke, R., Gage, D., Harris, K., Heaford, A., Howland, J., Kann, L., Lehoczky, J., LeVine, R., McEwan, P., et al. Initial sequencing and analysis of the human genome. *Nature* 409(6822): 860–921, 2001.
5. Venter, J.C., Adams, M.D., Myers, E.W., Li, P.W., Mural, R.J., Sutton, G.G., Smith, H.O., Yandell, M., Evans, C.A., Holt, R.A., Gocayne, J.D., Amanatides, P., Ballew, R.M., Huson, D.H., Wortman, J.R., Zhang, Q., Kodira, C.D., Zheng, X.H., Chen, L., Skupski, et al. The sequence of the human genome. *Science* 291(5507): 1304–1351, 2001.
6. Fodor, S.P., Read, J.L., Pirrung, M.C., Stryer, L., Lu, A.T., Solas, D. Light-directed, spatially addressable parallel chemical synthesis. *Science* 251(4995): 767–773, 1991.

7. Schena, M., Shalon, D., Davis, R.W., Brown, P.O. Quantitative monitoring of gene expression patterns with a complementary DNA microarray. *Science* 270(5235): 467–470, 1995.

8. Bernstein, B.E., Tong, J.K., Schreiber, S.L. Genomewide studies of histone deacetylase function in yeast. *Proc Natl Acad Sci U S A* 97(25): 13708–13713, 2000.

9. Hughes, T.R., Marton, M.J., Jones, A.R., Roberts, C.J., Stoughton, R., Armour, C.D., Bennett, H.A., Coffey, E., Dai, H., He, Y.D., Kidd, M.J., King, A.M., Meyer, M.R., Slade, D., Lum, P.Y., Stepaniants, S.B., Shoemaker, D.D., Gachotte, D., Chakraburtty, K., Simon, J., Bard, M., Friend, S.H. Functional discovery via a compendium of expression profiles. *Cell* 102(1): 109–126, 2000.

10. Kurdistani, S.K., Robyr, D., Tavazoie, S., Grunstein, M. Genomewide binding map of the histone deacetylase Rpd3 in yeast. *Nat Genet* 31(3): 248–254, 2002.

11. Bernstein, B.E., Humphrey, E.L., Erlich, R.L., Schneider, R., Bouman, P., Liu, J.S., Kouzarides, T., Schreiber, S.L. Methylation of histone H3 Lys 4 in coding regions of active genes. *Proc Natl Acad Sci U S A* 99(13): 8695–8700, 2002.

12. Robyr, D., Suka, Y., Xenarios, I., Kurdistani, S.K., Wang, A., Suka, N., Grunstein, M. Microarray deacetylation maps determine genomewide functions for yeast histone deacetylases. *Cell* 109(4): 437–446, 2002.

13. Kurdistani, S.K., Tavazoie, S., Grunstein, M. Mapping global histone acetylation patterns to gene expression. *Cell* 117: 721–733, 2004.

14. Schubeler, D., MacAlpine, D.M., Scalzo, D., Wirbelauer, C., Kooperberg, C., van Leeuwen, F., Gottschling, D.E., O'Neill, L.P., Turner, B.M., Delrow, J., Bell, S.P., Groudine, M. The histone modification pattern of active genes revealed through genomewide chromatin analysis of a higher eukaryote. *Genes Dev* 18(11): 1263–1271, 2004.

15. Solomon, M.J., Varshavsky A. Formaldehyde-mediated DNA-protein crosslinking: a probe for in vivo chromatin structures. *Proc Natl Acad Sci U S A* 82(19): 6470–6474, 1985.

16. Solomon, M.J., Larsen, P.L., Varshavsky, A. Mapping protein-DNA interactions in vivo with formaldehyde: evidence that histone H4 is retained on a highly transcribed gene. *Cell* 53(6): 937–947, 1988.

17. Iyer, V.R., Horak, C.E., Scafe, C.S., Botstein, D., Snyder, M., Brown, P.O. Genomic binding sites of the yeast cell-cycle transcription factors SBF and MBF. *Nature* 409(6819): 533–538, 2001.

18. Ren, B., Robert, F., Wyrick, J.J., Aparicio, O., Jennings, E.G., Simon, I., Zeitlinger, J., Schreiber, J., Hannett, N., Kanin, E., Volkert, T.L., Wilson, C.J., Bell, S.P., Young, R.A. Genomewide location and function of DNA binding proteins. *Science* 290(5500): 2306–2309, 2000.

19. Fujita, N., Jaye, D.L., Kajita, M., Geigerman, C., Moreno, C.S., Wade, P.A. MTA3, a Mi-2/NuRD complex subunit, regulates an invasive growth pathway in breast cancer. *Cell* 113(2): 207–219, 2003.

20. Gilbert, C., Kristjuhan, A., Winkler, G.S., Svejstrup, J.Q. Elongator interactions with nascent mRNA revealed by RNA immunoprecipitation. *Mol Cell* 14(4): 457–464, 2004.

21. Bernstein, B.E., Humphrey, E.L., Liu, C.L., Schreiber, S.L. The use of chromatin immunoprecipitation assays in genomewide analyses of histone modifications. *Methods Enzymol* 376: 349–360, 2004.

22. Horak, C.E., Snyder, M. ChIP-chip: a genomic approach for identifying transcription factor binding sites. *Methods Enzymol* 350: 469–483, 2002.

23. Ren, B., Dynlacht, B.D. Use of chromatin immunoprecipitation assays in genomewide location analysis of mammalian transcription factors. *Methods Enzymol* 376: 304–315, 2004.

24. Robyr, D., Grunstein, M. Genomewide histone acetylation microarrays. *Methods* 31(1): 83–89, 2003.

25. Robyr, D., Kurdistani, S.K., Grunstein, M. Analysis of genomewide histone acetylation state and enzyme binding using DNA microarrays. *Methods Enzymol* 376: 289–304, 2004.

26. Suka, N., Suka, Y., Carmen, A.A., Wu, J., Grunstein, M. Highly specific antibodies determine histone acetylation site usage in yeast heterochromatin and euchromatin. *Mol Cell* 8(2): 473–479, 2001.

27. Buck, M.J., Lieb, J.D. ChIP-chip: considerations for the design, analysis, and application of genomewide chromatin immunoprecipitation experiments. *Genomics* 83(3): 349–360, 2004.

28. Lieb, J.D., Liu, X., Botstein, D., Brown, P.O. Promoter-specific binding of Rap1 revealed by genomewide maps of protein-DNA association. *Nat Genet* 28(4): 327–334, 2001.

29. Lee, T.I., Rinaldi, N.J., Robert, F., Odom, D.T., Bar-Joseph, Z., Gerber, G.K., Hannett, N.M., Harbison, C.T., Thompson, C.M., Simon, I., Zeitlinger, J., Jennings, E.G., Murray, H.L., Gordon, D.B., Ren, B., Wyrick, J.J., Tagne, J.B., Volkert, T.L., Fraenkel, E., Gifford, D.K., Young, R.A. Transcriptional regulatory networks in *Saccharomyces cerevisiae*. *Science* 298(5594): 799–804, 2002.

30. Kadosh, D., Struhl, K. Repression by Ume6 involves recruitment of a complex containing Sin3 corepressor and Rpd3 histone deacetylase to target promoters. *Cell* 89(3): 365–371, 1997.

31. Wu, J., Suka, N., Carlson, M., Grunstein, M. TUP1 utilizes histone H3/H2B-specific HDA1 deacetylase to repress gene activity in yeast. *Mol Cell* 7(1): 117–126, 2001.

32. Vogelauer, M., Wu, J., Suka, N., Grunstein, M. Global histone acetylation and deacetylation in yeast. *Nature* 408(6811): 495–498, 2000.

33. Rohde, J.R., Cardenas, M.E. The tor pathway regulates gene expression by linking nutrient sensing to histone acetylation. *Mol Cell Biol* 23(2): 629–635, 2003.

34. Wang, A., Kurdistani, S.K., Grunstein, M. Requirement of Hos2 histone deacetylase for gene activity in yeast. *Science* 298(5597): 1412–1414, 2002.

35. Kimura, A., Umehara, T., Horikoshi, M. Chromosomal gradient of histone acetylation established by Sas2p and Sir2p functions as a shield against gene silencing. *Nat Genet* 32(3): 370–377, 2002.

36. Suka, N., Luo, K., Grunstein, M. Sir2p and Sas2p opposingly regulate acetylation of yeast histone H4 lysine16 and spreading of heterochromatin. *Nat Genet* 32(3): 378–383, 2002.

37. Weeraratna, A.T., Nagel, J.E., de Mello-Coelho, V., Taub, D.D. Gene expression profiling: from microarrays to medicine. *J Clin Immunol* 24(3): 213–224, 2004.

38. Hake, S.B., Xiao, A., Allis, C.D. Linking the epigenetic "language" of covalent histone modifications to cancer. *Br J Cancer* 90(4): 761–769, 2004.

39. Grignani, F., De Matteis, S., Nervi, C., Tomassoni, L., Gelmetti, V., Cioce, M., Fanelli, M., Ruthardt, M., Ferrara, F.F., Zamir, I., Seiser, C., Lazar, M.A., Minucci, S., Pelicci, P.G. Fusion proteins of the retinoic acid receptor-alpha recruit histone deacetylase in promyelocytic leukaemia. *Nature* 391(6669): 815–818, 1998.

40. Lin, R.J., Nagy, L., Inoue, S., Shao, W., Miller, W.H., Jr., Evans, R.M. Role of the histone deacetylase complex in acute promyelocytic leukaemia. *Nature* 391(6669): 811–814, 1998.

41. Mitsiades, C.S., Mitsiades, N.S., McMullan, C.J., Poulaki, V., Shringarpure, R., Hideshima, T., Akiyama, M., Chauhan, D., Munshi, N., Gu, X., Bailey, C., Joseph, M., Libermann, T.A., Richon, V.M., Marks, P.A., Anderson, K.C. Transcriptional signature

of histone deacetylase inhibition in multiple myeloma: biological and clinical implications. *Proc Natl Acad Sci U S A* 101(2): 540–545, 2004.

42. Horak, C.E., Mahajan, M.C., Luscombe, N.M., Gerstein, M., Weissman, S.M.,
Snyder, M. GATA-1 binding sites mapped in the beta-globin locus by using mammalian chip–chip analysis. *Proc Natl Acad Sci U S A* 99(5): 2924–2929, 2002.

43. Li, Z., Van Calcar, S., Qu, C., Cavenee, W.K., Zhang, M.Q., Ren, B. A global
transcriptional regulatory role for c-Myc in Burkitt's lymphoma cells. *Proc Natl Acad
Sci U S A* 100(14): 8164–8169, 2003.

9 DNA Microarrays as Functional Genomics Tools for Cancer Drug Discovery

Arindam Bhattacharjee

CONTENTS

ABSTRACT

Developments in DNA microarray technology and related "-omics" technologies have generated a vast amount of data that allows for a higher level of scrutiny of the human genome. This provides a tremendous opportunity for describing and understanding not only the wiring diagram and annotation of the human genome but also molecular defects in cancer genomes. These insights could significantly improve health-care management of human cancers by molecular targeted therapy, method development, and microarray assay-based screening techniques, among others. In this chapter, the usage of DNA microarrays in cancer biology and related approaches in drug discovery programs is reviewed, and recent approaches that have shown promise are discussed.

INTRODUCTION: DNA MICROARRAYS IN DRUG DISCOVERY

Development of the microarray technology in the last decade was aided by the interest of cancer drug manufacturers and biotechnology companies, as well as funding of cancer research by public groups (e.g., funding for cancer and genomics programs in the U.S. by the National Institutes of Health, NHGRI, and NCI programs such as the Directors Challenge Program). Public interest and funding resulted in a comprehensive body of knowledge, which, on a basic level, provided a catalog of expression signatures for human cancers (http://dc.nci.nih.gov/organization/publications, accessed February 2005). However, the efforts of this scientific community had broader implications beyond a cancer genome's afflicted genes or wiring diagram. The interest also led to a rapid commercialization of DNA microarray technology in recent years. Today, "tool maker" companies such as Agilent, Affymetrix, and Nimblegen have developed DNA microarray technologies, now routinely used in drug discovery and development efforts. The tool makers hope to prevail in their respective markets by making the drug discovery and development process successful, effective, efficient, and economical. Although, as many are quick to point out, these gene catalogs and parts lists are still in their infancy, a long way from curing cancer and filling up the drug discovery pipelines, the efforts have undoubtedly culminated in a solid understanding of microarray technology and molecular images of human cancers and syndromes. Without a thorough understanding of the molecular circuitry in cancer cells, finding new promising drugs will remain a formidable challenge. Until then, cancer patients will continue to be treated in a nonelegant fashion with conventional chemotherapy, which has nonspecific effects.

This chapter outlines the context in which DNA microarray technologies and applications are used in cancer research and the impact of this on drug discovery. Subsequently, new approaches in discovery and screens that utilize microarrays are reviewed. Finally, the impact of the technology in discovery efforts and possible uses of microarrays in the future are discussed.

FROM CANCER CLASSIFICATION
TO DRUG DISCOVERY

The early landmark paper, which unleashed the power of gene expression and microarray technologies, demonstrated accurate classification of hematologic malignancy-acute myeloid leukemia (AML) and acute lymphoid leukemia (ALL) [1]. This study demonstrated that molecular signatures could clearly classify patients at the clinic. The study led the way for microarrays to gain importance in several areas: tumor classification, prediction of tumor classes, molecular diagnostics, and revelation of the genetic defects, the Achilles' heel of cancers. Subsequently, many solid tumor studies were performed using DNA microarrays. For example, in breast and lung cancer, the focus moved from tumor classification based on unsupervised clustering of gene expression index [2–4] to dissection of solid tumors in the context of patient survival [5] and defining tumors by metastasis signatures from multiple data sets and cancer subtypes [6,7]. These cancer studies were able to reveal gene expression class-based molecular classification of cancer, and also to perform supervised analyses in which classes with different patient outcomes were associated with dysregulated genes or gene sets. In the breast cancer study [5], the set of genes that correlates with good or bad patient outcomes (metastasis or no metastasis) was determined, as were the marker genes for the estrogen receptor and BRCA1 pathway defects [5]. From this arose the notion that transcriptional profiling (which allowed monitoring of genes in a massive parallel fashion at a genomewide level) could help discover new tumor classes, pathway defects, patient stratification for treatment, and discovery of new drug candidates. Similarly, the discovery of fms-like tyrosine kinase 3 (flt3) receptor overexpression was observed in a new tumor class in ALL and AML patients, namely, MLL-rearranged acute lymphoblastic leukemia (MLL), which subsequently revealed mutations in flt3 receptor tyrosine kinase (RTK) [8,9]. Drugs that could target flt3, such as PKC-412, MLN-518, and CEP-701, could shut off signaling through flt3, especially in ft3 overexpressing cells or cells that harbored activating mutations in flt3. Phase I clinical studies with such drugs have shown modest tumor responses [10–12]. These results were contrasted with imatinib mesylate (Gleevec, Glivec, STI-571), a tyrosine kinase receptor inhibitor, which targets the ATP-binding portion of the kinase domain, shutting off kinase activity. The drug targets c-kit/PDGRF translocations or mutations in gastrointestinal stromal cancers (GIST) as well as bcr-abl mutant kinase fusion/translocation events in chronic myelogenous leukemia (CML), and the results in these tumors were impressive. The drug was recently approved by FDA and reduces tumor burden by several orders of magnitude in patients. More recently, mutations in the kinase domain of EGFR (another RTK) in a subclass of lung cancer (nonsmall cell lung cancer, NSCLC) correlate with sensitivity to inhibitors in the kinase domain of EGFR, providing a rational explanation for both the susceptibility of this class of cancer to such druggable compounds and the superior response in patients, albeit in less than 10% of NSCLC patients [13,14]. These three examples have therefore generated immense interest in looking at RTKs and their susceptibility to small molecules.

HUNTING AND SCREENING APPROACHES
IN DRUG DISCOVERY

Typically, there are two approaches to drug discovery. One method begins with screening for compounds from causative genes. The other approach is to first screen compounds for desirable effects, then pinpoint the causative genes. Following in the footsteps of the genome sequencing efforts, the early DNA microarray research community primarily aimed at selecting a gene list or testing a hypothesis by conducting partial genome screens. Because molecular classification of tumors could be achieved with fewer genes or features, and knowledge of molecular data was rudimentary, early efforts in microarrays were relegated to the molecular aspect of cancer classification. The recent availability of whole genome arrays has alleviated such constraints [15,16], making it possible to discover precise defects in pathways at the transcriptional level. In a recent talk at a scientific meeting on cancer research [17], it was emphasized that the low number of truly druggable targets (only a few hundred) was due to the poor annotation of the human genome (description or knowledge of gene function). The ability to create large repositories of data using microarray assays exists; however, the challenge is to make sense of the data and enable problem solving, generating new questions, and validating experiments that ultimately lead to generation of new drug compounds or therapies. One should not lose sight of the fact that proteins are the ultimate arbitrators of cellular function, and the genes encode the proteins. These networks and their role in tumorigenesis must be understood to determine effective therapies. Much to the dismay of Wall Street, the proposition that microarrays can quickly deliver new compounds has not proven to be true. In contrast, recent success with RTK inhibitors in cancers has renewed the hope of mechanism-based therapeutics.

Although all screening approaches cannot be reviewed here, a few deserve to be mentioned in the context of microarrays. The importance of drug screens using yeast or NCI-60 cancer cell lines was recognized very early in genomics-based screening [18–20]. Both specific targeting of pathways and wide screens are needed for drug discovery [21] and, in some cases, can be addressed by microarray-based assays. The insights that led to the discovery of Gleevec and its therapeutic impact are not readily replicated for the discovery of other drugs in which the same pathways are affected. Incomplete understanding of pathways and redundancies create challenges for screening approaches. It was noted that when lesions in RTK pathways were involved later in tumor progression, Gleevec was less effective, and thus for the concept of single-agent therapy, stage and state of tumor are important factors in determining clinical outcome [22]. Note that this statement does not indicate that microarrays cannot be used to discover compounds. Solving diseases such as cancer requires additional efforts far beyond whole genome analysis, a quick shortcut screen, and platform specifications such as sensitivity, specificity, and feature/spot density. It is accepted that cancer is plagued by complexity because of the etiology of the disease, patient response, heterogeneity and polygenic traits (rarely does cancer evolve from a single genetic event), tumor evolution, increased mutation rates, and chromosomal aberrations [22]. Thus, a battery of biological applications are attempted to precisely understand the problem and to test for specificity of the cancer drug in disparate disease states. DNA microarrays will invariably be included in the assortment of technological tools utilized.

Recently, Stegmaier and colleagues used microarrays to study AML differentiation, and were able to determine five predictive marker genes that reflected the change of AML cells to normal blood cell phenotype, as initiated by retinoic acid [23]. This showed that surrogate marker analysis of this desirable clinical end point could be achieved using microarray expression assays. The method involved identification of predictive markers, which were subsequently monitored by mass spectrometry, to screen thousands of drug candidates that elicited desirable responses following treatment. This approach bypassed the need to understand mechanisms of action or identification of compounds that interact with a single target and are therefore prone to poor responses because of issues of pathway redundancies or multiple aberrations of the cancer genome. This phenotypic screening approach holds considerable promise in pharmacology and discovery, because there are issues surrounding toxicity of targets, as well as the need to identify more compounds with desirable end points. However, the successful utilization of this method is dependent upon a priori knowledge and known outcomes that are desirable, such as phenotypes or markers. This approach also holds promise in screening additional compounds that help in creating robust drug classes or clinical end points.

Yet another approach that deserves to be mentioned was the discovery of RITA, an inhibitor that blocked interaction of p53 vs. hdm2, from a broad cell-based screen that sensitized tumors with functional p53 in in vitro and in vivo experiments [21]. In the future, additional compounds that elicit similar responses can be screened by approaches using microarrays to get detailed molecular views of drug activity or toxicity profiles. Additional approaches will be reviewed in application-specific methods (see the following text).

From a technological point of view, microarray assays have a broad impact on the dissection of cancer genomes and therapeutic intervention (see schema in Figure 9.1). The microarray technology is adaptable and flexible in the sense that one can query different molecular states in the cell by using the same basic platform for different microarray assays, such as array-based CGH, transcription profiling, or genomewide chromatin immunoprecipitation experiments [24]. Thus, instead of a single technology approach, questions relating to cancer drug discovery can be tailored to the most appropriate technology. The questions can also fit into mechanistic or phenotypic queries as described earlier or in the context of systems biology-level investigations. On occasion, DNA microarrays and a suitable platform, such as ChIP-on-chip or transcriptional profiling, would be the discovery tools of choice. The completion of the human genome sequencing project, as well as genomes of several model organisms, has enabled oligonucleotide DNA microarray-based research to fit into any of these assay modalities, thereby providing immense flexibility in querying the sequenced genomes for specific answers to address the overall goals of research and discovery projects. The methods described in the following paragraphs are powerful aids in understanding cancer genomes of patients and their response to particular drugs; how they can be developed or discovered is outlined in Figure 9.2.

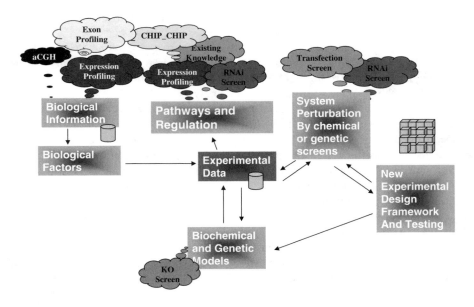

FIGURE 9.1 Cancer is a complex disease of impaired cellular process and occurs in diverse cell types. Many genes in the genome regulate cell cycle control and differentiation, thereby involving various genes and pathways that are causative and predictive and whose activities evolve with onset of cancer. In one view of systems biology, the biological knowledge and experimental data are used to define the relationships between various components of the system of interest and to extract biological variables. These relationships and variables can be used to build preliminary models to describe a particular biological process such as carcinogenesis or drug discovery. A comparison between the experimental data generated from perturbations of the system and predictions from various candidate models can be used to screen for the model or a set of parameters that best describes the phenomenon or observed data. One may also design new experiments eliciting different system responses between models to discriminate among candidate models. The outcome of this exercise can have far reaching effects on understanding the pathobiology of cancer at the molecular level and possible interventions. (Portions of the schema are adapted from Aggarwal, K., Lee, K.H. Functional genomics and proteomics as a foundation for systems biology. *Brief Funct Genomic Proteomic*. 2(3): 175–184, 2003).

THE ERA OF DNA MICROARRAYS

GENE EXPRESSION TECHNOLOGY AND THE FUTURE

Microarrays have been used for determining gene annotation, pathway dissection, tumor class prediction and discovery, drug marker identification, markers for drug toxicity, and associating tumor classes to clinical outcomes or drug responses [25]. However, as with any technology, there are caveats. For example, the susceptibility of NSCLC patients to gefitinib (Iressa) could not be determined by expression technology. Similarly, some tumor classes were tied to clinical outcomes [3,4], but it was difficult to pinpoint drug targets that would benefit these discovered tumor

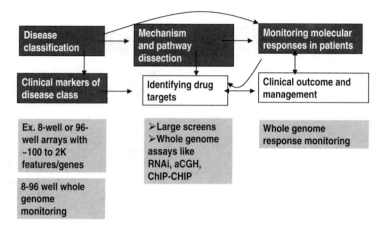

FIGURE 9.2 In diseases such as cancer, systems biology-level approaches are needed to understand basic gene regulation (i.e., a list of genes that are classifiers) or, perhaps, to predict class of distinct patient groups. Classification and its accuracy are very important for first-pass diagnosis of the disease in patient populations to better detect, diagnose, and describe the disease; these results have an impact on drug response and clinical trials. A tumor may look phenotypically identical by morphometric analysis, but could be very different at the molecular level as has been discovered for many solid tumors like lung cancer [3,4,6,13,14]. As evident from reviewing these studies, alterations in molecular circuitry that are causative of disease, or are markers in cohorts of patients, can also indicate druggable pathways. Sometimes, the pathway changes are not discernible or, perhaps, difficult to use as a target. A recent group has shown the importance of directly moving into marker responses that indicate drug activity match [23]. Complementary approaches may involve use of siRNA-based approaches to yield similar results or define drug function and mechanism of action [17]. The notion exists that profile-based, nonhypothesis-driven screen-based, and mechanism-based drug selection approaches are very powerful and will triumph over existing drug selection methodologies. Thus, from the point of view of finding a drug or intervention efforts, an approach using a combination of genomics and proteomics is very powerful and comprehensive. Diagnostics, discovery, and clinical management of patients can be addressed by microarrays, although all of them cannot be monitored by genomics approaches alone. Because of several legacy issues such as pathologists and their clinical practices, utilization of microarrays for determining cancer markers and their applicability for diagnostics and clinical trial evaluations has not become commonplace. Other usage needs for end users are as follows: density of features per slide, flexibility depending on assay type, need for more bases per feature (say, 60 [i.e., 90 mers] and beyond), a low-cost envelope for screens, high throughput without compromising density of features, adaptability of general assays and methods to array-based molecular diagnostics, and gene or pathway discovery.

classes from the list of genes that were deregulated and that initially characterized these tumor classes, even though the diagnostic benefit of classification existed. Although transcriptome analysis is a critical technology, other platforms could provide valuable information in conjunction or independently. The combined information needs to be effectively harnessed and orchestrated for a successful drug discovery program.

There are additional reasons why profiling alone cannot resolve all issues in drug discovery. First, cancer often involves changes in genetic programs and clonal expansion of a particular cell type. Although this causes widespread shifts in gene expression baseline measurements in the cancer tissue, the changes in specific genes may not be the cause of tumorigenesis. Second, transcription profiling is a snapshot of the overall steady-state level of mRNA and only a part of the gene expression program that involves synthesis and modifications in DNA, mRNA, and proteins. Multiple events before and after transcription regulate the final outcome of genetic expression. Some aspects of regulation may not be discovered because of poor sensitivity of the platform. Therefore, the need to perform diverse assays in addition to transcriptome analysis exists. Also, from a cell biology point of view, there are a large set of cellular conditions but only a few catalog gene expression compendiums to study such cellular states (see bottom part of Figure 9.2). Thus, the need for 96-well array formats has evolved. Finally, measurement of nucleic acids is typically noisy, so most users automatically use quantitative RT-PCR for validating results. However, transcriptome analysis can only uncover one piece of the story, and validating a biological artifact by an alternative technology or platform does not provide much value.

ALTERNATIVE DNA MICROARRAY-BASED TECHNOLOGIES IN CANCER RESEARCH AND INTEGRATIVE GENOMICS

Investigations in basic cancer biology have shown that many diverse biological assays can essentially be conceived and performed in conjunction with DNA microarrays. Microarrays can essentially handle a wide variety of assays, such as CGH, ChIP-on-chip, methylation, RNAi, and alternative splicing [24,26,27]. Today, most genome scientists in cancer biology intend to query multiple data sets, using diverse technologies to determine specific events that can form a foundation for therapeutic strategies. The ability to conduct such searches and analysis is no longer limited by the sequencing of genomes, but more by data analysis methods and data integration tools. Each of these data analysis tools allows scientists to look at specific, overlapping events to frame the overall portrait of a disease state and relevant therapeutic intervention strategies.

Alternative Splicing

A significant portion of the expressed genome exhibits exon rearrangement that leads to complexity of the proteome. Several groups have shown the ability to determine and monitor splice variants in various disease states via microarrays [27–29]. It is interesting that point mutations could affect splicing, in which the changes can be more readily discerned than from the mutation itself, especially in heterogeneous tumor populations. Given the mutation-based onset of several human cancers, this could be important in patient stratification and in understanding the mechanisms of action and drug discovery. In short, this technology reveals yet another layer of complexity in the expressed genome that was previously difficult to study, and holds considerable promise in revealing these events in the future.

RNA Interference

Although the RNAi phenomenon has been known for a while, only recently has it been possible to use it in mammalian systems [30]. RNAi technology can employ small interfering RNA (siRNA) or short hairpin RNAs (shRNA) to knock down specific expressed genes, thereby acting as a surrogate drug compound for testing genes and pathways that can halt cancer progression. The end points of these assays, such as drug effect or pathway effect, can be measured by DNA microarrays as readout systems [26,31]. Although earlier attempts in the creation of mammalian DNA libraries of shRNA vectors used off-line synthesized oligonucleotides, newer generation of libraries can be created by designing arrays that produce all possible target sequences in parallel on a microarray; these sequences are subsequently immortalized in cloning vectors [32]. The short hairpin target clones can be followed in cell populations by monitoring bar-code-tagged clones using a DNA microarray experiment [26]. Bernards and colleagues identified the affected NFkB pathway members in cylindromatosis disease (CYLD) using an RNAi screening approach and microarrays [17]. This led the team to identify and demonstrate the effectiveness of aspirin because of its known impact on blocking the NFκB pathway.

Array CGH

Comparative genomic hybridization (CGH) and fluorescent *in situ* hybridization techniques (FISH) have been well described in the cancer literature [33]. The implementation of CGH assay on microarrays can focus on genes and minimal regions of recurrent aberration. Early progress involved the use of BAC and cDNA arrays [33,34]. Recently, Lucito and colleagues published their work on oligonucleotide arrays using partial genome representation methods [35,36] to sample and label only a small portion of the genome via restriction enzymes. More recently, several groups have demonstrated the use of DNA oligonucleotide arrays using whole genome labeling and amplification methods [37,38]. This is significant, as the approaches can now be readily combined for genomewide analysis at the DNA level, such as methylation or ChIP-on-chip assays [24]. Array CGH has also been favored in diagnostic approaches because of its monotonic intensity across the genome, which allows for more simple normalization algorithms and accurate determination of aberrations against a normal genome. In combination with transcriptome analysis, the insights can be very powerful for determining gene dosage effects on pathways and chromosomal breakpoints. An alternative technology is the high-density single nucleotide polymorphism (SNP) screen for determining loss of heterozygosity (LOH) in cancer cells. LOH studies using SNP arrays can basically interrogate regions lost in chromosomes, which could harbor tumor suppressor genes [39,40]. However, this technique is dependent on SNP density [41] and may not be very valuable in studies in which such polymorphisms may be nonexistent at high density or in inbred model organisms such as mice.

Array-Based Sequencing

Sequencing technology has been used to discover activating mutations and develop treatment strategies. This has spurred interest in sequencing patient tumors for RTK mutations [13,14,42]. Gene-based resequencing is possible on oligonucleotide arrays and is perhaps more meaningful in a stepwise fashion as lesser quantities of genomic material are involved in the assay. The findings from several groups also indicate that, rather than tiling the whole gene, selecting areas of the gene that harbor mutations may be more effective [13,42]. If a typical gene is 1.2 knt in size, one could determine single base changes by querying $4 \times 1200 = 4800$ probe sequences. Thus, on a single array with a feature density of 50,000, roughly 10 distinct kinases and their targets could be sequenced. If only a portion of the region harboring mutations is used, almost all tyrosine kinase receptors could be screened for mutations, using a single DNA microarray of 50,000 features.

Chromatin IP and Regulome Analysis

Microarray methods have now been adapted to dissecting regulation of the genome [43]. The microarray-based chromatin immunoprecipitation (ChIP) experiments are well developed for understanding transcription factor function, chromatin modification, epigenetics (e.g., methylation), and its role in cancer. Genomewide analysis of transcription factors and their targets holds immense promise in cancer biology and drug discovery, and in teasing apart the function of proteins. The importance of protein interaction was demonstrated for the annotation of menin (a tumor suppressor protein whose dysregulation causes multiple endocrine neoplasia), which was solved by several protein-interaction experiments [44]. Microarray analysis involving ChIP demonstrated how RB pocket proteins and interactors are involved in regulating a cancer cell [45]. ChIP-on-chip assays have also demonstrated estrogen receptor alpha-mediated epigenetic silencing in breast cancer [46,47] and the regulatory role of c-myc in Burkitts lymphoma [48].

SUMMARY

Analyzing multiple assay data sets of different types, such as array CGH vs. gene expression, has helped resolve some of the issues in many cases, but such data sets are difficult to analyze simultaneously. Much of these types of studies are also dependent on informatics interfaces to synthesize and sort out important facts. DNA microarrays and similar technologies allow massive parallel assays. These assays are very important for drug discovery and development programs, especially those involving high-throughput screens. Although some continue to argue that microarray-based transcription profiling has generally proved to be unimportant in drug discovery, the sheer pace of discovery using microarrays and its role in gathering knowledge in the drug industry can no longer be ignored. Without this technology, the basic science and discovery processes involved in understanding disease and solving its problems would be extremely time-consuming and difficult.

IMPACT OF MECHANISM-BASED STUDIES IN CANCER DRUG DISCOVERY AND DEVELOPMENT

As with any disease, tackling cancer by drugs requires *a priori* knowledge of drug targets, the cell cycle, and an ability to carry out large-scale whole genome studies that provide biochemical and genetic evidence, as well as population studies such as population statistics and epidemiology. Such efforts are all dependent upon a series of tools, including informatics and platform tools such as sequencing, PCR, and DNA microarrays. In the discovery of cancer drugs such as Herceptin, EGFR inhibitor (Iressa and Tarceva) for lung cancer, and Imitanib (Gleevec) for CML and GIST, it has become common knowledge that a systematic dissection of molecular biology mechanisms is needed to understand the Achilles' heel of cancer and discover drugs that resolve the outcome of patients who suffer from cancers. Many of the findings reviewed here were dependent upon powerful genetic and biochemical approaches or assays. Poor understanding of tumor classes and effective stratification can also affect drug discovery outcomes and is perhaps the case for Iressa [13,14,49,50]. Microarray-based classification therefore has a tremendous role in classifying such cancer patients, as well as in the future of personalized medicine. Even findings from Stegmaier and colleagues [23] and RNAi studies [26] underscore not only the importance of gene manipulation, annotation, and affected pathways, but also the importance of microarray studies and the value of the information they provide for a mechanistic understanding of the underlying genetic lesions in human cancers.

It is also clear that many of the ideas to pursue the EGFR story came from questions pertaining to the dysregulated expression observed in tumors [13]. Similarly, gene expression dysregulation observed in MLL patients led to the discovery of a new tumor class, when researchers focused on flt3 overexpressed in this class of tumors [8]. Subsequently, the mutations in the flt3 and the drug efficacy response of a murine tumor model paved the way for drug testing [9]. This and other stories in the discovery of drugs against Erbb2 [42] are well documented in the literature and demonstrate how similar approaches and focuses on molecular studies can provide unexpected but exciting results.

Thus, proof-of-principle studies and new targets discovered in mechanism-based approaches to drug discovery have shown promise and have suggested ways of monitoring the genome and proteome at several levels, thereby providing a holistic view that should ultimately make drug discovery more effective. Also, these discoveries have suggested that, because of complexities of the disease, such as drug resistance, stem cell relapse, and clonal expansion of drug resistant tumor cells, understanding the disease in every patient is critical. Unfortunately, that cannot be determined by morphometric analysis alone, making microarray analysis of paramount importance.

FUTURE OF MICROARRAYS IN SYSTEM-LEVEL OVERVIEW OF CANCER AND THERAPEUTIC INTERVENTION

DNA microarrays were first used in the laboratory setting to answer specific questions. In general, data mining or reanalysis of archival data addressed by an "-omic" technology, such as DNA microarrays, addressed complex biological questions, but these laboratory-specific questions were neither data driven nor vast in scope. Today, biology in general, and even cancer biology, involves multiple array-based techniques and re-experimentation strategies (Figure 9.1). One can understand the response of a cancer cell to a drug by precisely monitoring the altered transcriptome on a DNA oligonucleotide microarray or choose from other nontranscriptional profiling technologies. The focus is ultimately on better understanding a wiring diagram of the cancer cell or tissue and the patient's response to a drug. This will require tools, and the tools will include microarrays among others. This big picture or "organismal" view of complex processes is often dubbed *systems biology* (Figure **9.1**).

Microarrays are not only used to understand biological questions, its use in a clinical setting with regard to drug responses and patient response predictions is being considered. Indications of side effects such as stroke and heart disease from prolonged treatment of osteoarthritis with cox2 inhibitor rofecoxib (Vioxx) has prompted the discovery process to utilize better markers for various chemical and clinical end points. As reviewed here, such clinical effects can be monitored by expression technology or be used to identify new compounds without such side effects. Microarrays have also gained the center stage in recent FDA approvals (Roche and cytochrome, p450 chip) and in diagnostics (http://www.agendia.com/, accessed February 2005). Microarrays have also risen to the challenge of consistency in interlaboratory measurements [51]. In the future, microarrays may also guide patient responses before and after treatment and/or patient recruitment for clinical trials based on expression signatures of tumors, which correlate with class identification and response rates.

CONCLUSIONS

Microarrays are now a fundamental tool in basic cancer biology and drug discovery and in patient-response evaluation. Although intense debates continue, microarray-based discoveries are used directly as clinical tools or for the development of secondary or diagnostic tools to measure clinical end points or therapeutic intervention methods.

ACKNOWLEDGMENTS

I thank Mel Kronick at Agilent Technologies for his review and critical comments.

ADDITIONAL NOTES

The opinions expressed herein are independent views of the author and not the official position of Agilent Technologies.

REFERENCES

1. Golub, T.R., Slonim, D.K., Tamayo, P., Huard, C., Gaasenbeek, M., Mesirov, J.P., Coller, H., Loh, M.L., Downing, J.R., Caligiuri, M.A., Bloomfield, C.D., Lander, E.S. Molecular classification of cancer: class discovery and class prediction by gene expression monitoring. *Science* 286(5439): 531–537, 1999.
2. Perou, C.M., Sorlie, T., Eisen, M.B., van de Rijn, M., Jeffrey, S.S., Rees, C.A., Pollack, J.R., Ross, D.T., Johnsen, H., Akslen, L.A., Fluge, O., Pergamenschikov,

A., Williams, C., Zhu, S.X., Lonning, P.E., Borresen-Dale, A.L., Brown, P.O., Botstein, D. Molecular portraits of human breast tumours. *Nature* 406(6797): 747–752, 2000.

3. Garber, M.E., Troyanskaya, O.G., Schluens, K., Petersen, S., Thaesler, Z., Pacyna-Gengelbach, M., van de Rijn, M., Rosen, G.D., Perou, C.M., Whyte, R.I., Altman, R.B., Brown, P.O., Botstein, D., Petersen, I. Diversity of gene expression in adenocarcinoma of the lung. *Proc Natl Acad Sci U S A.* 98(24): 13784–13789, 2001.

4. Bhattacharjee, A., Richards, W.G., Staunton, J., Li, C., Monti, S., Vasa, P., Ladd, C., Beheshti, J., Bueno, R., Gillette, M., Loda, M., Weber, G., Mark, E.J., Lander, E.S., Wong, W., Johnson, B.E., Golub, T.R., Sugarbaker, D.J., Meyerson, M. Classification of human lung carcinomas by mRNA expression profiling reveals distinct adenocarcinoma subclasses. *Proc Natl Acad Sci U S A.* 98(24): 13790–13795, 2001.

5. van't Veer, L.J., Dai, H., van de Vijver, M.J., He, Y.D., Hart, A.A., Mao, M., Petersen, H.L., van der Kooy, K., Marton, M.J., Witteveen, A.T., Schreiber, G.J., Kerkhoven, R.M., Roberts, C., Linsley, P.S., Bernards, R., Friend, S.H. Gene expression profiling predicts clinical outcome of breast cancer. *Nature* 415(6871): 530–536, 2002.

6. Beer, D.G., Kardia, S.L., Huang, C.C., Giordano, T.J., Levin, A.M., Misek, D.E., Lin, L., Chen, G., Gharib, T.G., Thomas, D.G., Lizyness, M.L., Kuick, R., Hayasaka, S., Taylor, J.M., Iannettoni, M.D., Orringer, M.B., Hanash, S. Gene-expression profiles predict survival of patients with lung adenocarcinoma. *Nat Med.* 8(8): 816–824, 2002.

7. Ramaswamy, S., Ross, K.N., Lander, E.S., Golub, T.R. A molecular signature of metastasis in primary solid tumors. *Nat Genet.* 33(1): 49–54, 2003.

8. Armstrong, S.A., Staunton, J.E., Silverman, L.B., Pieters, R., den Boer, M.L., Minden, M.D., Sallan, S.E., Lander, E.S., Golub, T.R., Korsmeyer, S.J. MLL translocations specify a distinct gene expression profile, distinguishing a unique leukemia. *Nat Genet.* 30: 41–47, 2002.

9. Armstrong, S.A., Kung, A.L., Mabon, M.E., Silverman, L.B., Stam, R.W., den Boer, M.L., Pieters, R., Sallan, S.E., Kersey, J.H., Fletcher, J.A., Golub, T.R., Griffin, J.D., Korsmeyer, S.J. Inhibition of FLT3 in MLL: Validation of a therapeutic target identified by gene expression based classification. *Cancer Cell* 3: 173–183, 2003.

10. Corbin, A.S., Griswold, I.J., La Rosee, P., Yee, K.W., Heinrich, M.C., Reimer, C.L., Druker, B.J., Deininger, M.W. Sensitivity of oncogenic KIT mutants to the kinase inhibitors MLN518 and PD180970. *Blood* 104(12): 3754–3775, 2004.

11. Stone, R.M., DeAngelo, D.J., Klimek, V., Galinsky, I., Estey, E., Nimer, S.D., Grandin, W., Lebwohl, D., Wang, Y., Cohen, P., Fox, E.A., Neuberg, D., Clark, J., Gilliland, D.G., Griffin, J.D. Patients with acute myeloid leukemia and an activating mutation in FLT3 respond to a small-molecule FLT3 tyrosine kinase inhibitor, PKC412. *Blood* 105(1): 54–60, 2005.

12. Smith, B.D., Levis, M., Beran, M., Giles, F., Kantarjian, H., Berg, K., Murphy, K.M., Dauses, T., Allebach, J., Small, D. Single-agent CEP-701, a novel FLT3 inhibitor, shows biologic and clinical activity in patients with relapsed or refractory acute myeloid leukemia. *Blood* 103(10): 3669–3676, 2004.

13. Paez, J.G., Janne, P.A., Lee, J.C., Tracy, S., Greulich, H., Gabriel, S., Herman, P., Kaye, F.J., Lindeman, N., Boggon, T.J., Naoki, K., Sasaki, H., Fujii, Y., Eck, M.J., Sellers, W.R., Johnson, B.E., Meyerson, M. EGFR mutations in lung cancer: correlation with clinical response to gefitinib therapy. *Science* 304(5676): 1497–1500, 2004.

14. Lynch, T.J., Bell, D.W., Sordella, R., Gurubhagavatula, S., Okimoto, R.A., Brannigan, B.W., Harris, P.L., Haserlat, S.M., Supko, J.G., Haluska, F.G., Louis, D.N., Christiani, D.C., Settleman, J., Haber, D.A. Activating mutations in the epidermal growth factor receptor underlying responsiveness of non-small-cell lung cancer to gefitinib. *N Engl J Med.* 350(21): 2129–2139, 2004.

15. Hardiman, G. Microarray Platforms-comparison and contrasts. *Pharmacogenomics* 5: 487–502, 2004.

16. Kronick, M.N. Creation of whole genome microarray. *Expert Rev. Proteomics* 1: 89–98, 2004.

17. Bernards, R. Functional identification of cancer-relevant genes. *95th American Association of Cancer Research*, Orlando, FL, March 2–5, 2004.

18. Hughes, T.R., Marton, M.J., Jones, A.R., Roberts, C.J., Stoughton, R., Armour, C.D., Bennett, H.A., Coffey, E., Dai, H., He, Y.D., Kidd, M.J., King, A.M., Meyer, M.R., Slade, D., Lum, P.Y., Stepaniants, S.B., Shoemaker, D.D., Gachotte, D., Chakraburtty, K., Simon, J., Bard, M., Friend, S.H. Functional discovery via a compendium of expression profiles. *Cell.* 102(1): 109–126, 2000.

19. Ross, D.T., Scherf, U., Eisen, M.B., Perou, C.M., Rees, C., Spellman, P., Iyer, V., Jeffrey, S.S., Van de Rijn, M., Waltham, M. et al. Systematic variation in gene expression patterns in human cancer cell lines. *Nat. Genet.* 24: 227–235, 2000.

20. Scherf, U., Ross, D.T., Waltham, M., Smith, L.H., Lee, J.K., Tanabe, L., Kohn, K.W., Reinhold, W.C., Myers, T.G., Andrews, D.T., Scudiero, D.A., Eisen, M.B., Sausville, E.A., Pommier, Y., Botstein, D., Brown, P.O., Weinstein, J.N. A gene expression database for the molecular pharmacology of cancer. *Nat Genet.* 24(3): 236–244, 2000.

21. Issaeva, N., Bozko, P., Enge, M., Protopopova, M., Verhoef, L.G., Masucci, M., Pramanik, A., Selivanova, G. Small molecule RITA binds to p53, blocks p53-HDM-2 interaction and activates p53 function in tumors. *Nat Med.* 10(12): 1321–1328, 2004.

22. Sawyers, C. Targeted cancer therapy. *Nature* 432(7015): 294–297, 2004.

23. Stegmaier, K., Ross, K.N., Colavito, S.A., O'Malley, S., Stockwell, B.R., Golub, T.R. Gene expression-based high-throughput screening(GE-HTS) and application to leukemia differentiation. *Nat Genet.* 36(3): 257–263, 2004.

24. Pollack, J.R., Iyer, V.R. Characterizing the physical genome. *Nat Genet.* 32(Suppl.): 515–521, 2002.

25. Oneill, G.M., Catchpoole, D.R., Golemis, E.A. From correlation to causality-microarrays, cancer, and cancer treatment. *Biotechniques* 34: S64–71, 2003.

26. Paddison, P.J., Silva, J.M., Conklin, D.S., Schlabach, M., Li, M., Aruleba, S., Balija, V., O'Shaughnessy, A., Gnoj, L., Scobie, K., Chang, K., Westbrook, T., Cleary, M., Sachidanandam, R., McCombie, W.R., Elledge, S.J., Hannon, G.J. A resource for large-scale RNA-interference-based screens in mammals. *Nature* 428(6981): 427–431, 2004.

27. Johnson, J.M., Castle, J., Garrett-Engele, P., Kan, Z., Loerch, P.M., Armour, C.D., Santos, R., Schadt, E.E., Stoughton, R., Shoemaker, D.D. Genomewide survey of human alternative pre-mRNA splicing with exon junction microarrays. *Science* 302(5653): 2141–2144, 2003.

28. Shoemaker, D.D., Schadt, E.E., Armour, C.D., He, Y.D., Garrett-Engele, P., McDonagh, P.D., Loerch, P.M., Leonardson, A., Lum, P.Y., Cavet, G., Wu, L.F., Altschuler, S.J., Edwards, S., King, J., Tsang, J.S., Schimmack, G., Schelter, J.M., Koch, J., Ziman, M., Marton, M.J., Li, B., Cundiff, P., Ward, T., Castle, J., Krolewski, M., Meyer, M.R., Mao, M., Burchard, J., Kidd, M.J., Dai, H., Phillips, J.W., Linsley, P.S., Stoughton, R., Scherer, S., Boguski, M.S. Experimental annotation of the human genome using microarray technology. *Nature* 409(6822): 922–927, February 15, 2001.

29. Le, K., Mitsouras, K., Roy, M., Wang, Q., Xu, Q., Nelson, S.F., Lee, C. Detecting tissue-specific regulation of alternative splicing as a qualitative change in microarray data. *Nucl Acids Res.* 32(22): e180, 2004.

30. Elbashir, S.M., Harborth, J., Lendeckel, W., Yalcin, A., Weber, K., Tuschl, T. Duplexes of 21-nucleotide RNAs mediate RNA interference in cultured mammalian cells. *Nature* 411(6836): 494–498, 2001.

31. Sordella, R., Bell, D.W., Haber, D.A., Settleman, J. Gefitinib-sensitizing EGFR mutations in lung cancer activate anti-apoptotic pathways. *Science* 305(5687): 1163–1167, 2004.

32. Cleary, M.A., Kilian, K., Wang, Y., Bradshaw, J., Cavet, G., Ge, W., Kulkarni, A., Paddison, P.J., Chang, K., Sheth, N., Leproust, E., Coffey, E.M., Burchard, J., McCombie, W.R., Linsley, P., Hannon, G.J. Production of complex nucleic acid libraries using highly parallel in situ oligonucleotide synthesis. *Nat Methods* 1: 241–248, 2004.

33. Albertson, D.G., Collins, C., McCormick, F., Gray, J.W. Chromosome aberrations in solid tumors. *Nat. Genet.* 34: 369–376, 2003.

34. Pollack, J.R., Perou, C.M., Alizadeh, A.A., Eisen, M.B., Pergamenschikov, A., Williams, C.F., Jeffrey, S.S., Botstein, D., Brown, P.O. Genomewide analysis of DNA copy-number changes using cDNA microarrays. *Nat Genet.* 23: 41–46, 1999.

35. Lucito, R., West, J., Reiner, A., Alexander, J., Esposito, D., Mishra, B., Powers, S., Norton, L., Wigler, M. Detecting gene copy number fluctuations in tumor cells by microarray analysis of genomic representations. *Genome Res.* 10(11): 1726–1736, November 2000.

36. Lisitsyn, N., Lisitsyn, N., Wigler, M. 1993. Cloning the differences between two complex genomes. *Science* 259(5097): 946–951.

37. Barrett, M.T., Scheffer, A., Ben-Dor, A., Sampas, N., Lipson, D., Kincaid, R., Tsang, P., Curry, B., Baird, K., Meltzer, P.S., Yakhini, Z., Bruhn, L., Laderman, S. Comparative genomic hybridization using oligonucleotide microarrays and total genomic DNA. *Proc Natl Acad Sci U S A.* 101(51): 17765–17770, 2004.

38. Brennan, C., Zhang, Y., Leo, C., Feng, B., Cauwels, C., Aguirre, A.J., Kim, M., Protopopov, A., Chin, L. High-resolution global profiling of genomic alterations with long oligonucleotide microarray. *Cancer Res.* 64(14): 4744–4748, 2004.

39. Lindblad-Toh, K., Tanenbaum, D.M., Daly, M.J., Winchester, E., Lui, W.O., Villapakkam, A., Stanton, S.E., Larsson, C., Hudson, T.J., Johnson, B.E., Lander, E.S., Meyerson, M. Loss-of-heterozygosity analysis of small-cell lung carcinomas using single-nucleotide polymorphism arrays. *Nat. Biotechnol.* 18(9): 1001–1005, 2000.

40. Bignell, G.R., Huang, J., Greshock, J., Watt, S., Butler, A., West, S., Grigorova, M., Jones, K.W., Wei, W., Stratton, M.R., Futreal, P.A., Weber, B., Shapero, M.H., Wooster R. High-resolution analysis of DNA copy number using oligonucleotide microarrays. *Genome Res.* 14(2): 287–295, 2004.

41. Janne, P.A., Li, C., Zhao, X., Girard, L., Chen, T.H., Minna, J., Christiani, D.C., Johnson, B.E., Meyerson, M. High-resolution single-nucleotide polymorphism array and clustering analysis of loss of heterozygosity in human lung cancer cell lines. *Oncogene* 23(15): 2716–2726, 2004.

42. Stephens, P., Hunter, C., Bignell, G., Edkins, S., Davies, H., Teague, J., Stevens, C., O'Meara, S., Smith, R., Parker, A., Barthorpe, A., Blow, M., Brackenbury, L., Butler, A., Clarke, O., Cole, J., Dicks, E., Dike, A., Drozd, A., Edwards, K., Forbes, S., Foster, R., Gray, K., Greenman, C., Halliday, K., Hills, K., Kosmidou, V., Lugg, R., Menzies, A., Perry, J., Petty, R., Raine, K., Ratford, L., Shepherd, R., Small, A., Stephens, Y., Tofts, C., Varian, J., West, S., Widaa, S., Yates, A., Brasseur, F., Cooper, C.S., Flanagan, A.M., Knowles, M., Leung, S.Y., Louis, D.N., Looijenga, L.H., Malkowicz, B., Pierotti, M.A., Teh, B., Chenevix-Trench, G., Weber, B.L., Yuen, S.T., Harris, G., Goldstraw, P., Nicholson, A.G, Futreal, P.A., Wooster, R., Stratton, M.R. Lung cancer: intragenic ERBB2 kinase mutations in tumours. *Nature* 431(7008): 525–526, 2004.

43. Odom, D.T., Zizlsperger, N., Gordon, D.B., Bell, G.W., Rinaldi, N.J., Murray, H.L., Volkert, T.L., Schreiber, J., Rolfe, P.A., Gifford, D.K., Fraenkel, E., Bell, G.I.,

Young, R.A. Control of pancreas and liver gene expression by HNF transcription factors. *Science* 303(5662): 1378–1381, 2004.

44. Hughes, C.M., Rozenblatt-Rosen, O., Milne, T.A., Copeland, T.D., Levine, S.S., Lee, J.C., Hayes, D.N., Shanmugam, K.S., Bhattacharjee, A., Biondi, C.A., Kay, G.F., Hayward, N.K., Hess, J.L., Meyerson, M. Menin associates with a trithorax family histone methyltransferase complex and with the hoxc8 locus. *Mol Cell.* 13(4): 587–597, 2004.

45. Cam, H., Balciunaite, E., Blais, A., Spektor, A., Scarpulla, R.C., Young, R., Kluger, Y., Dynlacht, B.D. A common set of gene regulatory networks links metabolism and growth inhibition. *Mol Cell.* 16(3): 399–411, November 5, 2004.

46. Jin, V.X., Leu, Y.W., Liyanarachchi, S., Sun, H., Fan, M., Nephew, K.P., Huang, T.H., Davuluri, R.V. Verifying estrogen receptor alpha target genes using integrated computational genomics and chromatin immunoprecipitation microarray. *Nucl Acids Res.* 32(22): 6627–6635, 2004.

47. Leu, Y.W., Yan, P.S., Fan, M., Jin, V.X., Liu, J.C., Curran, E.M., Welshons, W.V., Wei, S.H., Davuluri, R.V., Plass, C., Nephew, K.P., Huang, T.H. Loss of estrogen receptor signaling triggers epigenetic silencing of downstream targets in breast cancer. *Cancer Res.* 64(22): 8184–8192, 2004.

48. Li, Z., Van Calcar, S., Qu, C., Cavenee, W.K., Zhang, M.Q., Ren, B. A global transcriptional regulatory role for c-Myc in Burkitt's lymphoma cells. *Proc Natl Acad Sci U S A* 100(14): 8164–8169, 2003.

49. Druker, B.J. Molecularly targeted therapy: have the floodgates opened? *The Oncologist* 9: 357–360, 2004.

50. Demetri, G.D., von Mehren, M., Blanke, C.D., Van den Abbeele, A.D., Eisenberg, B., Roberts, P.J., Heinrich, M.C., Tuveson, D.A., Singer, S., Janicek, M., Fletcher, J.A., Silverman, S.G., Silberman, S.L., Capdeville, R., Kiese, B., Peng, B., Dimitrijevic, S., Druker, B.J., Corless, C., Fletcher, C.D., Joensuu, H. Efficacy and safety of imatinib mesylate in advanced gastrointestinal stromal tumors. *N Engl J Med.* 347(7): 472–480, August 15, 2002.

51. Dobbin, K.K., Beer, D.G., Meyerson, M., Yeatman, T.J., Gerald, W.L., Jacobson, J.W., Conley, B., Buetow, K.H., Heiskanen, M., Simon, R.M., Minna, J.D., Girard, L., Misek, D.E., Taylor, J.M.G., Hanash, S., Naoki, K., Hayes, D.N., Ladd-Acosta, C., Enkemann, S.A., Viale, A., Giordano, T.J. Interlaboratory comparability study of cancer gene expression analysis using oligonucleotide microarrays. *Clin Cancer Res.* 11: 565–572, 2005.

52. Aggarwal, K., Lee, K.H. Functional genomics and proteomics as a foundation for systems biology. *Brief Funct Genomic Proteomic* 2(3): 175–184, 2003.

10 High-Throughput Microarray Analysis

Sejal Sheth

CONTENTS

With the completion of the Human Genome Project 4 years ago came the hope and promise that the world's most ambitious sequencing effort would revolutionize pharmaceutical research and, ultimately, give us better therapies and improved patient care. However, during the decade-long project, scientists learned that the genome is far more complex than previously thought. The first estimate of 30,000 genes has given way to estimates of hundreds of thousands of splice variants, millions of newly discovered transcripts, and tens of millions of genetic polymorphisms. But the tools needed to understand this level of complexity simply did not exist.

The microarray, invented in 1989 by Stephen P. A. Fodor and colleagues [1–3], has emerged as a central technology that is helping to unravel much of the genome's complexity. Over the past 15 years, microarray information capacity has consistently increased, providing for a tool that allows meaningful whole-genome analysis, currently able to measure expression for nearly 50,000 transcripts or genotype more than 100,000 polymorphisms in a single experiment. This broadscale genetic analysis has not only helped to discover the underlying genetics for countless diseases but has fundamentally improved drug discovery and development research.

Before whole-genome microarray analysis, many drug development assays were typically limited to answering a very focused question, often generating a single data point. To perform comprehensive drug discovery, researchers must answer hundreds or even thousands of different questions, making the process slow, expensive, and prone to variability.

Microarrays have offered a significant improvement by measuring thousands of data points in a single assay, with the ability to analyze changes in gene expression and DNA sequence variation across the whole genome. However, microarray throughput and cost-efficiency have limited their application in pharmaceutical research, which requires analyzing far more samples than in biomedical research.

To enable industrialized microarray research, Affymetrix has recently developed an automated 96-array high-throughput (HT) system. The system automates the most labor-intensive steps in microarray processing — target preparation, washing, and staining — increasing productivity, throughput, and reducing the cost per assay. This decrease in cost, increase in throughput, and added reliability make the HT system ideally suited for drug discovery and development applications, including target identification and validation, compound profiling, and improved clinical trial outcome.

Genetic and Genomic Analyses Possible on a Single 96-Array Plate Containing Microarrays Manufactured at a Given Feature Size

Analysis Type		Feature Size	
		8 μm	5 μm
Transcripts/genes	96 well	25,500	65,000
	Per plate	2,448,000	6,240,000
SNPs	96 well	14,000	36,000
	Per plate	1,344,000	3,456,000
Base pairs	96 well	70,000	180,000
	Per plate	6,720,000	17,280,000

DEVELOPING THE HT ARRAY

The complexity of microarrays presented a challenge for engineering the HT system.

The Affymetrix HT array adapts the same GeneChip® technology and content to a standard 96-well plate footprint. Advances in feature size reduction have allowed significantly more content to be placed on smaller-sized arrays. And, by leveraging advanced automation methods, the HT system provides the consistency required to simultaneously analyze hundreds of high-content arrays.

The current HT microarray prototype contains 96 individual arrays mounted on a single plate, with each array containing the same genomic information as the company's Human Genome U133A array, but in approximately one fifth the surface area.

For each array of the 96-array plate, more than 500,000 probes are used to measure the expression of 18,400 human transcripts, meaning that each HT plate generates more than 48 million data points. By comparison, conventional HT screening may only generate a single data point per, well, a total of 96 data points per plate.

Each 96-array plate is processed and analyzed on a robotic Array Station that automates the microarray processing workflow. Figure 10.1 shows how this is done.

RNA Expression

	Day 1					Day 2			Day 3
Extract, Quantify, and Dilute RNA	Synthesize and Cleanup cDNA	IVT	cRNA Cleanup	Quantify and Dilute	Fragment	Hybridize		Wash and Stain	Scan
Customer Method	Array Station					48°C Incubator		Array Station	HT Scanner

FIGURE 10.1 Example of workflow.

This allows a high level of multiplexing in a single experiment and results in a significant decrease in sample-to-sample variation. To process an equivalent number of samples on GeneChip cartridges, a lab not only would have to dedicate extraordinary labor resources, but would require additional fluidics stations and multiple scanners as well. Modeling studies have determined that the automated system increases productivity by threefold or fourfold.

APPLICATIONS TO DRUG DISCOVERY AND DEVELOPMENT

Microarray technology has already revolutionized significant parts of the drug discovery process, but with the development of HT arrays, pharmaceutical companies can now more wholly implement and apply the technology. For example, at the beginning of the process, HT technology can play a role in disease pathway identification and validation, and later on, once a target has been identified, in compound screening and lead optimization. Researchers can then use the HT microarray system to manage clinical trials, potentially expediting the delivery of new drugs to market.

DISEASE PATHWAY IDENTIFICATION

HT array analysis provides researchers with a cost-efficient way to use genomewide expression profiling to generate hypotheses for complex disease mechanisms and to identify drug targets and their pathways. Additionally, GeneChip DNA analysis arrays have been used to discover the genetic basis of disease, by mapping disease genes with whole-genome single nucleotide polymorphism (SNP) assays [4–9]. The two array types complement each other: Gene expression arrays identify differentially regulated genes from related individuals, and DNA analysis arrays can validate those differences in fine mapping experiments.

DISEASE PATHWAY VALIDATION

Once a disease pathway is identified, researchers need to validate it, and verify that disrupting the pathway will actually affect the disease etiology. Using whole-genome expression profiling, scientists can understand a wide range of effects — desirable and undesirable — that result from disrupting a pathway; they are then able to better evaluate potential targets for drug design. Modern technologies, such as small interfering RNA (siRNA), are now being used to rapidly and specifically inhibit gene function, speeding up the exploratory process of validating useful drug targets. However, being able to affect many different genes quickly requires an equally efficient way to measure the downstream effects generated by those changes [10]. HT arrays enable researchers to simultaneously analyze the effects of nearly 100 different siRNA molecules on global gene expression.

Furthermore, the system will be used for microarray-based resequencing efforts to economically pinpoint disease-causing mutations and genetic variations in large clinical populations. The expression and sequence information generated from high-density microarrays gives researchers a more complete understanding of how a gene functions within a cell and adds significant value to the biological models used to validate gene targets.

COMPOUND SCREENING: MECHANISM OF ACTION

Following disease pathway identification and validation, whole-genome microarray analysis can be used to characterize lead compounds for selectivity and specificity and to identify compounds that disrupt expression of intended disease genes. Although existing technologies are well suited to measure the anticipated action of a development compound, these methods do not typically identify any additional or unexpected effects. Whole-genome expression analysis provides a complete and unbiased measure of both on-target and off-target effects for each compound tested. On-target effects are clearly desired; however, off-target changes in expression may help treat different diseases operating through a different mechanism. For example, despite their development to treat hypertension and depression, the respective block-buster successes of Viagra for erectile dysfunction and Wellbutrin for smoking cessation are prime examples of exploiting off-target drug action to serve other therapeutic markets. By developing large databases of information on the global activity for each member of a compound library, HT microarray expression analysis allows companies to ultimately create "smarter" compound libraries, with recorded and known effects for each member compound.

COMPOUND SCREENING: MECHANISMS OF TOXICITY

HT microarray gene expression screening not only helps to identify mechanisms of drug action, but also points to other off-target effects that may suggest the compound produces far too many side effects to be approved. For instance, if changes in gene expression match those of a known toxin, a compound can be eliminated from the screening process early in development, saving both time and money. Compound toxicity is typically not evaluated until later stages in the development pipeline, and this has become a major reason for the high attrition rate in drug development. In the past, the belief has been that once a compound is found to be active, it can be sufficiently modified to avoid toxic effects while retaining its specific activity [11]. However, a recent review of the literature demonstrates that, in general, successfully developed drugs undergo few modifications from their initial lead form [12]. Using HT microarrays to understand risk profiles for multiple compounds earlier in the development process allows for more efficient and cost-effective decision making regarding compound prioritization for future drug development.

MORE SUCCESSFUL CLINICAL TRIALS

By providing more complete genetic and genomic information, microarrays are helping researchers classify disease markers, predict drug efficacy, and more successfully manage clinical trials. The throughput and cost-efficiency of the HT system are the keys to industrializing microarray technology. There are already more than 40 examples of microarrays being used in large-scale trials.

For example, a recent Phase III clinical trial by Novartis Pharmaceuticals used expression profiles to predict the success or failure of Glivec/Gleevec treatment on chronic myelogenous leukemia [4]. Researchers analyzed gene expression patterns from patients prior to treatment and found a 31-gene "no response" signature, which predicts a 200-fold-higher probability of failed therapy.

Similarly, in a Phase II clinical trial conducted at the Dana Farber Cancer Research Institute for Millennium Pharmaceuticals' drug Velcade, researchers used GeneChip arrays to collect pharmacogenomic data from myeloma patients treated with the drug [5]. The scientists discovered a pattern consisting of 30 genes that correlate with response or lack of response to therapy. Clinical utility of biomarkers will be further assessed in a Phase III trial.

Though much progress has already been made using gene-expression analysis, studies to identify genes associated with drug response, efficacy, and toxicity may become one of the most promising applications for whole-genome DNA analysis. Tools like the GeneChip Mapping 100K Array Set (which can genotype more than 100,000 SNPs distributed across the genome) now allow researchers to readily genotype large populations of responders and nonresponders to a given drug for phenotypes including efficacy and toxicity.

With these kinds of genetic studies, scientists hope to elucidate the genes contributing to variable drug response. In key Phase III trials, microarray genotype analysis could be used to stratify patient populations to eliminate poor or toxic responders. Such stratification would help ensure maximum effectiveness through clearer statistical differentiation between drug and placebo, while also reducing trial size and costs, and improving the odds of drug approval.

Once a drug is on the market, patient stratification could also be used to accelerate drug expansion into new indications through faster, smaller, and more definitive Phase IV trials or to establish medical superiority of a late-to-market drug relative to entrenched competitors in an important class of patients. Genomewide genotype information will also fuel future research. By better understanding genetic mechanisms of drug response in patients, researchers will have made significant progress on finding next-generation drugs.

THE WAY AHEAD

As microarray technology advances and more content can be placed on smaller-sized arrays, the application of HT microarray systems to pharmaceutical development will become even more significant and extend beyond the traditional genetic and genomic experiments.

Though the ability to use HT microarrays representing the complete coding content of the human genome — more than 47,000 transcripts — will help accelerate discovery even further, HT analysis will also enable scientists to look beyond the coding content, and mine for drug targets and pathways in DNA sequences historically considered to be of inconsequential "junk." Human transcriptome analysis (i.e., the complete collection of transcribed elements of the genome) is made feasible by the HT system where an experiment can now be constructed to analyze an entire genome (coding and noncoding DNA) for structure–function relationships on a single plate. And the importance of noncoding DNA is only beginning to emerge; recent whole-genome association studies have implicated this "junk" in diseases like multiple sclerosis.

Similarly, advances in genotype analysis will be accelerated by microarrays that can analyze more SNPs, allowing scientists to more rapidly discover the genes associated with disease and drug response. Using next-generation HT SNP arrays promises to improve clinical trials and speed drugs to market by stratifying patients into more refined classes of disease and drug tolerance.

Efforts such as these are helping researchers use the genome sequence to improve pharmaceutical R&D and develop new therapies for improved disease management. Although the benefits of HT array analysis are only beginning to be realized, with the care taken to fit this technology into existing infrastructures, it offers the prospect of more efficient, cost-effective approaches to drug discovery and development.

REFERENCES

1. Fodor, S.P. et al. Light-directed, spatially addressable parallel chemical synthesis. *Science* 251, 767–773, 1991.
2. Fodor, S.P. et al. Multiplexed biochemical assays with biological chips. *Nature* 364, 555–556, 1993.
3. Pease, A.C. et al. *Proc. Natl. Acad. Sci. U.S.A. 91,* 5022–5026, 1994.
4. Gissen, P. et al. Mutations in VPS33B, encoding a regulator of SNARE-dependent membrane fusion, cause arthrogryposis-renal dysfunction-cholestasis (ARC) syndrome. *Nat Genet* 36: 400–404, 2004.
5. Middleton, F.A. et al. Genomewide linkage analysis of bipolar disorder by use of a high-density single-nucleotide-polymorphism (SNP) genotyping assay: a comparison with microsatellite marker assays and finding of significant linkage to chromosome 6q22. *Am J Hum Genet* 74: 886–897, 2004.
6. Puffenberger, E.G. et al. Mapping of sudden infant death with dysgenesis of the testes syndrome (SIDDT) by a SNP genome scan and identification of TSPYL loss of function. *Proc Natl Acad Sci U S A,* 2004.
7. Sellick, G.S., Garrett, C., and Houlston, R.S. A novel gene for neonatal diabetes maps to chromosome 10p12.1-p13. *Diabetes* 52: 2636–2638, 2003.
8. Shrimpton, A.E. et al. A HOX gene mutation in a family with isolated congenital vertical talus and Charcot-Marie-Tooth disease. *Am J Hum Genet* 75: 92–96, 2004.
9. Uhlenberg, B. et al. Mutations in the gene encoding gap junction protein alpha 12 (Connexin 46.6) cause Pelizaeus-Merzbacher-like disease. *Am J Hum Genet* 75: 2004.
10. Semizarov, D. et al. Specificity of short interfering RNA determined through gene expression signatures. *Proc Natl Acad Sci U S A* 100: 6347–6352, 2003.
11. Bleicher, K.H., Bohm, H.J., Muller, K., and Alanine, A.I. Hit and lead generation: beyond high-throughput screening. *Nat Rev Drug Discov* 2: 369–378, 2003.

12. Proudfoot, J.R. Drugs, leads, and drug-likeness: an analysis of some recently launched drugs. *Bioorg Med Chem Lett* 12, 1647–1650, 2002.
13. McLean, L.A., Gathmann, I., Capdeville, R., Polymeropoulos, M.H., and Dressman, M. *Clin. Cancer Res.* 10: 155–165, 2004.
14. Mulligan, G. et al. American Society of Hematology Annual Meeting, Philadelphia, 2002.

11 Laboratory Automation: Strategies for High-Volume Industrial Microarray Programs

Anton Bittner and Andrew A. Carmen

CONTENTS

ABSTRACT

Accelerating the rise of genomic research, microarrays have become an essential means for high-throughput gene expression profiling for both the pharmaceutical industry and academia. The diversity of microarray applications continues to broaden, experiment size is growing, and demands for higher throughput are being driven by the success of the technology. Similar to the rise of automated DNA sequencing technology in the late 1980s, specialized core labs, within industry and academia, are becoming the ideal environment to handle the requirements of the platform. Automation is a key component to achieving reproducibility and throughput in these environments. This chapter focuses on strategies for increasing throughput and decreasing variability through rational implementation of automation.

INTRODUCTION

With the advent of the microarray, the entire genome can now be represented on a single chip. The chips contain tens of thousands of individual elements each representing an individual gene. RNA can be extracted from experimental tissues and labeled with fluorescent probes and hybridized with these arrays. The relative fluorescence retained on each spot is an indication of how much each gene is expressed (in the case of gene expression analysis). However, to get consistent results with these complex systems, which can be cross compared in a meaningful manner, a number of steps must be taken to ensure consistency of data. This is particularly true in the industrial core lab setting, where many tens of thousands of arrays can be run every year, representing multiple billions of data points. With the advent of plate-based systems, presented in other chapters, one can consider this number moving to the hundreds of thousands of arrays per year. In such a setting, the management of variability in the laboratory (discussed here) and the management of data at an enterprise level (discussed here but more extensively in other chapters) is essential for success. Variability must be managed to leverage the maximal benefit from these systems over time. To face this task, we must break down these processes into their discrete elements and address each. Here we discuss one such strategy via the rational implementation of automation in the system to reduce costs and variability.

The array processing elements of a microarray experiment, hybridization, and wash, require expertise and specialized equipment for obtaining optimal results. As the initial scanned images have little meaning as is, data extraction and analysis software are necessary for these experiments and have similar specialization. Reproducibility and throughput, both omnipresent concerns for microarray experiments, are hallmarks of successful core labs, so the proliferation of these laboratories is no surprise. High volumes of focused tasks are performed repetitively, an ideal environment for automation. But the unique complications of microarray hybridization, washing, and staining provide challenges to effective implementation of such automation.

Even though its full potential is yet to be realized, microarray technology has already proven itself as a legitimate tool for gene expression research [20–22]. Published microarray experiments are now commonplace, and content has moved beyond discussing the technology to focus instead on the results of the experiment. Further maturity of the technology will increase the willingness of an institution to invest the time, money, and energy to adapt automation for microarray processes. Microarray substrates are already being adjusted to fit automation, moving away from the standard 1" × 3" slide to formats better suited for current automation. Many array manufacturers have already designed 96-well formats for oligonucleotide arrays, some of which are discussed in other chapters [2,17,19]. In this format, end-to-end automation of the protocol becomes more realistic, and justified as the number of samples increase.

BENEFIT OF AUTOMATION TO MICROARRAY EXPERIMENTS

Automation can benefit more than just the high-volume industrial core lab. The size of microarray experiments is growing throughout the field, while the expectations of quality — sensitivity and accuracy — continually rise as well [12]. Larger studies benefit in consistency from increasing the batch size of samples processed in parallel. Our own experience is that the most significant technical bias in a large data set is subset processing [unpublished data]. Not surprisingly, we have also observed that differences in the levels of skill and attention to detail between technicians have a profound effect on the quality of the data. Several published studies have shown that microarray experiments of the same design and the same materials conducted in different labs produce variable data [5–7]. Considering that core labs are often plagued with high turnover of entry-level staff due to the repetitive nature of the work, automation can reduce the bias normally afflicted by a newly trained technician. Well-designed automation should enable improvement upon the limitations of both small batch size and technician variability. Larger capacity also allows for greater replication, enhancing statistical significance and confidence in the data.

Cost savings are another benefit of automating processes in microarray experiments. Though there is an initial capital investment, automation lowers the costs associated with running large numbers of arrays. For management, the easiest factor to appreciate is the decrease in labor costs; whereas scaling up many of the manual processing methods can only be accomplished by adding more technicians. Automation also allows for extension of the productive workday for a lab, performing tasks overnight and during lunch. Fitting with the theory of constraints, the bottleneck should never be idled. Cost savings associated with economy of scale may also be realized.

Owing to the expense and time required to optimize and integrate automation, choosing which processes to automate requires careful thought. A targeted approach is a more rational and cost-effective means of implementing automation rather than indiscriminate end-to-end automation. The primary areas to target are those in which individual technique causes the greatest bias on the data. Quality and productivity should be the primary criteria for assessing automation needs, but cost savings and ergonomic risk mitigation are legitimate requirements as well. Identification of bottlenecks (which we define as the points at which capacity is limited) in the process can help focus automation efforts, but gains are only meaningful if the automation can realistically increase throughput without sacrificing quality. Automation usually has significant up-front costs and is time-consuming to optimize and integrate. Because the initial investment is large, the potential reward needs to be realistic and proportional to that investment.

AUTOMATION FOR ARRAY PROCESSING

Array processing is an area of particular concern, not only because of capacity-limiting constraints, but also because manual processing is subject to many variables. This makes array processing, the hybridization and washing of arrays, a particularly technique-dependent task, prone to inconsistency that automation can potentially improve. Overcoming this limitation has been the focus of many companies selling automation products specifically for microarray technology. Several tactics exist to approach these problems. Some use isolated sealed chambers, others are adapted from histology applications, and another combines both approaches.

The most common approach is to immobilize the slide in a sealed chamber plumbed to deliver target from a plate or tube as well as wash solutions pumped from bulk source containers. A metal plate beneath the slide controls hybridization temperature. The user loads the slides into the chamber, programs hybridization temperature and duration, orders and determines the volume of wash solutions, adds target (hybridization probes), and then returns when the cycle is complete. Benefits of this strategy are that the process is uninterrupted, and fluid delivery can be precise in volume and flow rate. It is also relatively easy to incorporate a feature to actively mix the target during hybridization by pumping small volumes of hybridization buffer in and out of the chamber. A weakness of this approach is that one unit can only process a limited number of slides, typically between 4 and 12. Daisy chaining the units improves upon the relatively low throughput, but not necessarily consistency, because each has its own fluid delivery apparatus, the performance of which can drift over time.

The design of histology instrumentation is predisposed to a 1" × 3" glass slide format, making them easily adaptable to standard-sized microarrays. An example of this approach is a robotic unit, currently marketed for microarray processing, which was originally designed for immunohistochemistry. On this instrument, up to 20 arrays are placed on the outside edge of a turntable. The instrument has a fluid delivery system that first adds target to the slide, followed by a layer of mineral oil that acts as a cover slip during the hybridization. The turntable, conducting heat directly to the slide, precisely controls hybridization temperature. Throughout hybridization, air jets accomplish mixing by gently circulating the mineral oil layer, creating turbulence in the target buffer underneath. Following hybridization, the instrument's fluid delivery system washes the slides according to a user-programmed protocol. Though this offers more slides per instrument than fixed-chamber units, detergents in wash buffers are required to clean the mineral oil may alter stringency.

We have also used other robotic systems designed specifically for histology applications for microarray prehybridization. Most cDNA chip production protocols require postdeposition modification to the arrays [23,24], a prehybridization step, or both. Performed manually in slide-staining dishes, the protocol is plagued with batch variation. Robotic systems can much better control the simple task of constant agitation for 5 min in a prehybridization buffer than the most attentive lab technician. This simple, inexpensive device merely moves a slide rack up and down, but with a uniform pace and motion that enables a drastic improvement in batch consistency for a seemingly minor section of the protocol.

A high-throughput robotic unit capable of processing 200 slides in a single run has also been developed [9], though not marketed. This instrument consists of a fluidics compartment separate from an incubator, both serviced by the same robotic transfer arm. Slides are stored in disposable cassettes in a carousel within the incubator. Cassettes are moved independently from the incubator to the fluidics compartment, where target is added to the array. The cassette is then returned to the incubator for hybridization. At a precise interval, the cassette is retrieved and placed back in the fluidics compartment for washing, then finally returned to the carousel for the user to scan. Precision of timing and fluid delivery, avoidance of intervention requirement, and unmatched throughput are all advantages of this system. Complexity of the system, large footprint, and costs that are prohibitive to most core labs are the disadvantages.

The benefits of automating target preparation are not as noteworthy as those for hybridization and wash. Because 96-well formats can be handled easily manually through multichannel pipetters, there is not an overwhelming advantage to throughput. Achieving consistency of manual target preparation is not unreasonable; in contrast to the array-processing steps, different labs using the same protocol can even achieve consistent results [5]. Automation will not necessarily improve reproducibility of the target prep, and may cause even more variable yields than manual prep [1]. Properly tuned robotic units can achieve great consistency, but the maintenance required for the system to continue operating at such a level can be more challenging than consistent manual pipetting. There are productivity benefits to well-designed automated protocols; however, the ability to run overnight decreases cycle time for an experiment. An obvious caveat to this is that the robotic units must be robust enough to be trusted to operate without anyone nearby. Great trustworthiness is required of an instrument, typically fairly complex, that is allowed to run unattended, and those deserving of that trust are rarely inexpensive. There are also strong ergonomic risk mitigation benefits that accompany reduction of manual pipetting steps, an attribute that is becoming increasingly valued in the workplace [18]. Greater reagent volumes are also required, and requirements for tips used on robotic systems add more expense compared to those used for manual pipetters, offsetting some of the labor cost savings.

DATA COLLECTION, ANALYSIS, AND SAMPLE TRACKING

Productivity gains from automating data collection, extraction, and quality control are perhaps the most important compared to all other parts of the microarray experiment. Scan time for a single microarray is widely variable and depends on scan resolution and array dimensions, but a typical run lasts 7 min. Without automation, a technician needs to spend less than 1 min, eight to nine times every hour unloading and loading slides. Although this accounts for less than 10% of someone's workday, the interruptions are frequent enough not to allow adequate concentration for other lab or analysis work, thus consuming an entire workday for one technician. Autoloader features are common options to many scanners now, and savings in labor costs quickly make up for the cost of the autoloader.

Data extraction in batch mode maintains autoloader productivity gains and is available in most feature extraction software packages. However, a massive bottleneck can form downstream in quality assessment, and software automating this process improves accuracy as well as productivity. Meaningful quality assessment should be much more than just determining the number of genes with intensities above noise. But manual determination and rejection of contaminated spots is almost entirely subjective and time-consuming. Outlier identification based on statistical analysis algorithms is a rapid, completely objective approach [10]. There are several flaws to manual outlier identification, most notably the inconsistency of the approach — not only between identically trained technicians, but also with the same technician over the course of a few hours. Accuracy is also problematic: discrimination of gray levels (white to black) for the human eye is 20–25 [11], hopeless in discriminating the 65,536-level resolution that a 16-bit image encodes.

Algorithms that compare three or more replicate features or replicate arrays can quickly identify which genes show a high coefficient of variation (CV). After

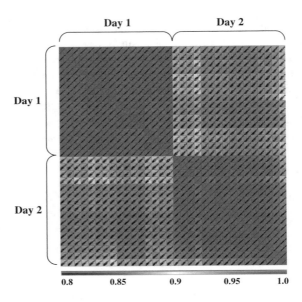

FIGURE 11.1 *(A color version follows page 204) Batch variability:* Two separate hybridizations of the same sample to cDNA arrays, replicated 16 times in each hybridization, run on consecutive days, are compared, with a color scale of correlation as indicated from 0.8 to 1.0. Average correlation between all arrays from Day 1 = 0.987, Day 2 = 0.981. Average correlation between arrays hybridized on different days = 0.945. Scatter plots within each individual square can be observed; less scatter will typically indicate a higher degree of correlation. Correlations calculated using OmniViz Desktop 3.9.0 using the CorScape tool.

identifying outlier genes, intensities from the individual features for those genes can be compared, and outlier spots then isolated and flagged. A more sophisticated algorithm can also examine distribution of signal within and around isolated spots to flag those with poor spot morphology or background problems. Using this approach with four replicate features per sample (two per array, two arrays per sample), we were able to successfully identify and remove outliers on our cDNA arrays to improve average CVs from 14% to less than 6%.

All of the posthybridization functions can be integrated through a Laboratory Information Management System (LIMS) that addresses each as separate modules. A bar coding system can be created to help track samples through this process with relative ease and little expense. Bar coding is an efficient means for tracking and identification, and robotic systems typically integrate bar code readers to accommodate tracking. Creating a LIMS sophisticated enough to recognize when to move images to feature extraction, and data to quality assessment, is not a trivial undertaking. We have integrated this type of system and it has proven to be a powerful way to enhance the productivity of those working in the lab, because they no longer have the pedestrian tasks of moving files and loading them into analysis software. A summary display is presented to an administrator after the data is processed, who makes the decision whether to upload to the database or fail the experiment. Because the system can operate unattended, data extraction can run overnight, dramatically improving throughput as well.

PITFALLS OF AUTOMATION

Poorly designed automation causes more problems than it solves, squanders budgets, wastes time and energy, and eventually lies fallow. Because it is useful only when operating, the dependability of a robotic system is as important as its capability. Equipment that breaks down disrupts the rhythm of workflow in the lab, and when it happens with frequency has an effect on throughput opposite to what was intended. Even with reliable systems, redundant instrumentation is desirable where cost is not prohibitive. Redundancy is almost compulsory in some cases. A microarray scanner, for example, is critical to the success of the experiment — no practical manual method for collecting data exists, and similar equipment is not likely to reside in neighboring labs.

There is a desire to stay on the cutting edge of technology, especially in a rapidly evolving field such as genomics, but obvious risks are involved with unproven systems. A mature product may be stable and well known, but also closer to obsolescence. Tolerance for risk, immediate need for production, and level of experience with automation should all factor into selection of robotic systems. The importance of intuitiveness of the software, flexibility of the instrument, and footprint are considerations that will also vary by the unique characteristics of each lab.

SUMMARY

Expectations on throughput and reproducibility steadily increase as microarray technology evolves. Automated workflows can increase batch size and consistency while decreasing ergonomic risk to lab technicians. For the sake of consistency, one is better off building a large capacity and using it with less frequency rather than processing smaller sets more frequently. Increased capacity associated with higher throughput allows for more data points, and in a carefully controlled experiment, better statistical significance. In this way, benefits of automation can extend to any microarray core lab, regardless of size, as long as the automation employed is tailored to the needs of the lab. Proper attention must be focused on selection of appropriate processes to automate as well as the robotic systems. Rational integration of automation must improve either efficiency, reproducibility, reduce costs, or alleviate ergonomic risks, but a gain in one of these areas must not come at the expense of another.

REFERENCES

1. Dorris, D.R., Ramakrishnan, R., Trakas, D., Dudzik, F., Belval, R., Zhao, C., Nguyen, A., Domanus, M., Mazumder, A. A highly reproducible, linear, and automated sample preparation method for DNA microarrays. *Genome Res.* 12(6): 976–984, 2002.
2. Zarrinkar, P.P., Mainquist, J.K., Zamora, M., Stern, D., Welsh, J.B., Sapinoso, L.M., Hampton, G.M., Lockhart, D.J. Arrays of arrays for high-throughput gene expression profiling. *Genome Res.* 11(7): 1256–1261, July 2001.

3. Hessner, M.J., Wang, X., Khan, S., Meyer, L., Schlicht, M., Tackes, J., Datta, M.W., Jacob, H.J., Ghosh, S. Use of a three-color cDNA microarray platform to measure and control support-bound probe for improved data quality and reproducibility. *Nucl Acids Res.* 31(11): e60, June 1, 2003.

4. Hessner, M.J., Meyer, L., Tackes, J., Muheisen, S., Wang, X. Immobilized probe and glass surface chemistry as variables in microarray fabrication. *BMC Genomics* 5(1), 53, 2004.

5. Piper, M.D.W., Daran-Lapujade, P., Bro, C., Regenberg, B., Knudsen, S., Nielsen, J., Pronk, J.T. Reproducibility of oligonucleotide microarray transcriptome analyses. An interlaboratory comparison using chemostat cultures of Saccharomyces cerevisiae. *J Biol Chem.* 277(40): 37001–37008, October 4, 2002.

6. Waring, J.F., Ulrich, R.G., Flint, N., Morfitt, D., Kalkuhl, A., Staedtler, F., Lawton, M., Beekman, J.M., Suter, L. Interlaboratory evaluation of rat hepatic gene expression changes induced by methapyrilene. *Environ Health Perspect.* 112(4): 439–448, 2004.

7. Irizarry, R.A., Warren, D., Spencer, F., Kim, I.F., Biswal, S., Frank, B.C., Gabrielson, E., Garcia, J.G.N., Geoghegan, J., Germino, G., Griffin, C., Hilmer, S.C., Hoffman, E., Jedlicka, A.E., Kawasaki, E., Martinez-Murillo, F., Morsberger, L., Lee, H., Petersen, D., Quackenbush, J., Scott, A., Wilson, M., Yang, Y., Ye, S.Q., Yu, W. Multiple-laboratory comparison of microarray platforms. *Nat Methods* 2(5): 345–350, 2005.

8. Yuen, T., Wurmbach, E., Pfeffer, R.L., Ebersole, B.J., Sealfon, S.C. Accuracy and calibration of commercial oligonucleotide and custom cDNA microarrays. *Nucl Acids Res.* 30(10): e48/1–e48/9, 2002.

9. Bittner, A., Carmen, A.A. Moving DNA microarray into high throughput, a rational approach. *Am Pharm Rev.* 6(2): 54, 56, 58, 60, 2003.

10. Bittner, A., Carmen, A.A., Leung, A., Liu, X., Smith, S., Sun, L., Wan, J., Xiao, H., Yao, X., Yieh, L. Methods and algorithms for performing quality control during gene expression profiling on DNA microarray technology. *U.S. Pat. Appl. Publ.* 2004, 25 pp.

11. www.intermed.dk/Digimage/Modul1.htm.

12. Hoffman, E.P., Awad, T., Palma, J., Webster, T., Hubbell, E., Warrington, J.A., Spira, A., Wright, G., Buckley, J., Triche, T., Davis, R., Tibshirani, R., Xiao, W., Jones, W., Tompkins, R., West, M. Guidelines: Expression profiling — best practices for data generation and interpretation in clinical trials. *Nat Rev Genet.* 5(3), 229–237, 2004.

13. Oh, M.-K., Scoles, D.R., Haipek, C., Strand, A.D., Gutmann, D.H., Olson, J.M., Pulst, S.-M. Genetic heterogeneity of stably transfected cell lines revealed by expression profiling with oligonucleotide microarrays. *J Cell Biochem.* 90(5), 1068–1078, 2003.

14. Jain, N., Thatte, J., Braciale, T., Ley, K., O'Connell, M., Lee, J.K. Local-pooled-error test for identifying differentially expressed genes with a small number of replicated microarrays. *Bioinformatics* 19(15): 1945–1951, 2003.

15. Martinsky, T. Microarrays: not just for genomics anymore. *PharmaGenomics* (Suppl.): 12–14, 16–24, 2004.

16. Brazma, A., Hingamp, P., Quackenbush, J., Sherlock, G., Spellman, P., Stoeckert, C., Aach, J., Ansorge, W., Ball, C.A., Causton, H.C., Gaasterland, T., Glenisson, P., Holstege, F.C., Kim, I.F., Markowitz, V., Matese, J.C., Parkinson, H., Robinson, A., Sarkans, U., Schulze-Kremer, S., Stewart, J., Taylor, R., Vilo, J., Vingron, M. Minimum information about a microarray experiment (MIAME)-toward standards for microarray data. *Nat Genet.* 29(4): 365–371, December 2001.

17. Lombardi, S. Industrializing microarrays (high-throughput microarray analysis) can help improve drug discovery and development. *Mod Drug Discov.* 7(12), 46–48, 2004.

18. Waters, T.R. National efforts to identify research issues related to prevention of work-related musculoskeletal disorders. *J Electromyogr Kinesiol* 14(1): 7–12, 2004.
19. Ferguson, J.A., Steemers, F.J., and Walt, D.R. High-density fiber-optic DNA random microsphere array. *Anal Chem.* 72: 5618, 2000.
20. Gershon, D. Microarrays go mainstream. *Nat Methods* 1(3): 263–269, 2004.
21. Petricoin, Emanuel F. III et al. Medical applications of microarray technologies: a regulatory science perspective. *Nat Genet* 32: 474–479, 2002.
22. Lord, Peter G., Papoian, T. Genomics and drug toxicity. *Science* 306(5696), 575, 2004.
23. Diehl, F., Grahlmann, S., Beier, M., Hoheisel, J. D. Manufacturing DNA microarrays of high spot homogeneity and reduced background signal. *Nucleic Acids Research* 29(7): e38, 2001.
24. Jiang, Y., Lueders, J., Glatfelter, A., Gooden, C., Bittner, M. Profiling human gene expression with cDNA microarrays, in *Current Protocols in Molecular Biology,* Ausubel, F. M., Brent, R., Kingston, R. E., Moore, D. D., Seidman, J. G., Smith, J. A., Struhl, K., (Eds.), Indianapolis IN: Wiley Interscience 2005, Unit 22.3.

12 Association Studies: Practical and Theoretical Considerations for Drug Discovery, Evaluation, and Beyond

Richard M. Kliman, Francis Barany, Nadine Cohen, and Reyna Favis

CONTENTS

INTRODUCTION

GAME PLAN FOR EARLY ADOPTERS

Some scientists quickly embrace new technologies with the desperate hope that a panacea for the unsolvable is at last at hand, but others in the field approach with greater caution. Historically, the pessimists have amassed more credible data and there is good cause to be skeptical. However, for a scientist to avoid the monikers of either Luddite or *naïf,* it is necessary to approach new technology with the intent to both expose the caveats and to engage one's creativity to overcome these limitations. The objective of this chapter is to provide a practical guide to understanding and, where possible, compensating for the caveats inherent in using microarray technology to facilitate association studies.

FOREWARNED IS FOREARMED: OVERVIEW OF THE CAVEATS

Caveats come in two basic flavors: those that arise owing to limitations in technology and those that are the consequence of biology. Although technical and biological limitations will largely be described as separate issues in the following, the reader will notice that there are frequent interactions between these two effects. Technical limitations may be ephemeral. Biological limitations are unchanging, although our understanding of these limitations changes.

Over time, technical limitations may eventually be overcome through the introduction of better processes; however, when assessing the "as is" environment, it is wise to take nothing for granted when exploring new technologies. For example, high-density microarrays for global gene expression analysis have been widely used in the scientific community over the last few years. Corrections in both design and/or fabrication have been implemented during this period. These improvements have been made possible because users have dutifully reported findings that impact performance (see *Lessons from Transcriptomics* in the following text). In contrast, although biological limitations are also being recognized, it is not always possible to adjust for this type of caveat.

Biological limitations are generally less amenable to engineering solutions. In the context of association studies, false positive associations can be generated because of, for example, unappreciated population substructure or sequence duplications in the genome. Population history with concomitant admixture and selection can create genetic architectures that affect one's ability to identify genes responsible for the phenotype of interest. In addition, variation contributing to a phenotype may not necessarily be in the form of a SNP, and may therefore be more difficult to identify using microarrays that detect mutations through direct hybridization of target and probe. Similarly, variation may reside in regions of the genome with vastly different GC-content, again posing difficulties for direct hybridization chips.

As with any high-throughput technology, implementation is a double-edged sword: users can produce large amounts of data with minimal effort compared to currently used methodologies; however, any error in design, manufacture, or process can have large-scale consequences for the quality of the data. In addition, biological realities impinge on data interpretation and must be considered if useful hypotheses are to be generated.

QUESTIONS TO CONSIDER BEFORE EMBARKING ON A STUDY

There are several questions that must be addressed by an investigator prior to launching a large-scale study. In practice, these questions include:

1. Is the presence or absence of the trait of interest really influenced mainly by genes, or does the environment play a more critical role?
2. Is it possible to develop a strong, testable hypothesis based on existing information?
3. Are DNA samples available from appropriate cases and controls that have richly detailed, supporting clinical information?
4. Is it economically feasible to rigorously test the genotyping platform of choice?

Although it is tempting for drug discovery scientists searching for new targets to attribute much of the variation observed in disease states to genetic causes, this is not necessarily the case. An example to consider is Parkinson's disease. Although there are well-documented familial cases with a strong genetic component, there are also cases where environmental chemicals created permanent parkinsonism, as well as cases where there appears to be genetic susceptibility and frank disease-only results upon exposure to environmental insult [1]. Before pursuing costly high-throughput genetic studies, scientists must have compelling evidence to support the claim that the trait of interest is influenced strongly enough by genetics to warrant pursuit. Once this question is addressed and a decision is made to go forward, the choice of methodology for the study should be considered.

Genetic studies can be conducted using either candidate gene or hypothesis-free approaches. If the existing literature supports the formulation of a strong, testable hypothesis, it is usually preferable to perform a candidate gene study. Similarly, candidate gene analysis may also be possible if data from global expression analysis yields information that suggests genes or signaling pathways that should be pursued to explain a phenotype. The guiding principle in this decision is generally the reduced cost compared to whole-genome approaches. An added benefit is that the statistical burden of multiple comparisons is also reduced for the investigator. If, on the other hand, no convincing evidence from existing data and literature are available to help formulate a hypothesis, whole-genome analysis should be pursued. An additional consideration for deciding the type of genetic study to pursue is the availability of appropriate DNA samples: fewer samples compel fewer comparisons.

Appropriate DNA samples are a critical component of an association study. In general, investigators aim for a balanced study design with equal numbers of cases and controls. To minimize genetic variation that is unrelated to the trait of interest, cases and controls are ethnically matched. In practice, it is generally only possible to do this crudely (e.g., Caucasian cases vs. Caucasian controls), and it is strongly advised that the cohort be assessed to verify that any associations do not result from unappreciated population substructure (see Reference 2 and the later subsection "Population Structure and Admixture"). Frequently, investigators will attempt to assemble a disease cohort from populations that are considered to be homogeneous and then to translate the findings from an association study to the general population. Another selection tool involves stratifying by clinical data. Because a single disease phenotype could be produced by multiple different genotypes, subtle clinical differences may help to select cases that most resemble each other in order to enrich data for subjects with the same underlying genotype. In addition to a well-defined phenotype, the magnitude of the clinical endpoint can be helpful in selecting subjects. To adequately power a study, it will be necessary to consult a statistician. Because each study is likely to possess unique characteristics, only general considerations will be mentioned in this section. To help guide the power analysis, the investigator should provide the statistician with as much information as possible. Useful information includes the frequency of the disease within the general population, the impact of candidate genes (i.e., odds ratios), whether there is any reason to believe certain alleles behave in a dominant or recessive manner, whether a candidate gene or a whole-genome scan is planned, etc. Once appropriate and sufficient samples have been obtained, the next step in developing a genetic study is to verify the performance of the platform.

The primary limiting factor in conducting genetic association studies is typically the available budget. A secondary consideration is the fact that DNA is a nonrenewable resource and future investigations will likely be planned around existing samples. Both factors require the investigator to be sure that the platform trusted to generate genotypes for the study is indeed trustworthy. When possible, it is advisable to pilot the platform in a positive control situation, including blinded replicate samples to ascertain the reproducibility of the data. An alternative approach is to test the platform for its ability to reproduce results from (now) classic examples that were successfully produced using traditional methods in select populations, e.g., mapping BRCA1 and BRCA2 [3] or BLM [4] using microsatellite markers. If neither approach is possible due to budgetary constraints, a long hard look at the quality control processes used during design, manufacture, and implementation of the platform is suggested before reaching a decision. If these processes are robust and convincing, it is still advisable to include blinded replicate samples during the actual study in order to verify consistency. Additional suggestions on assessing the functionality of a platform are presented in the next section.

We shall now discuss details of available technologies and potential pitfalls. This is followed by a critical consideration of the impact of biological variables on the ability to detect associations.

TECHNICAL CONSIDERATIONS

OVERVIEW OF SNP CHIP TECHNOLOGY

There are numerous technologies available for genotyping, with varying capacities for throughput, accuracy, and detection capabilities for different types of variation [5–7]. In terms of biochips that detect genetic variation, microarrays rely on two general strategies: direct hybridization of target and probe to detect mutations or liquid-phase enzymatic detection of mutations followed by hybridization to a solid support using nongenomic "zip-code" DNA sequences.

The strength of direct hybridization microarrays is the potential for high SNP content. For hypothesis-free whole-genome approaches, an important prerequisite for mapping genes associated with a trait of interest is sufficient SNP density across the genome (see the subsection "Frequencies and Spacing of Polymorphisms"). High-density SNP chips [8–10] are designed using quartets of probes that accommodate perfect match sequences for both possible alleles, as well as two mismatch probes. Quartets are created for both the sense and antisense versions of the sequence with the query SNP positioned at the center of the 25-base probe. By moving a 25-base sliding window along a reference sequence containing the SNP, the position of the query SNP within the probe is altered and additional quartets are created. In this way, 10 probe quartets with the best performance are selected for each SNP. Target is produced from genomic DNA by first digesting the DNA with a restriction enzyme. Following this, adapters compatible with the restriction site are ligated to the ends of the fragments and the DNA is subjected to PCR amplification using primers complementary to the adapters. The PCR product is then fragmented with DNase and terminal transferase is used to label the fragments with biotinylated nucleotides to produce the target that will be hybridized to the chip. Perfectly matched probes and targets are expected to hybridize efficiently, but mismatches should produce comparatively low signal or no signal. Visualization of hybridized target is made possible through a network formed by interactions between the biotinylated nucleotide, streptavidin, biotinylated antibodies directed against streptavidin and SAPE (streptavidin conjugated to the fluorescent group phycoerythrin). Complex algorithms and bioinformatics approaches are then required to analyze and interpret the information generated by the chip. In contrast to this approach, microarrays using zip-code tags completely separate the mutation detection step from hybridization.

Zip-code tag arrays are best suited for candidate gene analysis and ideal for analyzing haplotypes. The advantages of this approach include a reduction in the number of multiple comparisons, a straightforward analysis of the chip hybridization patterns and the ability of the investigator to determine the content of the chip without incurring a large expense. A simple, but powerful implementation of this strategy is PCR/LDR/Universal Array [11]. The SNP detection strategy for this approach relies on the polymerase chain reaction coupled to a ligase detection reaction. Following multiplex amplification of the target sequences of interest, multiplex SNP detection is accomplished by hybridizing two adjacent oligonucleotides to the target sequences. To detect a SNP, two discriminating primers with distinct fluorescent labels are designed with the two possible bases represented by the 3' terminal base on each primer. A common primer is designed to hybridize immediately adjacent to the discriminating primer and bears a unique 24-base sequence on its 3' end. If there is perfect complementarity to the target sequence at the junction between the discriminating and common primer, a thermostable ligase will join these two oligo-nucleotides. The unique 24-base sequence permits hybridization to the microarray because it is complementary to the zip-code sequences that are spotted on the chip. The identity of a particular SNP is revealed by the color of the fluorescence and the address to which it hybridizes. This approach has been validated in numerous cancer studies [12–16], where the sensitivity of the ligase-based mutation detection was exploited to detect low-level oncogenic mutations. The sensitivity has also been employed to facilitate sample pooling, which can enormously accelerate the progress of large-scale association studies [17]. Following publication of Gerry et al., this concept of separating mutation detection from hybridization to a microarray attained wide use in the scientific community, and this type of design was coupled with primer extension and other ligase-based approaches to detect SNPs using capillary, liquid, bead, and microchip arrays [18–37].

LESSONS FROM TRANSCRIPTOMICS: IS THE CHIP CONTENT TRUSTWORTHY?

Technical improvements continue to enhance the performance of microarrays assessing global gene expression. During the growing pains of early implementation, it was recognized that issues involving probes were a potential source of error. For cDNA microarrays, between 1% and 5% on average (and as high as 30%) of the clones used for spotting were incorrectly identified [38]. These cases of mistaken identity were generally traced back to errors in plate handling and cross-contamination of the bacterial cultures harboring the cDNA clones. Careful and systematic sequencing to verify the identities of the clones used for spotting has greatly reduced this source of error; however, sustained vigilance is necessary. For oligonucleotide arrays, errors involving probes were also encountered. In this case, owing to confusion in the public databases, sequences corresponding to the wrong strand of DNA were used as probe, resulting in erroneous sequences for up to one third of the addresses [38]. More recently, individual oligonucleotide probes on mammalian Affymetrix microarrays were assessed, and greater than 19% of the probes were found to deviate from their corresponding mRNAs in the RefSeq database [39]. To obtain better reproducibility between hybridization results, the authors of this study recommended eliminating these questionable probes from analysis. A permanent solution would involve redesigning probes based on the most current information available in the sequence information databases. Although the solutions and "work-arounds" to the above technical issues were relatively trivial, recognizing the presence of these errors *a priori* is a more formidable task, and one that will also be required to verify the quality of SNP chips.

Probe content and identity are caveats that readily translate from expression to variation microarrays. The greater the number of probes on an array, the more difficult it becomes to verify that all probes are correctly assigned. Because the probe verification method to check the content of Affymetrix expression chips used by Mecham et al. [39] (in the preceding text) is *in silico* and can be automated, this should perhaps be adapted for SNP chips. Probe verification can be made standard practice for every new version of a chip that is released and when sequence databases are updated. This action would provide peace of mind for the design aspect of a variation microarray, but the question of probe identity on the physical chip remains an open question. Addressing this question thoroughly requires a thorough consideration of the budget allocated. Although the most reliable approach would be to compare genotyping results for reference samples produced by the high-density SNP chip and another high-throughput technology, this is also the most expensive option and unlikely to be repeated with each new chip version. A more realistic alternative is to spot-check the chip results for a reference sample by sequencing. Probe sequence validation is considerably more straightforward for zip-code tag arrays.

A simple and effective method has been developed to rapidly verify the identity of probes on zip-code tag arrays [40]. Fluorescently labeled complements to the spotted sequences are combined such that the complements for each individual row and column of the array are pooled in separate batches. Sequential hybridization of femtomole quantities of each batch will allow the user to determine if the array lot has been fabricated without error. A flawless array lot will emit signal for each individual row or column without extraneous signal appearing anywhere else on the array. Stray signal would indicate cross-contamination of addresses or the presence of an illegitimate probe at a particular address. Quality analysis of universal arrays [11] is particularly economical, because these arrays can be stripped of target and reused; thus, only a few arrays from each lot need to be sacrificed to determine quality.

Besides confirming that the design and fabrication of the platform of choice is technically sound, another consideration is the platform's ability to detect the type of variation that underlies the disease phenotype.

THE INTERSECTION OF TECHNICAL AND BIOLOGICAL LIMITATIONS: VARIATION WITHIN GENETIC VARIATION

Due to the relative ease of detection, SNPs remain the focus of most drug discovery efforts. However, other types of mutation also have biological and pharmaceutical significance, e.g., insertion/deletions (BRCA1 and BRCA2 mutations in familial breast cancer), mononucleotide repeats (hereditary nonpolyposis colorectal cancer), translocations (formation of BCR-ABL oncoprotein in chronic myeloid leukemia), inversions (Hemophilia A) and duplications (CYP2D6 copy number and drug metabolism). Although no platform can detect all types of variation, it is essential to understand what platform to apply when variation other than (or in addition to) SNPs should be tested for association.

Direct hybridization arrays are designed to detect SNPs and are not well suited to detect most other types of variation. For example, several studies attempted to use direct hybridization arrays to identify all p53 mutations in enriched tumor tissue derived by selective microdissection. These studies succeeded in detecting only 81% [41], 84% [42], and 92% [43] of p53 mutations. In all cases, insertion/deletion mutations proved intractable to this detection scheme. In contrast, PCR/LDR/ universal array has demonstrated ability to detect insertion/deletion mutations well [17], length polymorphisms in mononucleotide [44], dinucleotide repeats [45], and even methylation of CpG islands [40]. In terms of large genomic rearrangements (translocations, inversions, and duplications), direct hybridization microarrays, PCR/LDR/universal array and other detection strategies using zip-code tag arrays could theoretically detect these aberrations. In all cases, if the breakpoints for these chromosomal rearrangements are perfectly consistent, probes can be designed across the junction that will capture sequence from both of the chromosomal regions involved. Depending on the sequence, direct hybridization arrays run the risk of producing false positives, because a portion of the sequence on the probe will exist regardless of whether a translocation exists. Enzymatic mutation detection based on ligation will not suffer from this constraint. If, on the other hand, the breakpoints are not predictable, none of these technologies can be used to detect chromosomal rearrangements and alternative, lower-throughput strategies must be employed.

Although the types of variation so far discussed have been well documented and frequently observed, additional forms of duplications within the genome are starting to be recognized. The origins and biological significance of these duplications are still a puzzle, but the implications for SNP-based association studies are worrying.

THE INTERSECTION OF TECHNICAL AND BIOLOGICAL LIMITATIONS: GENOMIC PECULIARITIES

The human genome project has provided researchers with the ability to examine the genome at a higher level than simple gene-by-gene analyses. As a result, patterns are emerging that could not have been visualized in the recent past. One surprising finding is the presence of numerous different types of duplications in the average genome. The extents to which these duplications arise by similar mechanisms and to what degree they will impact genotyping accuracy remain open questions.

Pseudogenes are a familiar form of duplication that can be described as complete or partial copies of genes that are unable to code for functional peptides. Two major types of pseudogenes exist. Duplicated pseudogenes arise from unequal crossover events, while processed pseudogenes result from retrotransposition, where cellular mRNA is reverse transcribed and inserted into DNA. Because pseudogenes do not generally possess *cis*-acting elements, these sequences are released from selective pressure and evolve neutrally. In a recent study that screened all intergenic regions in the human genome using homology searches and tests to identify functional nucleotide changes, 19,724 regions were revealed that appeared to evolve neutrally and were thus likely to encode pseudogenes [46]. A second study identified about 8,000 processed pseudogenes in the human genome [47]. Although pseudogenes are familiar forms of genomic duplications known to interfere with assay development, other types of duplications exist that do not necessarily mimic coding sequence.

Segmental duplications, or duplicons, are thought to comprise at least 5% of the genome [48,49] and many are yet to be annotated in the genome draft. Duplicons are defined as being greater than 1 kb in length and having greater than 90% similarity between copies. Recently, a careful analysis of duplicon SNPs used monospermic complete hydatidiform moles (CHMs), which contain fully homozygous genomes [50]. This clever experiment allowed the investigators to distinguish between true SNP alleles at a single genome locus and SNPs that arise from paralogous sequence variants (PSVs). Complex patterns of genotypes were detected that could be explained by various models: homozygosity in one copy of a duplicon with heterozygosity in the other; two different homozygotes represented in either copy of the duplicon; multisite variation represented by heterozygosity in both copies of the duplicon; and multisite variation with additional copies bearing variously heterozygous and both homozygous versions. Of note, when the SNP data was reanalyzed without consideration for allele copy number, only half of the apparent SNPs that were actually complex multisite variants deviated from Hardy–Weinberg equilibrium (HWE). Clearly, complex multisite variants of this type create distortions in the interpretation of genotype frequencies at the true SNP locus. Interfering duplications are, however, not limited to long sequences. More recently, additional forms of duplications have been identified that involve short sequence lengths.

Short duplications between 25 and 100 bp have been found that have peculiar and consistent spacing patterns [51]. There appear to be three populations of doublets: those that are separated by at most 100 bp, those with distances between 100 to 10 kb and doublets that are separated by greater than 10 kb or are found on distinct chromosomes. Full (both doublets and the spacer region) and partial conservation (single core sequence with or without spacer sequence) of these doublets have been observed between human and chimpanzees, indicating a relatively ancient evolutionary origin.

In contrast to the duplications discussed above, large-scale copy number polymorphisms of about 100 kb and greater (dubbed CNPs) [52] appear to be of more recent vintage. Using representational oligonucleotide microarray analysis, 20 individuals were compared. The investigators found that on average, individuals differed by 11 CNPs and the average length of a CNP interval was 465 kb. Because the tissues used for these studies (blood and sperm) have high mitotic indices, it is possible that this observation is the result of mitotic recombination. If this mechanism is responsible, this implies that analyses of different tissues with different propensities for mitosis may result in different genotype calls for the same individual. In addition, duplication for some chromosomes may be mirrored by deletions in others [53], also complicating genotype analysis by increasing the count for homozygotes due to the presence of hemizygotes. Further analysis of these CNP phenomena is warranted. Indeed, improved information for all types of duplications would provide for better assay development.

Although mutations that can distinguish duplicated regions from true genes accumulate over time, similarity to paralogous functional genes can still interfere with PCR or chromosomal *in situ* hybridization experiments [54]. Based on this observation and the sheer volume and complexity of potential duplications, it seems likely that SNP detection could be influenced by the presence of these types of sequences. If the duplication is relatively new evolutionarily, and has not had time to accumulate many differences from the true gene, microarray probes designed to query the true gene will cross-hybridize to both sequences.

Probes and PCR primers can be designed to avoid duplications in many instances; however, it is necessary to have *a priori* knowledge of these sequences during the design phase to achieve specificity. Without prior knowledge, interference from duplications could go unnoticed during microarray analysis because the algorithms used to call SNPs are not designed to detect copy number differences. If the true gene is homozygous at a particular base and some fraction of duplications (complex or otherwise) within a population is heterozygous, the microarray will wrongly report that the position can be both heterozygous and homozygous when the population is queried. This would go unnoticed as long as the genotype frequencies did not deviate from HWE.

Although probes producing genotype frequencies that deviate from HWE values are routinely eliminated during the design phase of a platform, comparatively small numbers of samples are used for development, compared to the number of samples required for the numerous association studies that will use that platform. Put simply, design phase samples only a small fraction of the variation that exists and probes that will prove faulty in the long run will be retained. Further, HWE as a criterion for eliminating probes is not foolproof if the resulting signal from the duplication is consistent with HWE in the overall population. (Recall that this was the finding for half of the complex multisite variants when the CHM data described above was analyzed without regard to copy number.) In summary, it is highly unlikely that all faulty probes on an array have been eliminated.

To eliminate the potential for confounding due to cross-reaction of probes, Fredman et al. [50] recommended the use of fully homozygous CHMs or haploid genomes in upstream assay validation during platform development. This recommendation has merit if genomic duplication with contrary and complex genotypes is truly shown to be a widespread and significant phenomenon within human populations. Given that the generation of labeled target for high-density SNP chips involves a nonspecific process of amplification of genomic representations, known regions of duplication cannot be filtered out at this stage. Eliminating duplicated regions can only be achieved when quartets are selected during development of the platform. In contrast, PCR/LDR/Universal array can avoid duplicated regions through PCR and/or LDR selection [55,56], provided that information is available to identify duplications and to highlight any variation found in duplicates. Other enzymatic detection strategies may employ similar approaches for selective analysis; however, as stated, the first step is to know which regions are affected. In time and with further research, it should be possible to determine the degree to which genomic duplications may cause confounding in SNP-based association studies.

In addition to the confounding potentiated by various forms of genomic duplication, the history of the population used to investigate a trait of interest and complex gene-by-environment interactions will influence the likelihood of identifying genes associated with a trait of interest. As stated earlier in this chapter, technical limitations are relatively easy to overcome; accommodating biological constraints poses more of a challenge.

BIOLOGICAL CONSIDERATIONS

GENES, ENVIRONMENT, AND COMPLEXITY

The most important assumption underlying identification of genes that contribute to variation in medically relevant traits is that such genes exist. Given that this assumption is valid, one is faced with a broad range of genotype–phenotype relationships. Is most phenotypic variation associated with highly penetrant variation at a single gene of major effect? Or are many genes involved, with epitasis possibly reducing penetrance? Is penetrance influenced by environmental factors, and do gene $\times$ environment interactions complicate matters further?

From a public health standpoint, it is apparent that diseases that affect a large number of individuals may not be associated with single genes of major effect [57]. The same cannot be said for genetic variation that influences the major routes of metabolization of pharmaceuticals. Thus, functional and pharmacogenomics (involving many loci, e.g., relating to etiology of disease) and pharmacogenetics (involving one or few loci, e.g., relating to drug metabolism) may benefit from different approaches. Regardless, because identification of relevant genes usually requires initial correlation between the character states of the *trait of interest* and a *mapped marker*, it is essential to appreciate the processes responsible for the origin and eventual loss of such correlations.

LINKAGE DISEQUILIBRIUM

Definitions

Fundamental concepts intrinsic to the ability to apply association studies to identify genes involved in a trait of interest include "linkage" and "linkage disequilibrium" (LD). *Linkage* refers to the tendency of two or more loci to be inherited together because of their location near one another on the same chromosome. Tightly linked loci do not assort independently during the creation of germ cells (meiosis). This results in a state of disequilibrium, because the two-locus haplotype frequency for a pair of alleles is not equal to the product of the individual allele frequencies. This difference reflects the degree of linkage disequilibrium. Because two loci in close proximity to one another will tend to associate nonrandomly, this will lead to a higher degree of LD for the two loci. The underlying hope when performing association studies is that mapped markers will be able to identify genes contributing to the trait of interest due to LD.

Frequencies and Spacing of Polymorphisms

Both linkage and LD mapping require that the character states of the trait of interest and the mapped marker reflect genetic variation. In many studies today, the mapped marker is itself a genetically polymorphic site (e.g., a SNP). Genetic polymorphism is the result of at least one mutation, such that a given genetic locus has an ancestral state and at least one derived state. Most mutations will be quickly lost to genetic drift, but a few will rise to measurable frequency over time. For this reason, when a polymorphism is due to a single mutation, low-frequency derived states tend to be younger than high-frequency derived states.

If a polymorphism is due to multiple mutations at the same position, the frequency of the derived state will be higher. Unfortunately, LD with nearby polymorphisms will be lower, thus decreasing the likelihood that susceptibility loci will be detected through association with a mapped marker. Pritchard [58] shows that susceptibility alleles can reach appreciable frequencies by recurrent mutation: essentially mutation balances weak selection against the deleterious allele. Interestingly, such independent mutations in the same gene, detected by family-based linkage analysis, can provide very strong evidence for disease causation (e.g., DJ1 and early-onset parkinsonism [59]).

Figure 12.1a shows the expected frequency spectrum of derived states at poly-morphic sites (assuming one mutation per site) for a sample of 1000 individuals in a population of constant size; the y-axis is on a log scale for ease of visualization. Figure 12.1b shows the probability that the frequency of the derived state will be at least the value on the x-axis. For example, only about 21.5% of polymorphic sites will have a derived state frequency at or above 0.2 (see highlighted point in Figure 12.1b). For an expanding population, such as that of humans, the frequency distribution of derived states, d, is left-skewed. Thus, for a given level of pairwise polymorphism, relative to a constant population, an expanding population will have a greater density of polymorphic sites, but with lower average frequency of the derived state.

It should be noted that, without an outgroup, the ancestral/derived state polarity of a polymorphic site is unknown. However, because low frequency derived states are much more common than the corresponding high frequency derived states (i.e., those with frequency of $1-d/k$, where k is the number of haplotypes screened), the derived state is usually the rare state.

The actual density of polymorphic sites in any sample depends, in large part, on the overall level of polymorphism. For k haplotypes sampled from a population of constant size N over time, the expected density of polymorphic sites with derived state frequency at least d/k is given by the formula

$$\text{site density} = \left(\theta \sum_{i=d}^{k-1} i^{-1} \right)^{-1},$$

where θ is the expected pairwise polymorphism, given the population size and the mutation rate. This parameter is generally estimated from the observed pairwise polymorphism [60] or the observed number of polymorphic sites [61]. Figure 12.1c shows the expected number of base pairs between polymorphic sites under a constant N model and $\theta = 0.0008$ (slightly greater than the value of 0.0007 estimated by Reich et al. [62]) for sites with derived state frequencies up to 0.5. For example, in this constant N model, we expect sites with derived frequencies at or above 0.2 to be spaced, on average, every 775 bp (see highlighted point in Figure 12.1c).

Correlating a Mapped Marker and a Trait of Interest

We are now in a position to consider the correlation between character states of a mapped marker (locus A) and a trait of interest (locus B) — i.e., linkage disequilib-rium [63]. The simplest measure of LD is D, the difference between the observed frequency of a given haplotype (e.g., A_0B_0) and its expected frequency under the assumption of independent assortment (i.e., $D = f(A_0B_0) - f(A_0)f(B_0)$). Note that D can be positive or negative; in a two-gene, two-allele model, $|D|$ is the same for any of the four haplotypes. A limitation to the use of $|D|$ as a measure of LD is that it is sensitive to allele frequencies. It reaches its maximum ($|D| = 0.25$) when all alleles are at a frequency of 0.5. An alternative index, D', scales D by the maximum possible value given the observed allele frequencies. However, a more intuitive index is simply the product–moment correlation of character states (0 or 1) of the two loci.

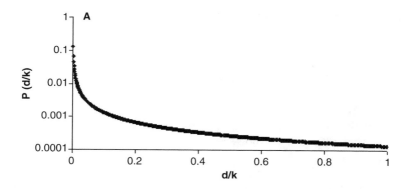

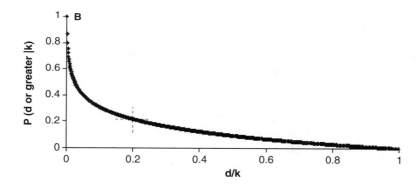

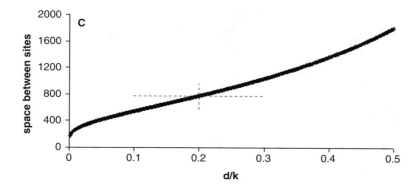

FIGURE 12.1 Frequency of derived states and the number of polymorphic sites with a given frequency of derived states. (A) The expected frequency spectrum for a sample of 1000 individuals in a population of constant size; the y-axis is on a log scale for ease of visualization. (B) The probability that the frequency of the derived state will be at least the value on the x-axis. (C) Expected number of base pairs between polymorphic sites with a derived state frequency at or above d/k under a constant N model and $\theta = 0.0008$.

It can be shown that the squared correlation coefficient of the allelic states of the two genes is

$$r^2 = \frac{(f(A_0 B_0) - f(A_0)f(B_0))^2}{f(A_0)f(A_1)f(B_0)f(B_1)} = \frac{D^2}{f(A_0)f(A_1)f(B_0)f(B_1)}.$$

It can also be shown by algebraic rearrangement that the value of the test statistic for a chi-square test of independence of the character states of the two loci is kr^2. One can then easily test the hypothesis that two loci are in linkage equilibrium, because kr^2 is chi-square distributed with one degree of freedom [64]. The critical value of a chi-square test with one d.f. is 3.841. Thus, for a sample size of $k = 100$ haplotypes, r^2 must exceed 0.03841 to infer nonzero linkage disequilibrium. This drops to 0.00384 if the sample size is increased to $k = 1000$.

High LD When Frequencies Match for Derived States of the Mapped Marker and Trait of Interest

It is reasonable to ask whether a statistically significant, but low, value of r^2 is useful for mapping a locus that influences the trait of interest. The answer to this question is a cautionary "yes." Consider a derived state at a polymorphic mapped marker that has been sampled at a frequency of d/k. For r^2 to be at its maximum value (1.0), certain conditions must be met. First, the sampled frequencies of the two character states at the trait of interest must also be d/k and $1 - d/k$. Second, the character states of the two loci must be in phase: individuals with the rare character state at the mapped marker must have the rare character state for the trait of interest. Figure 12.2 shows a gene tree for two completely linked sites

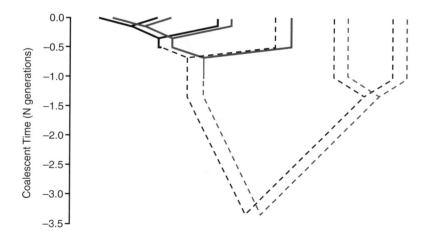

FIGURE 12.2 Gene tree. Black lines represent the tree for one site; gray lines represent the tree for a second, completely linked site. Lines are dashed prior to, and solid subsequent to, the mutation responsible for polymorphism. When the two mutations occur on different branches of the tree, three haplotypes must be present, even without recombination between loci.

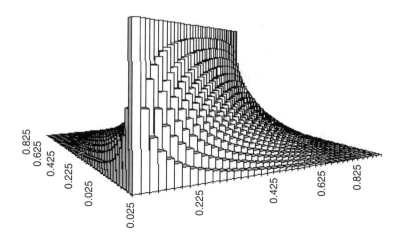

FIGURE 12.3 Maximum possible r^2 for all combinations of derived state frequencies ranging from 0.025 to 0.975 for two loci. As the derived state frequencies become increasingly different, the maximum possible value of r^2 goes down. The figure is symmetrical with respect to the ridge.

(one depicted in black, the other in gray). It shows that, when the two frequencies are different because of historical contingency, three haplotypes must be present even in the absence of recombination between the sites. It should be clear that the strongest association will occur when the two mutations occur on the same branch of the tree. A gene tree relating k haplotypes will have $2(k-1)$ branches of varying lengths (in generations), and the probability that a mutation will occur on a given branch is proportional to the branch length. The probability that both mutations will occur on the same branch — a requirement for perfect LD — is, therefore, exceedingly small. Only a small fraction of polymorphisms will be in strong LD, even in the absence of recombination, because of this historical contingency requirement.

Because derived-state frequencies will usually be different, even for a pair of completely linked polymorphic sites, it is worth considering the effect this has on LD. Figure 12.3 shows the maximum possible r^2 for combinations of derived-state frequencies ranging from 0.025 to 0.975 for two loci. Note that the r^2 along the diagonal (where the frequencies of derived states are equal) is 1.0. The remainder of the plot is saddle shaped, showing that the maximum possible value of r^2 decreases as the frequencies of the derived states at the two sites diverge.

The Consequences of Genetic Recombination

Of course, the assumption of no recombination between sites is unrealistic. The gene tree that underlies the frequencies of the character states of the trait of interest will be correlated to that of the mapped marker if recombination subsequent to the causative mutation has not effectively randomized the character states

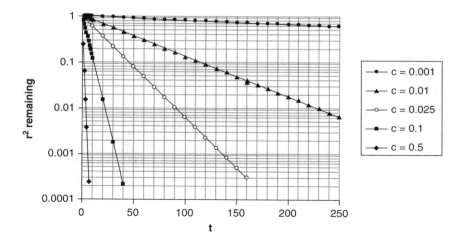

FIGURE 12.4 r^2 remaining over time for given recombination rates. The variable c in the figure corresponds to *rec* in the text; t is time in generations. Values are calculated from the formula $r^2_t \cong r^2_0(1 - c)^{2t}$.

of the two loci. Intuitively, the two trees will be less congruent if there has been greater opportunity for recombination. The consequence of this, from the standpoint of LD, is that, in the absence of any change over time in allele frequencies, D must decrease as time passes and recombination occurs. In an infinite population (i.e., one with no genetic drift), this LD *decay* process can be described as follows: D at generation t can be related to D at generation 0 by the simple formula $D_t = D_0(1 - rec)^t$, where *rec* is the recombination probability per generation between the polymorphic sites [65]. Because r^2 is proportional to D^2, $r^2_t \cong r^2_0(1 - rec)^{2t}$, assuming no change in character state frequencies over time. Figure 12.4 shows the expected decay of r^2 for different recombination rates. The take-home message is that r^2 will be lower today than it was in the past, but it may remain significant for some time.

A paradox should be apparent from the above discussion: if LD decays over time, how does it arise in the first place? For association mapping, the assumption is that LD between two polymorphic sites originates at the time of the second mutation. However, this is not equivalent to $t = 0$ in the decay formula. Rather, if the appropriate haplotypes (those bearing the initially rare derived states) become sufficiently common soon thereafter, then D will initially rise. (Recall that D reaches its maximum when allele frequencies equal 0.5). This initial change in SNP frequencies may be due to drift or it may be driven by natural selection (e.g., by hitchhiking to selectively favored linked mutations). We can think of $t = 0$ as corresponding to that arbitrary point in time when D peaks and begins its expected decay. What really matters is that generating and maintaining LD becomes more difficult as recombination rate increases.

Population Structure and Admixture

It is important to note that LD can arise for reasons unrelated to physical linkage between a mapped marker and a gene associated with a trait of interest. Natural selection (including selection *in utero*) can make certain combinations of character states appear at different frequencies than expected on the basis of the product rule. More problematic for association mapping in humans is population structure followed by admixture. Consider a population in which a character state of the trait of interest has risen to an unusually high frequency, for example, by genetic drift. Presumably, frequencies of character states at various mapped markers have also changed in frequency. There need be no linkage disequilibrium between the loci *within* populations. However, consider the possibility that the combination A_1B_1 has become common in one population. That population now mixes with another, in which A_0B_0 is common. Even if the two loci are independently assorting (i.e., $rec = 0.5$), it could take multiple generations for the character states to become uncorrelated. As shown in Figure 12.4, r^2 is expected to decay to 6.25% of its starting value in two generations. The generation time in humans is about 25 years. If r^2 was 0.1 as a consequence of admixture 50 years ago, it would be about 0.00625 today. For a sample of $k = 1000$ haplotypes, this would be a statistically significant level of LD. Fortunately, there are association mapping strategies that ameliorate the problem of population structure/admixture [66], but these require special sampling protocols (such as genotyping both parents of all cases and controls) that may be unfeasible for many studies.

Correcting for Multiple Comparisons

The final consideration, before turning to mapping strategies, is correction for multiple tests. As noted earlier, the statistical significance of LD can be estimated on the basis of a simple chi-square test. With large sample sizes, even very small values of r^2 can indicate significant LD between a mapped marker and a trait of interest. But what happens if the trait of interest is compared to several mapped markers? One simple definition of the p-value of a statistical test is the probability that the strength of the signal (e.g., r^2) used to reject a null hypothesis could arise just by chance. This means that, for x tests and a single-test significance cutoff of α (traditionally 0.05), we expect $x\alpha$ tests to be spuriously significant. A conservative correction is the Bonferroni correction [67], which sets the cutoff for statistical significance at α/x. That the Bonferroni correction is overly conservative is a common criticism, but it is clear that the multiple testing problem is a significant problem, particularly for whole-genome analyses [68]. Thus, association analysis involving many mapped markers requires some form of correction for multiple tests. If this correction is conservative (for example, the use of the Bonferroni correction when some mapped markers are not independently assorting), some useful information will be lost. If this correction is liberal, for example, by allowing an arbitrarily acceptable false-discovery rate [69], researchers will be misled into searching for a relevant gene when none exists. When possible, replication of association studies with independent samples can decrease the likelihood of a false positive association; however, systematic biases, such as those involving population structure, may allow replication of false positive associations.

MAPPING STRATEGIES

If there were a simple way to map all genes that contribute to variation in a trait of interest, the point of this chapter would be moot. The reality is that tradeoffs are unavoidable. On the one hand, whole-genome analysis would allow one to map all relevant genes. However, this would require screening huge numbers of mapped markers, making false positive correlations a virtual guarantee. Assuming an average recombination rate of 0.5 cM/mb in the human genome, a marker density of 0.5 cM would require 3,000 markers; a density of 0.1 cM would require 15,000 markers. Figure 12.5 shows the number of generations for loss of statistical significance (following Bonferroni correction) in a sample of $k = 1,000$ haplotypes for markers starting at a given r^2 with the trait of interest. Along the x-axis are markers that are progressively farther from the gene of interest (i.e., in increments of 0.1 or 0.5 cM). Because the closest marker may, for reasons of historical contingency, never have been in strong LD with the gene of interest, it may be necessary to rely on more

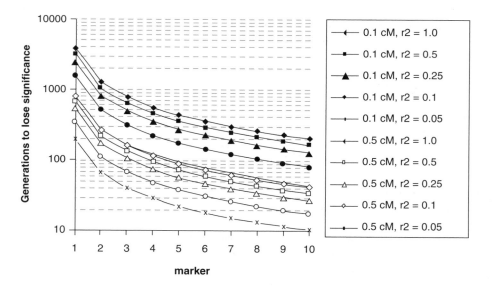

FIGURE 12.5 Number of generations for loss of statistical significance (following Bonferroni correction) of markers starting at a given r^2. Marker density is set to 0.1 or 0.5 cM. It is assumed that the first marker is 0.05 or 0.25 cM from the gene of interest (i.e., half the marker density). Subsequent markers are farther from the gene interest, in successive increments of 0.1 or 0.5 cM along the x-axis. Sample size is $k = 1,000$ haplotypes. Statistical significance of kr^2 is determined by a chi-square test with one degree of freedom. Following Bonferroni correction for 15,000 tests (0.1-cM density), a critical value of r^2 of 0.021614 is required for statistical significance. For 3,000 tests (0.5-cM density), the critical value of r^2 is 0.018536. In contrast, the critical value of r^2 for a single test is 0.003841.

distant markers. The problem is that more distant markers will recombine more rapidly with the gene of interest. It is clear that, unless r^2 is initially very high, it is unlikely that LD would be statistically significant today.

Would a "block-like" structure of the genome improve whole-genome mapping efforts? Variation across the genome in recombination rate, as well as gene density, should lead to increased LD among polymorphisms located between recombination hot spots. McVean et al. [70] have suggested that there may be about 10,000 recombination hot spots in the genome, which would divide it into blocks of 300 kb. Taken to its extreme, there would be no recombination within blocks, such that, aside from cases of recurrent mutation, all polymorphisms would be in maximum LD for their relative frequencies (see Figure 12.3). If one were to select a handful of markers based on a frequency match with the trait of interest, one could presumably find one that is in very high LD. The catch, however, is that one could only infer that polymorphism of interest is somewhere in the block. If a limited amount of recombination is allowed in the block, more careful mapping might be possible, but more sites would have to be screened. Further, there is no guarantee that a site in stronger LD with the gene of interest will be available, given the limitations in marker density due to low polymorphism levels in general. Also, LD with a gene of interest is not necessarily greater for closer markers, due to the historical contingency effect. Thus, if there is very low recombination within a block, fine-scale association mapping may not work. As it stands, the "block-like" structure of the genome is debatable. Wall and Pritchard [71,72] find that, though a uniform recombination rate model is unsuitable for the human genome, data are not fully consistent with a strict haplotype block model.

The alternative to whole-genome analysis is to take some form of candidate gene (or candidate region) approach. One could begin with an *a priori* list of candidate genes, and determine if nearby SNPs are in LD with the trait of interest. The obvious drawback is that the analysis is limited by the imagination of the investigators, insofar as the subjective decision as to what constitutes a candidate gene. Linkage analysis may be used to narrow the region to a megabase scale, and association analysis could then focus on SNPs within the region. By narrowing the regions of the genome to be screened, one could either assay fewer SNPs (decreasing the cost of correction for multiple tests) or screen SNPs at a higher density (increasing the likelihood that one will be in high LD with the polymorphism of interest). As recently noted by Blangero [73], progress in identifying relevant loci has generally begun with a more traditional linkage-based approach involving extended pedigrees. The example of *DJ-1* and early-onset parkinsonism is especially interesting, as the investigators were forced to rule out all of the "obvious" candidate genes within a 5.6-Mb region of chromosome 1 [59]. However, because their search had been narrowed to this smaller region, they were able to assay other transcribed regions to find mutations that were perfectly associated (as homozygotes) with disease. Of course, one could argue that this study did not rely at all on association mapping, *per se*. However, other efforts have used LD, in conjunction with linkage analysis, to pinpoint medically important genes [74]. Recent reviews by Terwilliger and Weiss [75] and Blangero [73] merit attention by those interested in further exploration of these issues.

CONCLUDING REMARKS

The intent of this chapter is to highlight some of the caveats that must be addressed during association studies and to serve as a resource for discovery by scientists. By bringing the issues of interest to the attention of researchers, we are hopeful that many of the problems encountered will be resolved prior to commencing a study. As success stories for association studies reach publication, it will be possible to more accurately assess the factors that drive success, redirect efforts to avoid ineffective approaches, and refine strategies to further improve the state of the art.

REFERENCES

1. Greenamyre, J. and T. Hastings. Parkinson's — divergent causes, convergent mechanisms. *Science* 304: 1120–1122, 2004.
2. Pritchard, J.K. and N.A. Rosenberg. Use of unlinked genetic markers to detect population stratification in association studies. *Am J Hum Genet* 65: 220–228, 1999.
3. Miki, Y., J. Swensen, D. Shattuck-Eidens, P.A. Futreal, K. Harshman, S. Tavtigian, Q. Liu, C. Cochran, L.M. Bennett, W. Ding, et al. A strong candidate for the breast and ovarian cancer susceptibility gene BRCA1. *Science* 266: 66–71, 1994.
4. Ellis, N.A., A.M. Roe, J. Kozloski, M. Proytcheva, C. Falk, and J. German. Linkage disequilibrium between the FES, D15S127, and BLM loci in Ashkenazi Jews with Bloom syndrome. *Am J Hum Genet* 55: 453–460, 1994.
5. Kirk, B.W., M. Feinsod, R. Favis, R.M. Kliman, and F. Barany. Single nucleotide polymorphism seeking long term association with complex disease. *Nucl Acids Res* 30: 3295–3311, 2002.
6. Chen, X. and P.F. Sullivan. Single nucleotide polymorphism genotyping: biochemistry, protocol, cost and throughput. *Pharmacogenomics J* 3: 77–96, 2003.
7. Kwok, P.-Y. Methods for genotyping single nucleotide polymorphisms. *Annu. Rev. Genomics Hum. Genet.* 2: 235–258, 2001.
8. Kennedy, G.C., H. Matsuzaki, S. Dong, W.M. Liu, J. Huang, G. Liu, X. Su, M. Cao, W. Chen, J. Zhang, W. Liu, G. Yang, X. Di, T. Ryder, Z. He, U. Surti, M.S. Phillips, M.T. Boyce-Jacino, S.P. Fodor, and K.W. Jones. Large-scale genotyping of complex DNA. *Nat Biotechnol* 21: 1233–1237, 2003. Epub September 7, 2003.
9. Lipshutz, R.J., S.P. Fodor, T.R. Gingeras, and D.J. Lockhart. High density synthetic oligonucleotide arrays. *Nat Genet* 21: 20–24, 1999.
10. Hacia, J.G. Resequencing and mutational analysis using oligonucleotide microarrays. *Nat Genet* 21: 42–47, 1999.
11. Gerry, N.P., N.E. Witowski, J. Day, R.P. Hammer, G. Barany, and F. Barany. Universal DNA microarray method for multiplex detection of low abundance point mutations. *J Mol Biol* 292: 251–262, 1999.
12. Favis, R. and F. Barany. Mutation detection in K-ras, BRCA1, BRCA2, and p53 using PCR/LDR and a universal DNA microarray. *Ann N Y Acad Sci* 906: 39–43, 2000.
13. Favis, R., J. Huang, N. Gerry, A. Culliford, P. Paty, S.T., and F. Barany. Harmonized microarray/mutation scanning analysis of TP53 mutations in undissected colorectal tumors. *Hum. Mutation* 24: 63–75, 2004.
14. Dong, S.M., G. Traverso, C. Johnson, L. Geng, R. Favis, K. Boynton, K. Hibi, S.N. Goodman, M. D'Allessio, P. Paty, S.R. Hamilton, D. Sidransky, F. Barany, B. Levin, A. Shuber, K.W. Kinzler, B. Vogelstein, and J. Jen. Detecting colorectal cancer in stool with the use of multiple genetic targets. *J Natl Cancer Inst* 93: 858–865, 2001.

15. Fouquet, C., M. Antoine, P. Tisserand, R. Favis, M. Wislez, F. Como, N. Rabbe, M.F. Carette, B. Milleron, F. Barany, J. Cadranel, G. Zalcman, and T. Soussi. Rapid and sensitive p53 alteration analysis in biopsies from lung cancer patients using a functional assay and a universal oligonucleotide array: a prospective study. *Clin Cancer Res.* 10: 3479–3489, 2003.

16. Overholtzer, M., P.H. Rao, R. Favis, X.Y. Lu, M.B. Elowitz, F. Barany, M. Ladanyi, R. Gorlick, and A.J. Levine. The presence of p53 mutations in human osteosarcomas correlates with high levels of genomic instability. *Proc Natl Acad Sci U S A* 100: 11547–11552, 2003.

17. Favis, R., J.P. Day, N.P. Gerry, C. Phelan, S. Narod, and F. Barany. Universal DNA array detection of small insertions and deletions in BRCA1 and BRCA2. *Nat Biotechnol* 18: 561–564, 2000.

18. Bell, P.A., S. Chaturvedi, C.A. Gelfand, C.Y. Huang, M. Kochersperger, R. Kopla, F. Modica, M. Pohl, S. Varde, R. Zhao, X. Zhao, M.T. Boyce-Jacino, and A. Yassen. SNPstream UHT: ultra-high throughput SNP genotyping for pharmacogenomics and drug discovery. *Biotechniques* (Suppl.): 70–72, 74, 76–77, 2002.

19. Day, J.P. A multiplexed ultra-high throughput genotyping platform based on the oligonucleotide ligation assay: Design pipeline, conversion rate, and genotyping accuracy. In *The 53rd American Society of Human Genetics Annual Meeting.* 2003.

20. Baner, J., A. Isaksson, E. Waldenstrom, J. Jarvius, U. Landegren, and M. Nilsson. Parallel gene analysis with allele-specific padlock probes and tag microarrays. *Nucl Acids Res* 31: e103, 2003.

21. Consolandi, C., E. Busti, C. Pera, L. Delfino, G.B. Ferrara, R. Bordoni, B. Castiglioni, L.R. Bernardi, C. Battaglia, and G. De Bellis. Detection of HLA polymorphisms by ligase detection reaction and a universal array format: a pilot study for low resolution genotyping. *Hum Immunol* 64: 168–178, 2003.

22. Hardenbol, P., J. Baner, M. Jain, M. Nilsson, E.A. Namsaraev, G.A. Karlin-Neumann, H. Fakhrai-Rad, M. Ronaghi, T.D. Willis, U. Landegren, and R.W. Davis. Multiplexed genotyping with sequence-tagged molecular inversion probes. *Nat Biotechnol* 21: 673–678, 2003.

23. Iannone, M.A., J.D. Taylor, J. Chen, M.S. Li, P. Rivers, K.A. Slentz-Kesler, and M.P. Weiner. Multiplexed single nucleotide polymorphism genotyping by oligonucleotide ligation and flow cytometry. *Cytometry* 39: 131–140, 2000.

24. Oliphant, A., D.L., Barker, J.R. Stuelpnagel, and M.S. Chee, BeadArray technology: enabling an accurate, cost-effective approach to high-throughput genotyping. *Biotechniques* (Suppl.): 56–80, 60–61, 2002.

25. Han, M., X. Gao, J.Z. Su, and S. Nie. Quantum-dot-tagged microbeads for multiplexed optical coding of biomolecules. *Nat Biotechnol* 19: 631–635, 2001.

26. Busti, E., R. Bordoni, B. Castiglioni, P. Monciardini, M. Sosio, S. Donadio, C. Consolandi, L. Rossi Bernardi, C. Battaglia, and G. De Bellis. Bacterial discrimination by means of a universal array approach mediated by LDR (ligase detection reaction). *BMC Microbiol* 2: 27, 2002.

27. Chen, J., M.A. Iannone, M.S. Li, J.D. Taylor, P. Rivers, A.J. Nelsen, K.A. Slentz-Kesler, A. Roses, and M.P. Weiner. A microsphere-based assay for multiplexed single nucleotide polymorphism analysis using single base chain extension. *Genome Res* 10: 549–557, 2000.

28. Epstein, J.R., J.A. Ferguson, K.H. Lee, and D.R. Walt. Combinatorial decoding: an approach for universal DNA array fabrication. *J Am Chem Soc* 125: 13753–13759, 2003.

29. Hirschhorn, J.N., P. Sklar, K. Lindblad-Toh, Y.M. Lim, M. Ruiz-Gutierrez, S. Bolk, B. Langhorst, S. Schaffner, E. Winchester, and E.S. Lander. SBE-TAGS: an array-based

method for efficient single-nucleotide polymorphism genotyping. *Proc Natl Acad Sci U S A* 97: 12164–12169, 2000.

30. Jarvius, J., M. Nilsson, and U. Landegren. Oligonucleotide ligation assay. *Methods Mol Biol* 212: 215–228, 2003.

31. Ladner, D.P., J.H. Leamon, S. Hamann, G. Tarafa, T. Strugnell, D. Dillon, P. Lizardi, and J. Costa. Multiplex detection of hotspot mutations by rolling circle-enabled universal microarrays. *Lab Invest* 81: 1079–1086, 2001.

32. Mikhailovich, V., S. Lapa, D. Gryadunov, A. Sobolev, B. Strizhkov, N. Chernyh, O. Skotnikova, O. Irtuganova, A. Moroz, V. Litvinov, M. Vladimirskii, M.C. Perelman, V. Erokhin, A. Zasedatelev, and A. Mirzabekov. Identification of rifampin-resistant Mycobacterium tuberculosis strains by hybridization, PCR, and ligase detection reaction on oligonucleotide microchips. *J Clin Microbiol* 39: 2531–2540, 2001.

33. Taylor, J.D., D. Briley, Q. Nguyen, K. Long, M.A. Iannone, M.S. Li, F. Ye, A. Afshari, E. Lai, M. Wagner, J. Chen, and M.P. Weiner. Flow cytometric platform for high-throughput single nucleotide polymorphism analysis. *Biotechniques* 30: 661–666, 668–669, 2001.

34. Zhong, X.B., P.M. Lizardi, X.H. Huang, P.L. Bray-Ward, and D.C. Ward. Visualization of oligonucleotide probes and point mutations in interphase nuclei and DNA fibers using rolling circle DNA amplification. *Proc Natl Acad Sci U S A* 98: 3940–3945, 2001.

35. Zhong, X.B., R. Reynolds, J.R. Kidd, K.K. Kidd, R. Jenison, R.A. Marlar, and D.C. Ward. Single-nucleotide polymorphism genotyping on optical thin-film biosensor chips. *Proc Natl Acad Sci U S A* 100: 11559–11564, 2003.

36. Pastinen, T., M. Raitio, K. Lindroos, P. Tainola, L. Peltonen, and A.C. Syvanen. A system for specific, high-throughput genotyping by allele-specific primer extension on microarrays. *Genome Res* 10: 1031–1042, 2000.

37. Fan, J.B., X. Chen, M.K. Halushka, A. Berno, X. Huang, T. Ryder, R.J. Lipshutz, D.J. Lockhart, and A. Chakravarti. Parallel genotyping of human SNPs using generic high-density oligonucleotide tag arrays. *Genome Res* 10: 853–860, 2000.

38. Knight, J. When the chips are down. *Nature* 410: 860–861, 2001.

39. Mecham, B.H., D.Z. Wetmore, Z. Szallasi, Y. Sadovsky, I. Kohane, and T.J. Mariani. Increased measurement accuracy for sequence-verified microarray probes. *Physiol Genomics*, 18: 308–315, 2004.

40. Favis, R., N.P. Gerry, Y.W. Cheng, and F. Barany. Applications of the universal DNA microarray in molecular medicine, in *Methods in Molecular Medicine: Microarrays in Clinical Diagnostics*, T.O. Joos and P. Fortina, Eds. The Humana Press Inc: 2004.

41. Ahrendt, S., S. Halachmi, J. Chow, L. Wu, N. Halachmi, S. Yang, S. Wehage, J. Jen, and D. Sidransky. Rapid p53 sequence analysis in primary lung cancer using an oligonucleotide probe array. *Proc Natl Acad Sci U S A* 96: 7382–7387, 1999.

42. Wikman, F.P., M.L. Lu, T. Thykjaer, S.H. Olesen, L.D. Andersen, C. Cordon-Cardo, and T.F. Orntoft. Evaluation of the performance of a p53 sequencing microarray chip using 140 previously sequenced bladder tumor samples. *Clin Chem* 46: 1555–1561, 2000.

43. Wen, W.H., L. Bernstein, J. Lescallett, Y. Beazer-Barclay, J. Sullivan-Halley, M. White, and M.F. Press. Comparison of TP53 mutations identified by oligonucleotide microarray and conventional DNA sequence analysis. *Cancer Res* 60: 2716–2722, 2000.

44. Zirvi, M., T. Nakayama, G. Newman, T. McCaffrey, P. Paty, and F. Barany. Ligase-based detection of mononucleotide repeat sequences. *Nucl Acids Res* 27: e40, 1999.

45. Zirvi, M., D.E. Bergstrom, A.S. Saurage, R.P. Hammer, and F. Barany. Improved fidelity of thermostable ligases for detection of microsatellite repeat sequences using nucleoside analogs. *Nucl Acids Res* 27: e41, 1999.

46. Torrents, D., M. Suyama, E. Zdobnov, and P. Bork. A genome-wide survey of human pseudogenes. *Genome Res* 13: 2559–2567, 2003.

47. Zhang, A., P.M. Harrison, Y. Liu, and M. Gerstein. Millions of years of evolution preserved: a comprehensive catalog of the processed pseudogenes in the human genome. *Genome Res* 13: 2541–2558, 2003.

48. Lander, E.S. et al. Initial sequencing and analysis of the human genome. *Nature* 409: 860-921, 2001.

49. Venter, J.C. et al. The sequence of the human genome. *Science* 291: 1304–1351, 2001.

50. Fredman, D., S.J. White, S. Potter, E.E. Eichler, J.T.D. Dunnen, and A.J. Brookes. Complex SNP-related sequence variation in segmental genome duplications. 2004. advanced online publication, *Nature Genetics*.

51. Thomas, E.E., N. Srebro, J. Sebat, N. Navin, J. Healy, B. Mishra, and M. Wigler. Distribution of short paired duplications in mammalian genomes. *Proc Natl Acad Sci USA* 101: 10349–10354, 2004.

52. Sebat, J., B. Lakshmi, J. Troge, J. Alexander, J. Young, P. Lundin, S. Maner, H. Massa, M. Walker, M. Chi, N. Navin, R. Lucito, J. Healy, J. Hicks, K. Ye, A. Reiner, T.C. Gilliam, B. Trask, N. Patterson, A. Zetterber, and M. Wigler. Large-scale copy number polymorphism in the human genome. *Science* 305: 525–528, 2004.

53. Tischfield, J.A. Loss of heterozygosity or: how I learned to stop worrying and love mitotic recombination. *Am J Hum Genet* 61: 995–999, 1997.

54. Hurteau, G.J. and S.D. Spivack. mRNA-specific reverse transcription-polymerase chain reaction from human tissue extracts. *Anal. Biochem.* 307: 304–315, 2002.

55. Day, D.J., P.W. Speiser, E. Schulze, M. Bettendorf, J. Fitness, F. Barany, and P.C. White. Identification of non-amplifying CYP21 genes when using PCR-based diagnosis of 21-hydroxylase deficiency in congenital adrenal hyperplasia (CAH) affected pedigrees. *Hum Mol Genet* 5: 2039–2048, 1996.

56. Day, D.J., P.W. Speiser, P.C. White, and F. Barany. Detection of steroid 21-hydroxylase alleles using gene-specific PCR and a multiplexed ligation detection reaction. *Genomics* 29: 152–162, 1995.

57. Pritchard, J.K. and N.J. Cox. The allelic architecture of human disease genes: common disease-common variant or not? *Hum Mol Genet* 11: 2417–2423, 2002.

58. Pritchard, J.K. Are rare variants responsible for susceptibility to complex diseases? *Am J Hum Genet* 69: 124–137, 2001. Epub June 12, 2001.

59. Bonifati, V., P. Rizzu, M.J. van Baren, O. Schaap, G.J. Breedveld, E. Krieger, M.C. Dekker, F. Squitieri, P. Ibanez, M. Joosse, J.W. van Dongen, N. Vanacore, J.C. van Swieten, A. Brice, G. Meco, C.M. van Duijn, B.A. Oostra, and P. Heutink. Mutations in the DJ-1 gene associated with autosomal recessive early-onset parkinsonism. *Science* 299: 256–259, 2003. Epub November 21, 2002.

60. Nei, M. and W.H. Li. Mathematical model for studying genetic variation in terms of restriction endonucleases. *Proc Natl Acad Sci U S A* 76: 5269–5273, 1979.

61. Watterson, G.A. On the number of segregating sites in genetical models without recombination. *Theor Pop Biol* 7: 256–276, 1975.

62. Reich, D.E., M. Cargill, S. Bolk, J. Ireland, P.C. Sabeti, D.J. Richter, T. Lavery, R. Kouyoumjian, S.F. Farhadian, R. Ward, and E.S. Lander. Linkage disequilibrium in the human genome. *Nature* 411: 199–204, 2001.

63. Lewontin, R.C. On measures of gametic disequilibrium. *Genetics* 120: 849–852, 1988.

64. Weiss, K.M. and A.G. Clark. Linkage disequilibrium and the mapping of complex human traits. *Trends Genet* 18: 19–24, 2002.

65. Lynch, M. and B. Walsh. *Genetics and Analysis of Quantitative Traits*. Sunderland, MA: Sinauer Associates Inc., 1998.

66. Spielman, R.S., R.E. McGinnis, and W.J. Ewens. Transmission test for linkage disequilibrium: the insulin gene region and insulin-dependent diabetes mellitus (IDDM). *Am J Hum Genet* 52: 506–516, 1993.

67. Rice, W.R. Analyzing statistical tests. *Evolution* 43: 223–225, 1989.

68. Carlson, C.S., M.A. Eberle, L. Kruglyak, and D.A. Nickerson. Mapping complex disease loci in whole-genome association studies. *Nature* 429: 446–452, 2004.

69. Sabatti, C., S. Service, and N. Freimer. False discovery rate in linkage and association genome screens for complex disorders. *Genetics* 164: 829–833, 2003.

70. McVean, G.A., S.R. Myers, S. Hunt, P. Deloukas, D.R. Bentley, and P. Donnelly. The fine-scale structure of recombination rate variation in the human genome. *Science* 304: 581–584, 2004.

71. Wall, J.D. and J.K. Pritchard. Assessing the performance of the haplotype block model of linkage disequilibrium. *Am J Hum Genet* 73: 502–515, 2003.

72. Wall, J.D. and J.K. Pritchard. Haplotype blocks and linkage disequilibrium in the human genome. *Nat Rev Genet* 4: 587–597, 2003.

73. Blangero, J. Localization and identification of human quantitative trait loci: king harvest has surely come. *Curr Opin Genet Dev* 14: 233–240, 2004.

74. Boutin, P., C. Dina, F. Vasseur, S.S. Dubois, L. Corset, K. Seron, L. Bekris, J. Cabellon, B. Neve, V. Vasseur-Delannoy, M. Chikri, M.A. Charles, K. Clement, A. Lernmark, and P. Froguel. GAD2 on chromosome 10p12 is a candidate gene for human obesity. *PLoS Biol* 1: E68, 2003.

75. Terwilliger, J.D. and K.M. Weiss. Confounding, ascertainment bias, and the blind quest for a genetic "fountain of youth." *Ann Med* 35: 532–544, 2003.

13 Approaches for Microarray Data Validation

Sergey E. Ilyin

Microarray data has numerous scientific applications and, in many instances, investigators may not need to go beyond the statistical analysis of an expressional profile obtained on array. An example would be a situation in which a certain microarray signature is proven to have a diagnostic validity. For example, RNA samples from 74 patients with Dukes' B colon cancer were analyzed using an Affymetrix U133a GeneChip [1]. Class prediction approaches were used to identify gene markers that can best discriminate between patients who would experience relapse and patients who would remain disease-free. Gene expression profiling identified a 23-gene signature that predicts recurrence in Dukes' B patients [1]. In a number of other instances, however, validation of microarray data may become an integral part of research projects. Validation means answering one or several of the following questions:

1. Are the observed statistically significant changes for certain genes real (relates to false positive)?
2. Are the observed statistically insignificant values for certain genes real (false negative findings)?
3. An additional aspect of validation relates to functional characterization. Microarray experiments are often carried out to identify the potential function of certain genes, and, as such, these modeling experiments require functional validation.

Prior to validation of microarray findings, microarray data needs to undergo a rigorous quality control, normalization, and statistical analysis to identify candidate genes. Microarray data analysis is a well-studied area, described elsewhere [2–5]. Confirmation of microarray data using some independent methods is the most straightforward task (see Figure 13.1). There are a number of independent technologies, which can be used for expressional profiling (i.e., the Northern blot, RNase protection assay [6]). Despite being quite old, these technologies still offer

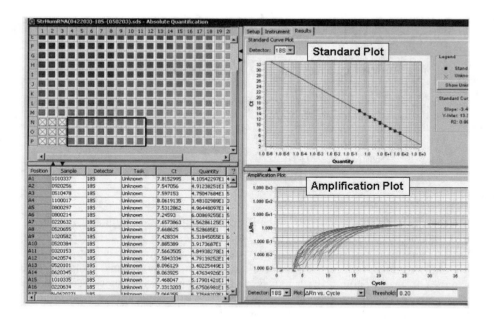

FIGURE 13.1 *(A color version follows page 204)* A representative amplification plot showing amplification of reference samples. Linear relationship between amount of input material and CT values was observed across all reference samples suggesting a broad dynamic range of detection.

some unique advantages. For example, Northern blot analysis enables one to detect the sizes of mRNA. TaqMan qRT–PCR is currently a method of choice for microarray data validation, because of the relative simplicity of the procedure and availability of validated reagents for a significant number of genes. Applied Biosystems (http://www.appliedbiosystems.com) offers a compelling choice of redesigned human (over 40,000 ready-to-use human gene detection reagents), rat, and mouse gene detection reagents. TaqMan qRT–PCR is a kinetic PCR [7], which means that the accumulation of product is observed in real time. A liquid hybridization assay is incorporated into the reaction. Standards are generally incorporated in each plate run, and these may include: standards with a known amount of material to estimate unknowns, negative controls (contamination control), and RT control to verify absence of genomic DNA. The type of standards used in the experiment will determine whether it is a quantitative or semiquantitative characterization. Most experimental designs also incorporate the analysis of a housekeeping gene to verify that an equal amount of material is used for all samples. Data analysis and interpretation of these types of experiments was previously described. General observations made in the course of microarray data confirmation are:

1. It is highly advisable to design PCR primers for the same sequence as the one represented on microarray as it significantly increases chances of successful validation.
2. Microarray data exhibits a so-called compression effect [8] or smaller fold change compared to the real one. Compression may significantly affect the observed fold change for a low-level expressed gene and, in some cases, could make these changes even undetectable, thus leading to false negatives.

False negatives are a more complicated problem to investigate. False negatives may arise from different technical issues (for example, poor probe design, a problem with a specific spot, or a very low level of expression leading to extreme compression). In general, much less attention is paid to the existence of false negatives. A problem may be studied by performing large-scale parallel microarray/qPCR experiments with genes representing different ranges of expression. TaqMan qPCR confirmation is facilitated by the advent of novel methodologies for sample preparation; such as the novel poly-A mRNA capture plate-based method [9]. This technology eliminates all traditional steps of sample preparation and all steps including RNA extraction, RT, and PCR are performed in the same tube, thus significantly enhancing throughput of this method and making it automation friendly [10]. As the sample is lysed in poly-T-containing plates, mRNA gets captured, wells are washed, and subsequent steps of RT and PCR are performed. Figure 13.2 shows the results of an experiment designed to determine if mRNA binding to the GenePlate HT would

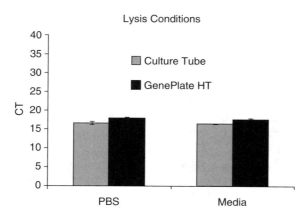

FIGURE 13.2 Ten thousand K562 cells were suspended in either PBS or growth media containing 10% FCS. The cells were either lysed in separate culture tubes by mixing equal parts of 2X lysis buffer with the medium or PBS, then added to the GenePlate HT, or were lysed directly in the GenePlate HT using 2X lysis buffer. Samples were reverse transcribed and amplified as described. The results represent the mean of four replicates. (From Maley, D., Mei, J., Lu, H., Johnson, D.L., and Ilyin, S.E. Multiplexed RT- PCR for high-throughput screening applications. *Comb Chem High Throughput Screen* 7(8), 727–732, 2004. With permission.)

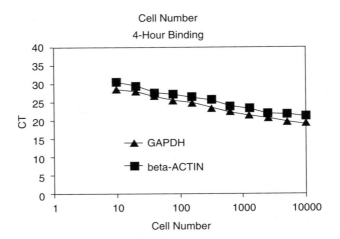

FIGURE 13.3 Increasing numbers of cells were plated in the GenePlate HT, to determine the limits of RNA capture. 1:2 dilutions from 10,000 cells/well were extracted with 2X lysis buffer, then the poly-A RNA allowed to capture for 4 h to the oligo-T on the plate. The wells were washed, and cDNA reverse transcribed for 2 h at 37°C. GAPDH or beta-Actin cDNA were quantified using TaqMan directly in the GenePlate HT as described in Methods. The results were expressed as the number of amplifications needed to reach a threshold of detection (CT). (From Maley, D., Mei, J., Lu, H., Johnson, D.L., and Ilyin, S.E. Multiplexed RT- PCR for high-throughput screening applications. *Comb Chem High Throughput Screen* 7(8), 727–732, 2004. With permission.)

be influenced by serum proteins or other media components, and if the cellular mRNA can be extracted directly in the GenePlate HT. There was a minimal difference between cells lysed in PBS or in growth media containing 10% FCS. There was no statistically significant difference between lysis directly in the GenePlate HT and lysis in separate culture tubes with subsequent transfer to GenePlate HT. Figure 13.3 demonstrates the range of cell numbers that can be used in the GenePlate HT. The reduction was linear through the entire range of cells from 10,000 to approximately 10.

Figure 13.4 is a time course for binding of lysates to a GenePlate HT. The oligo-dT plate requires time for the polyA RNA to bind. Allowing the lysate to sit in the plate 1 h was sufficient to capture the gene product. Figure 13.5 demonstrates a dose response of a certain gene AOD (assay on demand), and the changes in both the target gene (AOD) and a nontarget gene (GAPDH). Induction and amplification of the target gene occurred when the primer/probe set was used alone, or multiplexed with GAPDH primers/probe. There was no observed difference between the induction curves. At the same time, signals from the second target (GAPDH) did not vary with increasing compound concentrations.

The BioTrove qPCR microfluidics device (http://www.biotrove.com/) is another approach to microarray data validation. The OpenArray™ transcript analysis system performs real-time PCR reactions achieving a throughput of over 3000 reactions per array (Figure 13.6). This technology is based on a sandwich of hydrophobic layers encapsulating a hydrophilic inner part of the well. When this sandwich is immersing

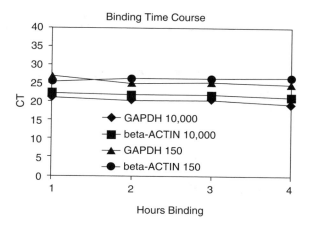

FIGURE 13.4 The time required for optimum binding of poly-A RNA to the GenePlate HT was evaluated by lysis of cells in the plate, followed by increasing binding times. Ten thousand and 150 cells were plated, in replicate wells. 2X lysis buffer was added directly to the culture medium, and then the plates incubated at 37°C until the next time point. Lysis buffer was added to the next set of wells, and the plates returned to the incubator. Following the last hour of binding, the final wells were lysed, and the plates mixed at room temperature. Following lysis and binding, all of the wells were washed 3X with wash buffer, then cDNA reverse transcribed, and DNA amplified by TaqMan as described. Results were expressed as CT. (From Maley, D., Mei, J., Lu, H., Johnson, D.L., and Ilyin, S.E. Multiplexed RT- PCR for high-throughput screening applications. *Comb Chem High Throughput Screen* 7(8), 727–732, 2004. With permission.)

in the water-based solution, capillary force fills each capillary with a fairly constant amount of solvent. Each well contains a primer pair probe encapsulated in the thermo labile material so the release of reagents occurs during the first cycle of the PCR reaction. Each chip contains a number of negative and positive controls in addition to experimental samples. A significant advantage of OpenArray is in the system scalability. Different manipulations could be performed in the sandwiched chips and materials could be transferred using capillary forces (Figure 13.7).

A note worthy of mentioning with regard to the comparison between microarray and other approaches relates to informatics support of such exercises. Although it may be a relatively straightforward task to compare the expression of a couple of genes between microarray and qPCR data, large data sets require a completely different way of consideration. Guided applications incorporating statistical and visualization tools offer convenient and timesaving solutions. Example of such solution is Spotfire/R application for microarray and PCR analysis described by Stephen Prouty and his colleagues [11]. Solution enables to deploy application on a server and distribute guided statistical routines across organization. In this integrated approach, Spotfire is used as an interface with end user, and statistical analysis is performed on R or/and S-PLUS server (Figure 13.8).

The functional characterization of genes identified in a microarray experiment may be achieved using functional genomics tools. If a straightforward test system

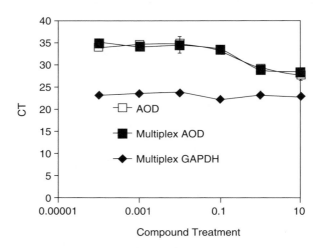

FIGURE 13.5 A multiplex experiment testing the induction of a gene by a compound compares the RT-PCR signal generated in a reaction containing a single primer/probe set with the same reaction containing a second, housekeeping primer/probe set. Ten thousand K562 cells were treated with increasing concentrations of an inducer compound for 2 h in a GenePlate HT at 37°C. The cells were lysed and RT-PCR was conducted as described in Methods. An assay on demand (AOD) primer/probe set targeting the induced gene was employed to quantify the changes in message induced by the compound. A duplicate set of inductions was performed using an additional primer/probe set targeting GAPDH in a multiplex reaction with the AOD. The results were expressed as the threshold CT as described. The results of the multiplex reaction are described as two curves, one for the AOD target gene, and a GAPDH curve from the same multiplex wells. The AOD utilized a FAM TaqMan signal, while GAPDH uses the fluorescent signal VIC. Compound, tested in doses ranging from 0.001 μM to 10.00 μM, was used as indicated. Compound treatment induced significant upregulation of AOD (over 50-fold, as data expressed in threshold cycle numbers). (From Maley, D., Mei, J., Lu, H., Johnson, D.L., and Ilyin, S.E. Multiplexed RT-PCR for high-throughput screening applications. *Comb Chem High Throughput Screen* 7(8), 727–732, 2004. With permission.)

for functional validation could be created *in vitro*, then a battery of functional genomics tools could be used for this purpose: viral vectors [12], antisense, aptamers [13], and siRNA [14,15]. *In vivo* validation generally requires the development of animal models and is fairly time-consuming. An exception to this is a situation in which the localized delivery of a gene-expression-modifying agent is used to create an animal model [12]. *In vitro* functional validation could be achieved by delivering a gene expression-modulating reagent to the cell in culture and subjecting modified and control cell populations to a battery of tests to interrogate gene function. This experimentation can be automated [14]; in fact, a novel experimental paradigm, *functional informatics*, which is based on the integration of automation and bio-informatics, was described and validated. The integrative functional informatics

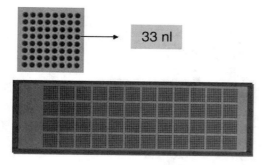

chip	channel	sub_array_x	sub_array_y	row	column	UserSamp	Concentration	Ct	Ct Confide	Tm	Expected	Tm Confid	RefSeq ID	Functional	UniGene ID
1	A1a1	A	1	a	1	TK011-5ca	0	13.2926	0.883428	85.17088	84.46553	0.999589	NM_01455	ion channe	272287
1	A1a2	A	1	a	2	TK011-5ca	0	15.75315	0.99408	85.09245	84.402	0.999962	NM_00526	connexon	198249
1	A1a3	A	1	a	3	TK011-5ca	0	15.92889	0.992607	79.7646	79.69856	1.36E-05	NM_02243	channel pc	80395
1	A1a4	A	1	a	4	TK011-5ca	0	15.48377	0.897317	86.93428	86.80764	0.71678	NM_00443	potassium	163936
1	A1a5	A	1	a	5	TK011-5ca	0	15.5239	0.967388	0	85.52528	0	NM_00421	purine rec	167076
1	A1a6	A	1	a	6	TK011-5ca	0	15.18785	0.915991	84.44491	84.13346	0.999969	NM_00112	transporte	331602
1	A1a7	A	1	a	7	TK011-5ca	0	15.50167	0.935743	78.53833	79.40003	0.999969	NM_001402		0
1	A1a8	A	1	a	8	TK011-5ca	0	15.40244	0.998706	0	78.54879	0	Note		0
1	A2a1	A	2	a	1	TK011-5ca	0	15.1687	0.952318	79.35796	79.41258	0.083301	NM_00707	transporte	293411
1	A2a2	A	2	a	2	TK011-5ca	0	14.8668	0.967509	80.82127	81.11592	0.22395	NM_01474	transporte	79305
1	A2a3	A	2	a	3	TK011-5ca	0	17.19327	0.97529	84.98974	85.55853	0.9502	NM_13363	transporte	125878
1	A2a4	A	2	a	4	TK011-5ca	0	15.93712	0.913997	79.35796	79.36179	0.999969	NM_15292	transporte	166887

FIGURE 13.6 Microfluidics OpenArray from BioTrove performs uHT Real Time PCR in 33 nl total reaction volume. Solution achieves throughput of 3000 reactions per chip. Top: OpenArray overview. Bottom: Example of data generated in the course of real-time PCR experiment.

definition has been included in the Cambridge Healthtech Institute Bioinformatics Glossary: (http://www.genomicglossaries.com/content/Bioinformatics_gloss.asp).

Functional informatics takes advantage of automated assays and integrates them with machine learning techniques for analysis and hypothesis generation. This integration enables target identification, validation, lead generation, and optimization to be performed seamlessly in the same system. Information obtained in the first rounds of testing will be processed, integrated, and used for hypothesis generation using neuronal networks or other types of methodologies. Input from this modeling can be fed back directly into the integrated robotic system for testing, thus enabling fully automated operations to define novel knowledge space. Functional Informatics also facilitates a novel approach to personalized medicine by enabling economically viable testing of individual patients' biology [17,18]. A current limitation of this approach is in the fairly high cost of large-scale experimentation using industrial robotics. Just as an example, in a 96-well format, each data point in an expression profiling assay is between $1.00 to $3.00, cost drops substantially by transitioning to a 384-well format (i.e., $0.33 to $1.00), but still remains prohibitively high for large-scale studies. Microfluidics, however, opens a totally new dimension by potentially reducing cost to several cents per well. Reduced cost, in turn, enables the investigator to perform an increased number of testing and refine knowledge about

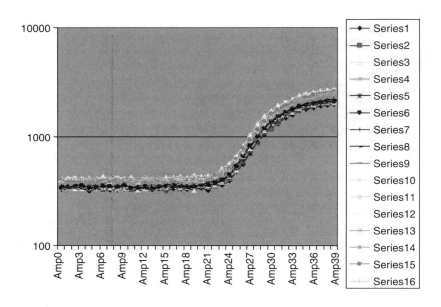

FIGURE 13.7 *(A color version follows page 204)* Example of control amplification (all wells equally loaded with positive control) performed in OpenArray on BioTrove system.

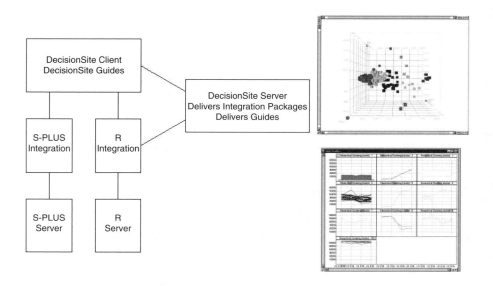

FIGURE 13.8 *(A color version follows page 204)* Integrated solutions for data analysis facilitate data processing, integration, and reporting. Spotfire DecisionSite is used as an interface. The user communicates directly with the R and S-PLUS servers, and uses the DecisionSite server for software updates.

the biological role of the molecule and potential side effects of a new treatment, thus derisking patients and enhancing value to pharmaceutical and biomedical organizations. Fairly interesting studies can be performed to adapt microarray technology for functional characterization. First, cDNAs or siRNA are printed on the chip, cells are then plated, and then image-based analysis of transfected cells can be performed. For example, a cell-based microarray system was described [19,20]. cDNAs or siRNA are printed on the chip, cells are plated and image-based analysis of transfected cells is performed. The system offers significant throughput as at least 10,000 different sequences can be simultaneously tested on a single chip. The lack of physical boundaries between spots imposes certain limitations on this technology and limits the types of assays that can be performed. The possibility to incorporate physical boundaries to limit diffusion, while maintaining throughput capability represents an exciting and highly practical opportunity.

ACKNOWLEDGMENTS

The authors would like to thank Daniel Horowitz, Albert Pinhasov, Stephen Prouty, and Carlos R. Plata-Salamán for their valuable comments and examples.

REFERENCES

1. Wang, Y., Jatkoe, T., Zhang, Y., Mutch, M.G., Talantov, D., Jiang, J., McLeod, H.L., and Atkins, D. Gene expression profiles and molecular markers to predict recurrence of Dukes' B colon cancer. *J Clin Oncol* 22(9), 1564–1571, 2004.
2. Hackl, H., Cabo, F.S., Sturn, A., Wolkenhauer, O., and Trajanoski, Z. Analysis of DNA microarray data. *Curr Top Med Chem* 4(13), 1357–1370, 2004.
3. Krajewski, P. and Bocianowski, J. Statistical methods for microarray assays. *J Appl Genet* 43(3), 269–278, 2002.
4. Planet, P.J., DeSalle, R., Siddall, M., Bael, T., Sarkar, I.N., and Stanley, S.E. Systematic analysis of DNA microarray data: ordering and interpreting patterns of gene expression. *Genome Res* 11(7), 1149–1155, 2001.
5. Quackenbush, J. Computational analysis of microarray data. *Nat Rev Genet* 2(6), 418–427, 2001.
6. Ilyin, S.E., Gayle, D., and Plata-Salaman, C.R. Modifications of RNase protection assay for neuroscience applications. *J Neurosci Methods* 84(1–2), 139–141, 1998.
7. Giulietti, A., Overbergh, L., Valckx, D., Decallonne, B., Bouillon, R., and Mathieu, C. An overview of real-time quantitative PCR: applications to quantify cytokine gene expression. *Methods* 25(4), 386–401, 2001.
8. Yue, H., Eastman, P.S., Wang, B.B., Minor, J., Doctolero, M.H., Nuttall, R.L., Stack, R., Becker, J.W., Montgomery, J.R., Vainer, M., and Johnston, R. An evaluation of the performance of cDNA microarrays for detecting changes in global mRNA expression. *Nucl Acids Res* 29(8), E41–41, 2001.
9. Maley, D., Mei, J., Lu, H., Johnson, D.L., and Ilyin, S.E. Multiplexed RT- PCR for high-throughput screening applications. *Comb Chem High Throughput Screen* 7(8), 727–732, 2004.
10. Pinhasov, A., Mei, J., Amaratunga, D., Amato, F.A., Lu, H., Kauffman, J., Xin, H., Brenneman, D.E., Johnson, D.L., Andrade-Gordon, P., and Ilyin, S.E. Gene

expression analysis for high-throughput screening applications. *Comb Chem High Throughput Screen* 7(2), 133–140, 2004.

11. Prouty, S., Nathan, D., Ledwith, J., Salisbury, M., Lyon, G., Messer, A., Amaratunga, D., Go, O., J., W. and Ilyin, S.E. Integrative tools for data analysis in pharmaceutical R&D. *PharmaGenomics.* In press. 2004.

12. Darrow, A.L., Conway, K.A., Vaidya, A.H., Rosenthal, D., Wildey, M.J., Minor, L., Itkin, Z., Kong, Y., Piesvaux, J., Qi, J., Mercken, M., Andrade-Gordon, P., Plata-Salaman, C., and Ilyin, S.E. Virus-based expression systems facilitate rapid target *in vivo* functionality validation and high-throughput screening. *J Biomol Screen* 8(1), 65–71, 2003.

13. Burgstaller, P., Girod, A., and Blind, M. Aptamers as tools for target prioritization and lead identification. *Drug Discov Today* 7(24), 1221–1228, 2002.

14. Elbashir, S.M., Harborth, J., Lendeckel, W., Yalcin, A., Weber, K., and Tuschl, T. Duplexes of 21-nucleotide RNAs mediate RNA interference in cultured mammalian cells. *Nature* 411(6836), 494–498, 2001.

15. Hammond, S.M., Bernstein, E., Beach, D., and Hannon, G.J. An RNA-directed nuclease mediates post-transcriptional gene silencing in Drosophila cells. *Nature* 404(6775), 293–296, 2000.

16. Xin, H., Bernal, A., Amato, F.A., Pinhasov, A., Kauffman, J., Brenneman, D.E., Derian, C.K., Andrade-Gordon, P., Plata-Salamán, C.R., and Ilyin, S.E. High-throughput siRNA-based functional target validation. *J Biomol Screen* 9(4), 286–293, 2004.

17. Ilyin, S.E., Belkowski, S.M., and Plata-Salaman, C.R. Biomarker discovery and validation: technologies and integrative approaches. *Trends Biotechnol* 22(8), 411–416, 2004.

18. Ilyin, S.E., Pinhasov, A., Vaidya, A.H., Amato, F.A., et al. Emerging paradigms in applied bioinformatics. *BioSilico* 3: 86–88, 2003.

19. Wu, R.Z., Bailey, S.N., and Sabatini, D.M. Cell-biological applications of transfected-cell microarrays. *Trends Cell Biol* 12(10), 485–488, 2002.

20. Ziauddin, J. and Sabatini, D.M. Microarrays of cells expressing defined cDNAs. *Nature* 411(6833), 107–110, 2001.

14 Microarray Enterprise Information Management: What Is It and Why Is It Important?

Peter Morrison, Pankaj Prakash, and Soheil Shams

CONTENTS

ABSTRACT

A microarray experiment is conceptually simple — the expression levels of many genes (many thousands, in most cases) in a particular biological sample are measured [1,2]. However, to place the results of an array experiment in proper context, a great deal of data must be collected and properly organized and interrelated with respect to how the expression data were collected and later how these data were transformed into biologically relevant information. In this chapter, we will outline the microarray experimental workflow for both self-spotted as well as commercial array platforms.

203

We will concentrate on the data generated at each step of this process and the various approaches that can be taken to store and manage such data. We will highlight the importance of structured data management in ensuring result integrity through the use of an example. We will then provide an examination of how experimental process data can be used for performing quality control checks, facilitating collaborative research, meeting publication requirements, and encapsulating knowledge. An essential goal of this discussion is to highlight the importance of proper collection, management, and integration of experimental data when conducting microarray experiments. Throughout this chapter, we will use the GeneDirector® product from BioDiscovery, Inc. as an example of a comprehensive data management solution.

INTRODUCTION

Microarray experimentation is unlike most studies undertaken by biologists. The issues that make this process unique are as follows: (1) it takes days to generate all the raw data, (2) the large expense necessitates collaboration and sharing of resources, (3) the large volume of data requires active management, (4) results are generated as large tables and complex images, and (5) there is a large diversity and heterogeneity in the types of data generated at different stages of the experimental process, which include detailed notes in laboratory notebooks, instrumentation log files, images, etc. The complexity of the different file formats and handwritten notes and their complex relationships are inherently a substantial challenge for even the most basic microarray experiment. In the end, the user must be able to correctly associate and integrate these different files and associated data to achieve a meaningful analytical understanding.

The microarray experimental process can be broken into a set of logical steps. This chapter will present the process in terms of the following distinct components along the experimental data flow: (1) designing the array, (2) processing samples, (3) hybridizing arrays, (4) scanning array images, (5) generating and summarizing the expression measurements, and (6) discovering biological insights by analysis and data mining. Our goal in this chapter is to demonstrate the value of a data management system in ensuring the accuracy of the results obtained from array experiments or at least one's confidence in the data. Furthermore, we aim to demonstrate the importance of collecting and archiving all the experimental information in a structured fashion to both aid in enhancing the technical quality control capability, which may be applied to future experiments, as well as simplify the data submission for peer-reviewed publications. Figure 14.1 shows the inherent relationships among these processes.

In the remainder of this chapter, we will discuss the importance of data management for assurance of data integrity, including an example in which conventional methods for data management have resulted in data corruption and eventual loss of value of an important large-scale microarray experiment. Then we will discuss in detail all aspects of array experiment data management, from array design to quantified arrays. Finally, we provide some concluding remarks and future directions.

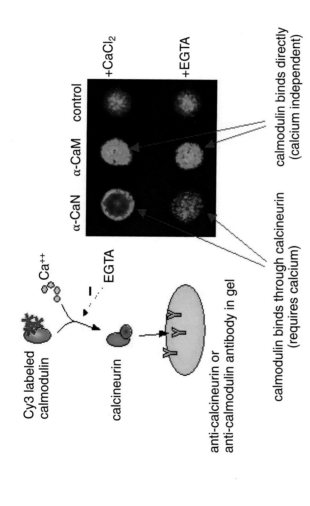

COLOR FIGURE 2.11 Sequence-specific protein–DNA recognition in three-dimensional HydroArray. (From Gurevich, D., et al., A novel three-dimensional hydrogel-based microarray platform, *JALA*, 6(4): 87–91, 2001. With permission.)

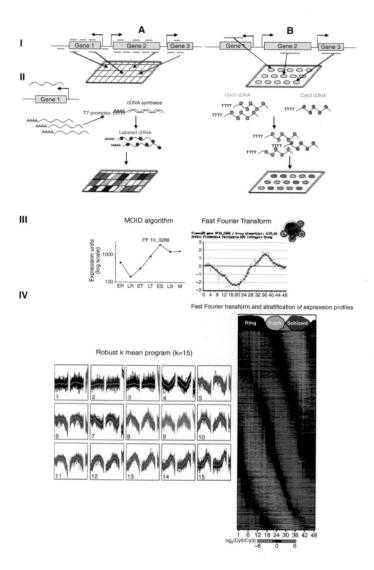

COLOR FIGURE 3.2 Comparison of the two microarray methods used for the malaria parasite gene expression's life cycle (A, short oligonucleotides by *in situ* synthesis; Affymetrix, Santa Clara, CA) or malaria cell cycle (B, long oligonucleotides by robotic deposition of nucleic acids onto a glass slide). I — Probe Design: For the high-density 25-mer oligonucleotide array, multiple probes per gene are placed on the array (A). In the case of robotic deposition, a single (75-mer) probe is generally used for each gene. II — Preparation of labeled material for measurement of gene expression using a cRNA labeled protocol (A) or a cDNA labeling protocol using the Cy3 (or Cy5) for a two-color strategy (B). III — Experimental design and expression level using different algorithms: MOID algorithm (A) and Fourier transform (B). III — Cluster analysis using the robust k-mean algorithm (A) or the fast Fourier transform (FTT) (B).

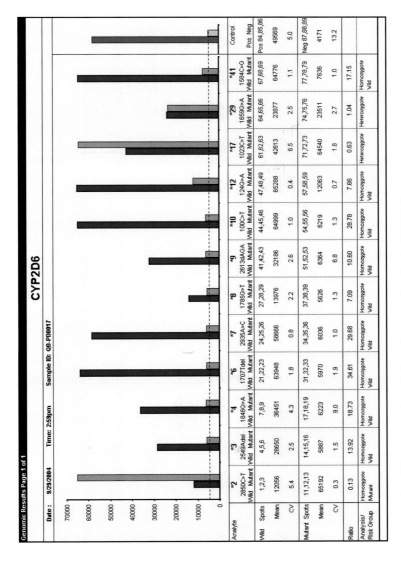

COLOR FIGURE 6.4 Result analysis and presentation for the CYP2D6 assay.

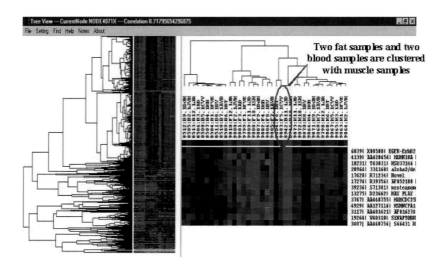

COLOR FIGURE 7.5 Clustering on 36 samples reveals 4 spurious samples. Gene number (at least twofold change in one sample and geo-mean): 4303.

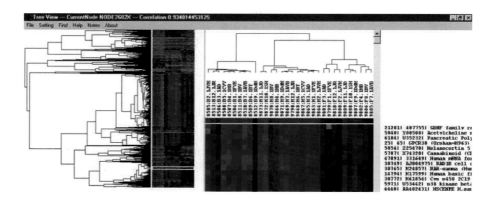

COLOR FIGURE 7.6 Clustering on the remaining 32 samples after excluding 4 spurious samples. Gene number (at least twofold change in one sample and geo-mean): 3709.

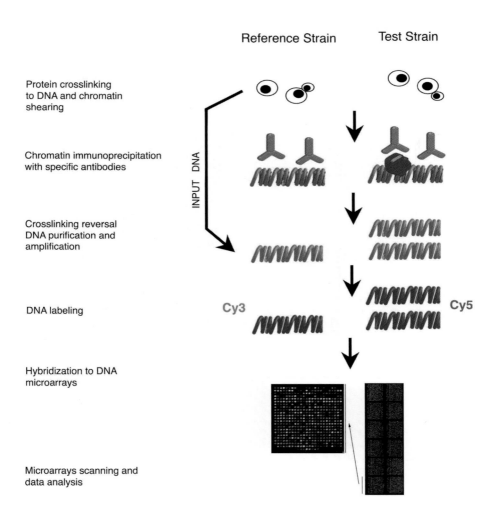

COLOR FIGURE 8.1 General ChIP-on-chip method overview. The "test" and "reference" experiments are run in parallel. The test strain contains the protein of interest (or its epitope-tagged version), whereas the reference strain has neither the protein nor the tag. Chromatin from formaldehyde-treated cells is sheared by sonication to an average size of 500 bp and immuno-precipitated using highly specific antibodies raised against the analyzed protein. After immu-noprecipitation and crosslinking reversal, DNA is purified and amplified by PCR. INPUT DNA from the reference strain (crosslinked and sonicated) is used as an alternative in case immuno-precipitation does not yield enough DNA. INPUT DNA is amplified by PCR directly, bypassing the immunoprecipitation step. This approach, however, does not control for any potential weak antibody specificity. Enriched DNA from the test and reference strains is then labeled with the fluorophores Cy3 and Cy5, respectively. The fluorescent DNA probes are subsequently com-bined and hybridized to a DNA microarray. For any given DNA fragment on the microarray, the ratio of the normalized fluorescent intensities between the two probes reflects the protein enrichment in the test relative to the reference experiment.

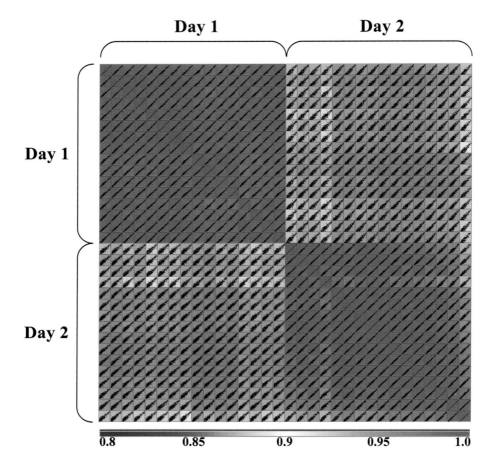

COLOR FIGURE 11.1 *Batch variability:* Two separate hybridizations of the same sample to cDNA arrays, replicated 16 times in each hybridization, run on consecutive days, are compared, with a color scale of correlation as indicated from 0.8 to 1.0. Average correlation between all arrays from Day 1 = 0.987, Day 2 = 0.981. Average correlation between arrays hybridized on different days = 0.945. Scatter plots within each individual square can be observed; less scatter will typically indicate a higher degree of correlation. Correlations calculated using OmniViz Desktop 3.9.0 using the CorScape tool.

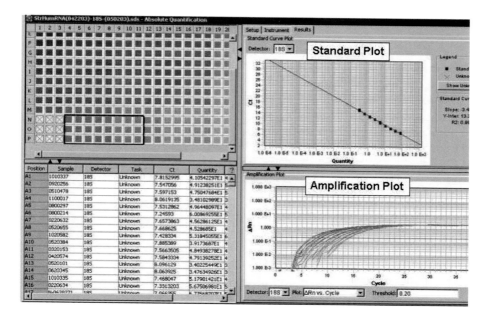

COLOR FIGURE 13.1 A representative amplification plot showing amplification of reference samples. Linear relationship between amount of input material and CT values was observed across all reference samples suggesting a broad dynamic range of detection.

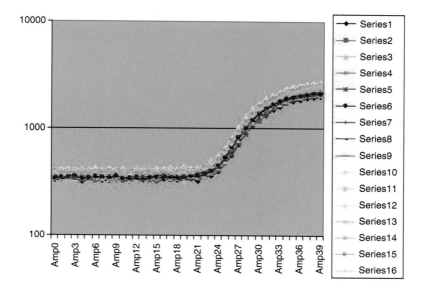

COLOR FIGURE 13.7 Example of control amplification (all wells equally loaded with positive control) performed in OpenArray on BioTrove system.

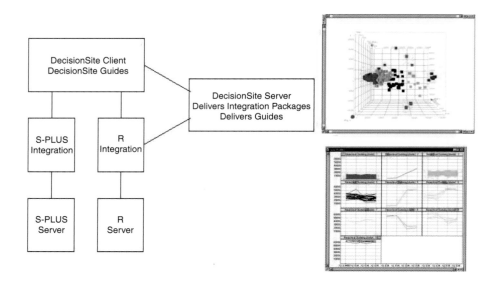

COLOR FIGURE 13.8 Integrated solutions for data analysis facilitate data processing, integration, and reporting. Spotfire DecisionSite is used as an interface. The user communicates directly with the R and S-PLUS servers and uses the DecisionSite server for software updates.

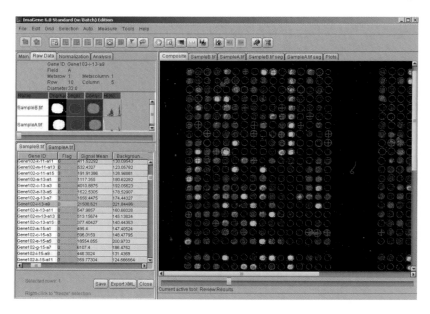

COLOR FIGURE 14.10 GeneDirector's image analysis module ImaGene for the spot quantification, flagging, and removal of contamination.

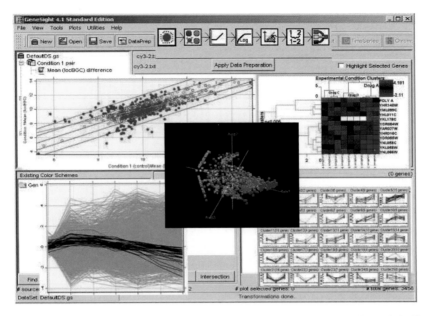

COLOR FIGURE 14.12 GeneDirector's data mining and statistical analysis module Gene-Sight for knowledge discovery from the microarray data.

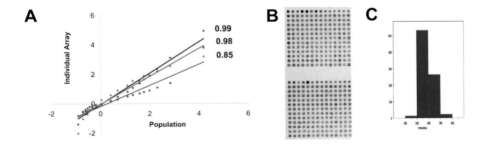

COLOR FIGURE 15.2 The concordance correlation coefficient (CCC) detects many target-labeling failures. (A) CCC of array with a good quality (dark blue), poor array (pink), and the array shown in B (light blue). (B) A representative area of the array with low overall hybridization to probes. (C) A histogram of intensities for the array shown in B.

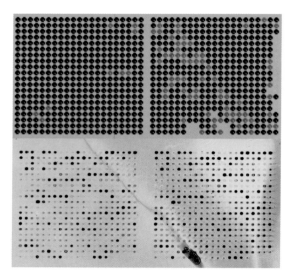

COLOR FIGURE 15.5 The outlier algorithm identifies spots affected by hybridization. The top panel is an applet display of regions in the bottom panel. Red-labeled outliers were based on high replicates. Yellow spots were outliers by both variance and morphology.

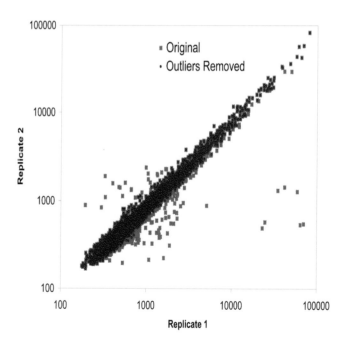

COLOR FIGURE 15.6 QC2 identifies and removes outlier intensities from the data set. Replicate arrays were compared with each other in an intensity plot on a log scale. Data are shown before (pink) and after (blue) outlier removal.

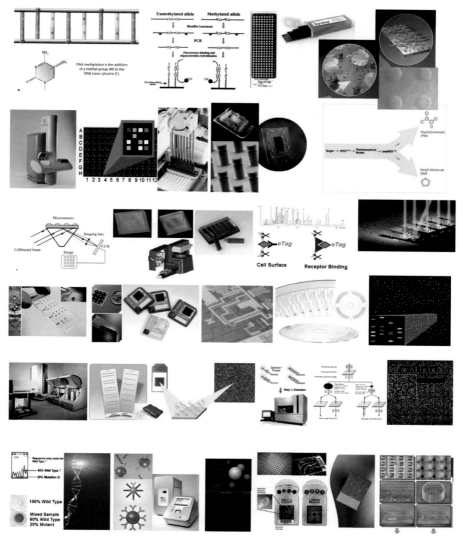

COLOR FIGURE 19.2 A: DMH (differential methylation hybridization). B: DNA Methylation. C: Glycominds, Ltd. D: Biocept. E: Ciphergen. F: High-Throughput Genomics. G: Protagen. H: Zyomyx. I: Jerini. J: Genoptics. K: HTS Biosystems. L: Zeptosens. M: Aclara/Virologic. N: Protiveris. O: Advalytix. P: Calipertech. Q: HandyLab. R: Gyros Systems. S: Xeotron. T: Sequenom. U: Illumina. V: Affymetrix. W: Applied Biosystems. X: PerkinElmer. Y: Parallele. Z: Tebu-Bio. AA: Orchid. BB: Nanosphere. CC: Lynx Therapeutics. DD: Nanogen. EE: CMS. FF: Combimatrix. GG: GeneOhm. HH: Nanoplex. II: BioArray Solutions. JJ: Luminex. KK: PamGene. LL: Metrigenix. MM: Solexa. NN: Graffinity. OO: Febit. PP: Genospectra. QQ: Epoch Biosciences. RR: Exiqon. SS: Nimblegen. TT: Perlegen. UU: ArrayIt. VV: Genomic Solutions. WW: Idaho Technology. XX: Asper Biotech. YY: Agilent Technologies.

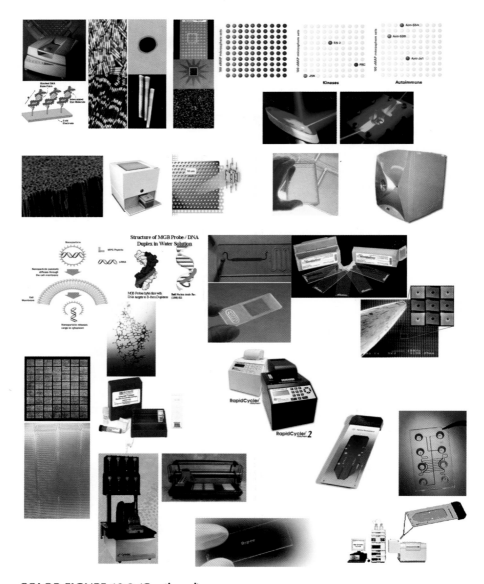

COLOR FIGURE 19.2 (Continued).

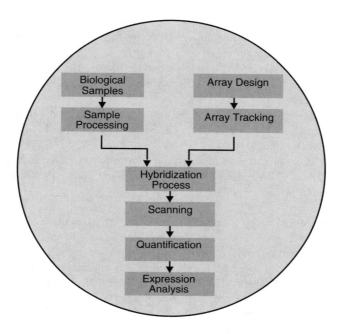

FIGURE 14.1 Microarray experimental process.

IMPORTANCE OF SYSTEMATIC DATA
MANAGEMENT FOR QUALITY ASSURANCE

Because a microarray experiment involves several different data types, ranging from nucleotide sequences of probes to analyzed expression results, it is important to have a microarray data management system that organizes this wide range of data and enforces logical relationships between the data items to maintain data integrity. Traditionally, a combination of computer file system and the laboratory notebook has been used to "organize" all data associated with an experiment. For example, a scientist might use the file name to specify a number of diverse data attributes, such as array type, sample information and ID. A case in point would be the file name 7093–7085#659astro133a.cel, which might indicate the following information: It is a HG-U133A Affymetrix GeneChip CEL file and the Array ID is 7093–7085 and sample information includes the patient ID 659 and tissue diagnosis of the brain tissue sample as an astrocytoma. To locate information about how the labeled extract was generated, such as the RNA extraction protocol, who did the extraction and when this was done, and what label lot number was used, the scientist must locate and refer to files on many computers in a laboratory as well as through laboratory notebooks in the hope that such information was indeed noted. There are several inherent problems with this traditional process:

1. It is very time-consuming: Especially when the number of experiments increases, it will take a significant amount of time and personnel to find the relevant information.
2. Data can get lost: If the technician that performed the experiment leaves the organization, it might be very difficult to search for the information in the notebooks or file system.
3. Difficult to enforce policies: If the site wants to capture a particular piece of information, e.g., RNA quality metrics, it is very hard to enforce this policy with a manual system.
4. Highly error prone: Any typos in the file name can make a drastic error in the result. Being able to enforce logical constraints will alleviate many sources of possible errors.

The last point just stated is one of the most important aspects of having a robust data management system. It is very troublesome to find out that the results of an expensive experiment might have been compromised because of very simple errors. A real-world example of this was highlighted as part of a recent Critical Assessment of Microarray Data Analysis (CAMDA) conference [3]. The participants of this conference competition were provided with publicly available published data sets to analyze. What the winning team discovered was not a scientific breakthrough, but rather evidence that the data set provided had been corrupted by a simple manual error in file manipulation [4]. This occurred through an altered mapping of genes to the coordinate locations on a subset of the arrays in the data set. Such a corruption of data is highly improbable when robust data management is used to track and relate data across an experiment. For example because all the arrays shared the same "array design," it would have been impossible, in this case, for a subset of the arrays to have incorrectly mapped features, as this variation would be constrained to force consistency. This type of rational consistency enforcement is a significant feature of a robust data management solution. It is informative to quote from the abstract of the winning analysis:

> One of the greatest challenges in dealing with microarray data is keeping track of the auxiliary information that surrounds the gene expression measurements extracted from scanned images of microarray experiments. Unlike the sequence data collected for the human genome project, microarray data is highly structured. Any number of factors, including sample sources, sample preparation protocols, hybridization conditions, and microarray lot numbers, can critically affect our ability to interpret the results of a set of microarray experiments.

It should be clear that for almost all array experiments, having a data management system in place is becoming a necessity rather than a luxury. It is becoming simply too expensive to correct these mistakes after the fact, when they can be readily and cost effectively avoided by implementing an effective data management system. Today, there are both academic/shareware solutions as well as more robust and better-supported commercial data management solutions. Examples of these options are BASE [5], GeneTraffic [6], Affymetrix GCOS Server [7], and BioDiscovery's GeneDirector [8]. All these systems have advantages and disadvantages that must be weighed carefully before any group commits to a single solution. In this chapter we will primarily describe the data management capabilities of BioDiscovery's GeneDirector product, as it is one of the more comprehensive systems.

DATA MANAGEMENT NEEDS AT EACH STEP OF ARRAY EXPERIMENT

In this section, we will provide detailed discussion of the need and type of data management required at each step for the microarray experiment.

ARRAY DESIGN

The first issue that arises when one wants to conduct a microarray experiment is designing or selecting the array that will address the experimenter's questions. An array design is the association of a selection of genes that are of interest to the investigator and the mapping of the probe sequences to array elements. Usually, this array is in the form of a glass slide, but other types or platforms exist. In general, arrays can be divided in two major groups: home-spotted and commercial catalog arrays. The home-spotted variety is produced by the experimenter (or a core facility of an institution). These arrays are often produced using robotic arrayers that transfer the probes from microtiter plates to the surface of a glass slide to create the microarray [9]. These arrays are used more commonly among those studying less-characterized organisms or in more focused studies in which the investigators want to probe only a small portion of the expressed genome. The commercial catalog arrays can also be subdivided into two categories. The first consists of arrays of oligonucleotides produced *in situ* by synthesis-based methods using photolithography [7]. The second is material deposited using industrial methods, which involve electromechanical equipment and high-performance inkjet technology [10,11]. The probes themselves can be either full-length cDNAs, PCR fragments, or chemically synthesized short oligonucleotides of variable length, selected either experimentally from clones found in interesting cDNA libraries or computationally from nonredundant sequence databases. The array design process also consists of mapping these gene sequences (probes) to a physical location on the array, often referred to as a feature.

The probe design aspect can be computationally demanding and experimentally challenging. Initial microarrays were simply based on clones from cDNA libraries, and the issue of concern was tracking these cDNAs across wells and plates and their subsequent mapping to microarray features [12]. This information did require data management, as not only were the plate well positions mapped to the probes, but also the cDNA clones or their PCR products were validated by sequencing and assessed for quality before printing. This is still done in custom spotted cDNA arrays, but as the technology has moved quickly to oligo-based probes for numerous reasons, this mapping is not the only issue, but instead, much interest has revolved around optimizing the criteria for the representative oligonucleotide set. Furthermore, microarray manufacturers have expended much effort to empirically validate the efficacy of their selected probes, demonstrating the limitations of computation alone in identifying ideal probes. Fortunately, there are many commercial sources one can turn to nowadays to address this problem. Nonetheless, as the impact of the array design on experimental results remains poorly understood by the community, the primary reason that many select an array or manufacturer remains cost.

ARRAY TRACKING

Arrays are usually purchased, to cover the anticipated needs of an experiment. This set of arrays typically shares many properties provided by the manufacturer, such as manufacturing lot number and "use by" dates, based on the half-life of the probe stability or some other metric. The user laboratory personnel will track additional information for the arrays, which may include an array bar code, their storage location and condition, and date received. These arrays will typically be stored until the experiment requires them.

When these arrays are consumed, new arrays must be purchased, and tracking their usage may become a much more complicated task. This task may be further complicated if, in the interval between ordering, either new lots are manufactured or, potentially, the array design itself has been revised. This has ramifications likely to need addressing during the analysis phase of the experiment as well, and thus, tracking these details is crucial to interpreting the data. It has been observed that different array lots may generate different signal values because of variations intrinsic to printing of individual probes and variation in array fabrication. Although data generated from the arrays within a particular lot are generally considered consistent and highly reproducible by most manufacturers, they can vary significantly across different lots or print runs. Such systematic changes, sometimes referred to as *lot effects*, can often be effectively addressed and corrected using statistical processing, but may be insignificant depending on the experimental analysis undertaken. However, in order to perform this task, the user must have readily available information associating a given quantification to a particular array's lot number. Therefore, to ensure proper treatment of data downstream, it is vital to collect, manage, and associate this information early in the experimental process. Figure 14.2 shows, as an example, the array data capture view in BioDiscovery's GeneDirector product. Figure 14.3 shows how, using the relationship viewer, the association between a particular quantification and a particular array is kept and can be quickly retrieved.

SAMPLE HANDLING/PROCESSING

A variety of sample types are used in the microarray process, ranging from biomaterials, i.e., animal or plant tissues and cultured cell lines, to RNA extracted from the biomaterial, amplified, and then labeled with a dye or conjugate. The samples at each stage are related to one another by a series of particular laboratory protocols. For instance, investigators use protocols for tissue handling, cell treatment and collection, extraction of the RNA, amplification, and labeling. Furthermore, the microarray experimental process relies upon a variety of laboratory materials purchased from many manufacturers, all of which may be important for the technical reproduction of the resulting data.

It is a challenge to keep track of samples through a microarray experiment. Besides information about the samples themselves, including the location where

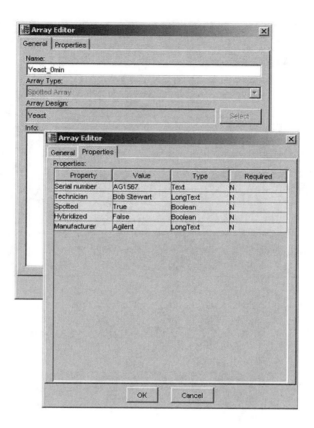

FIGURE 14.2 GeneDirector stores the arrays in the database, indicating the array design that they belong to. In addition, the array editor also displays the properties belonging to the arrays.

they are stored, the protocols used to create them should also be tracked, which is often done across the pages of laboratory notebooks and the documents kept in various computer folders. This organization challenges the curious investigator when they wish to retrospectively verify either interesting or disappointing experimental results. They may be left wondering whether the results might have been different if they had better controlled for factors such as RNA quality, experimental protocols, or the reagents used. However, without effective data management and organization, it becomes quite a challenge to confidently verify these and other factors in the experimental process.

The Sample Tracker module of GeneDirector, for example, offers a solution to all the challenges mentioned previously. This module manages the data related to biomaterials, RNA, amplified RNA, and labeled extract; e.g., RNA labeled with Cy3 and Cy5 dyes. In addition to the sample information, sample protocols can also be imported into GeneDirector and associated with their respective samples (Figure 14.4).

The sample tracker also allows users to create or import sample properties (Figure 14.5). The properties could be text, numeric, Boolean, image, or binary files

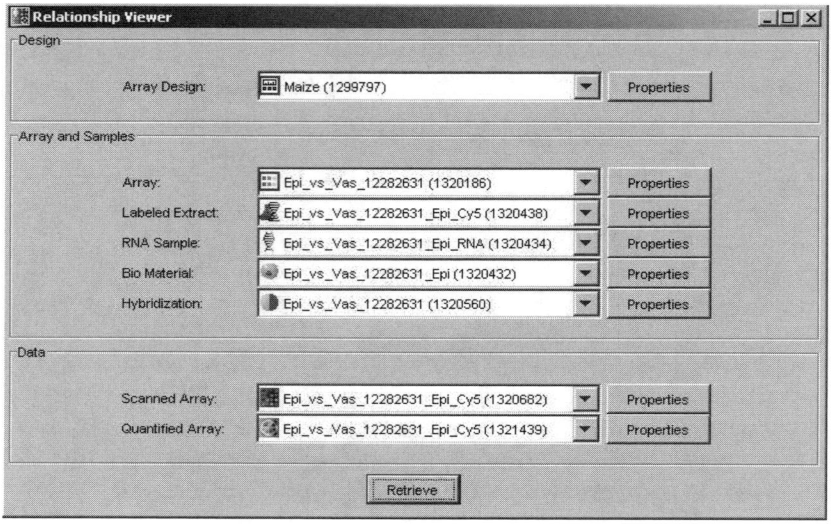

FIGURE 14.3 The relationship viewer in GeneDirector retrieves and displays how any selected entity of a project is related to the other entities. For example, this figure of relationship viewer shows that "Epi_vs_Vas_12282631_Epi_Cy5" quantification belongs to "Epi_vs_Vas_12282631" array and "Maize" array design.

FIGURE 14.4 This figure displays the GeneDirector sample tracker for RNA along with the source biological material it is derived from and the RNA extraction protocol.

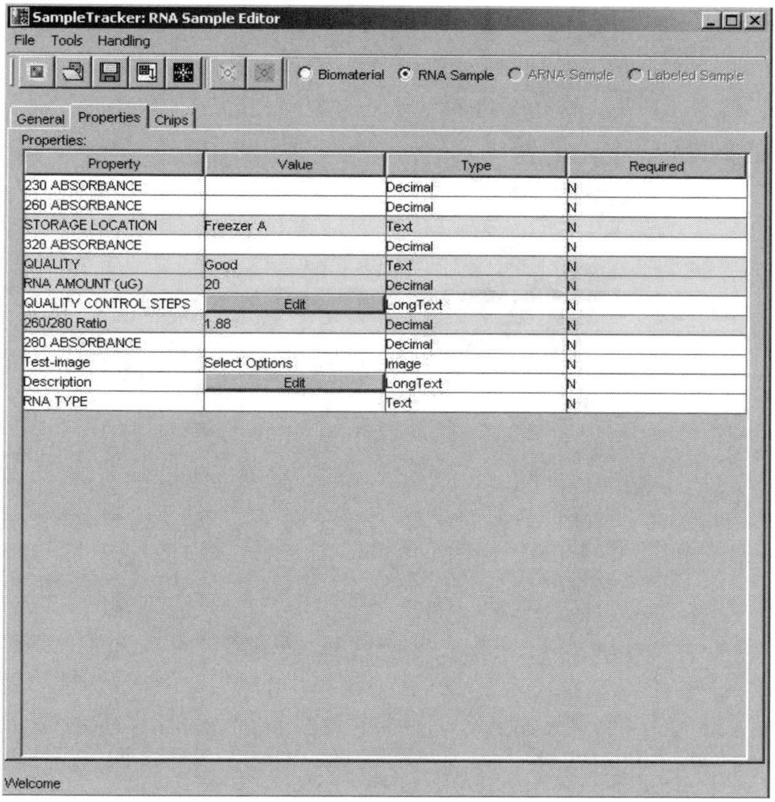

FIGURE 14.5 Sample properties stored in the GeneDirector sample tracker.

and could be set by the administrator as required; there could also be optional properties. Additionally, the experimental factors, those sample properties of interest for the experiment, can be specified. Because experimental factors can be quite different depending on the type of experiment, the data management software must provide the capability to define the type of study and, depending on the type, present the corresponding fields. For example, if the experiment is a drug dosage study, experimental factors would include such items as the drug, dosage level, treatment time, etc. More details about this requirement can be found at the MIAME Web site [13].

The importance of good quality RNA can never be overemphasized to ensure good quality expression results. Many scientists now routinely use accepted methods to check RNA quality before committing to spend time and resources associated with an experiment. One of the most popular tools in this area is the bioanalyzer product marketed by Agilent Technologies [14]. This product can simultaneously measure the RNA quality of multiple samples on a microfluidic chip. It is important to associate the results of this analysis with the biological sample that is to be hybridized (cDNA, RNA, or labeled RNA) in an array experiment. For example, in GeneDirector, the sample tracker offers direct importation of BioAnalyzer data files

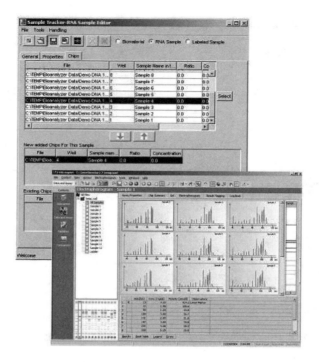

FIGURE 14.6 This figure shows the association of an RNA sample in GeneDirector with the corresponding BioAnalyzer data. After selecting a sample in the sample tracker, it can directly be launched into the BioAnalyzer image viewer for RNA quality assessment.

as well as the capability of directly launching the image viewer, enabling easy tracking and quality assessment of target samples (Figure 14.6). These data are associated with the proper sample type in GeneDirector. This comes handy in a situation in which when, after months of research, the user finds interesting results and wonders about the RNA quality.

Hybridizing

Hybridization is an important part of the microarray process in which probes on the arrays are hybridized with RNA samples labeled, typically, with dyes (e.g., Cy3 or Cy5). This step joins the information regarding an array (probes on the array surface) with information from the sample. Tracking various aspects of this process and organizing them in a common framework greatly facilitates process validation and results verification. As an example, we show how GeneDirector provides a system for tracking these experimental details about hybridizations (Figure 14.7). The hybridization editor also stores properties related to a particular hybridized array (Figure 14.8). In addition, the hybridization protocol can be readily viewed (Figure 14.9).

FIGURE 14.7 GeneDirector hybridization editor showing the labeled samples and array used for hybridization along with its protocol.

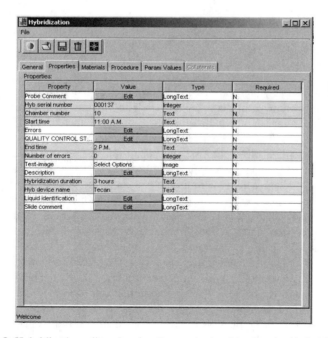

FIGURE 14.8 Hybridization editor showing the required and nonrequired hybridization properties. The properties could be text, numeric, Boolean, or binary files.

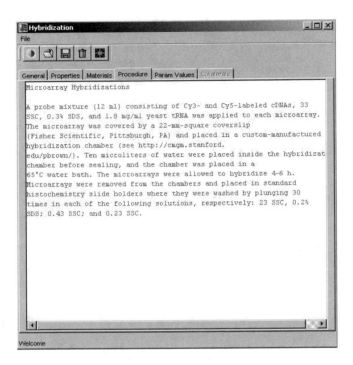

FIGURE 14.9 Hybridization procedure panel showing the actual protocol used for array hybridization in the project.

ARRAY SCANNING

Hybridization occurs when the detectable target is mixed with the surface-affixed probes, enabling changes in fluorescence levels to be converted to expression measurements. The process in which this label is first detected (or visualized) and measured (or quantified) at the probe level is called *imaging* or *scanning*. The microarray image data file is the first true result that the experimentalist can interrogate or simply look at and assess. The display generally consists of a pseudo color image showing the target fluorescing in the pattern of the fixed arrayed probes. This image is subsequently converted to a spreadsheet, representing the intensities of the pixels of every feature or probe. It is at this stage of the experiment that data volume first becomes a real concern and handling data management issues assumes importance.

The image file itself is generated through the scanner hardware and imaging software system. This technology is usually laser based, enabling the selective capture of the labeled sample by its emission spectra. Array scanners offer their own unique and competitive features, which enable the user to individually optimize the image acquisition of their particular hybridized array. It is advisable that along with the resultant image file, typically in TIFF (tagged image file format), the selected instrument parameter settings should also be saved. These parameters or properties may include model of scanner, serial number, software (or version), sensitivity setting (PMT, voltage), laser power, pixel resolution (commonly 10 μM), dynamic range (commonly 16 bit) as well as operator, and an output scanner log. If such settings are not noted, it is possible that apparent overexpression measures from a sample may be confounded by altered settings, and the experimental analytical results corrupted.

IMAGE QUANTIFICATION

As outlined in the previous section, all array experiments eventually result in an image file. This image file needs to be computationally processed to generate gene expression values. This process is referred to as the *image quantification* step. At this step, the location of each feature (spots in spotted arrays and square elements in Affymetrix arrays) must be found, the signal intensity values measured and, potentially, spot-quality measures generated. There are numerous software tools in the market that provide the necessary algorithms to accomplish this task. The major differences between various available tools are their ability to accurately locate the features in the array, to correctly segment signal pixels from background and potential contaminants, and finally, to generate various quality measures on the quantified values. One of the most complete tools for spotted-array image analysis is BioDiscovery's ImaGene product [8]. The reader is referred to a recent publication on details of image analysis and quality measurements process for spotted arrays [15].

For a comprehensive data management system to track the entire array experiment workflow, it must be able to either directly process scanned arrays in the database to generate quantified arrays within the database, or provide tools for importing quantified array data and properly associating this information to existing arrays and labeled samples within the database. In the case of GeneDirector, this tool provides both options. GeneDirector is directly integrated with the ImaGene software, which offers automated quantification and quality assessment of the spots and removal of contaminants from the spots. Furthermore, there are batch import facilities in GeneDirector, allowing the user to load expression data from the file system into the database and provide the necessary data relationships. Figure 14.10 shows a screenshot of the ImaGene tool implemented within GeneDirector, and Figure 14.11 shows the automated batch image analysis in action.

PROCESSED DATA

Once array images have been quantified, the database contains the raw material that is needed to perform statistical analysis and visualization of expression data. Here again, there is a large and growing variety of software products and open-source tools, such as the Bioconductor package [16], which can be used to perform the analysis. It is important to note that currently few of these products utilize the valuable experimental process information that can and should be used to improve data quality [17]. Importantly, the results of an analysis should directly be linked to the source quantification values. In a highly structured database implementation, such as GeneDirector, this process is automatic and also enables the user to "drill down" to any point in the experimental workflow. For example, if a particular gene is deemed to be very significant in a disease condition based on differential expression data, before

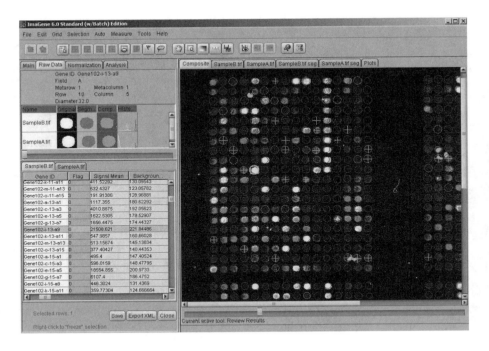

FIGURE 14.10 *(A color version follows page 204)* GeneDirector's image analysis module ImaGene for the spot quantification, flagging, and removal of contamination.

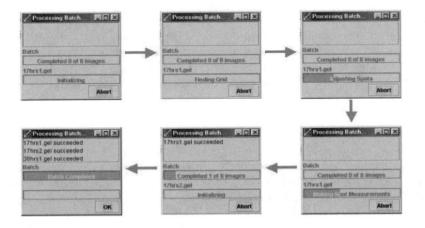

FIGURE 14.11 The batch image processing module of GeneDirector's ImaGene showing the automated batch processing of multiple scanned arrays without human supervision.

undertaking expensive validation experiments (e.g., running RT-PCR experiments), the scientist can quickly traverse backward to find the quality of RNA used in the experiment, source of the biomaterial, protocols associated with hybridization or labeling, and so forth. With a relational database designed to hold such information at basic atomic level, the task of performing such a process review would take considerable time and resources.

The GeneDirector product is a good example of a data management system that can provide functionality as described previously. In addition to managing all the data as already discussed, expression data can be analyzed using the integrated GeneSight analysis tool or exported as text files to be analyzed using any other commercial software package. The data analysis module GeneSight®, embedded within GeneDirector, is a comprehensive bioinformatics software solution that offers exploratory data mining and confirmatory statistical analyses tools to obtain biological insights from the complex and high-dimensional microarray experiments.

The quantified arrays can be uploaded into GeneSight for sophisticated data analysis (Figure 14.12). The data transformation methods embedded in GeneSight include the following steps: background adjustment, flagged spot removal, combining replicates, replacing missing values, flooring, log transformation, pairing, and several normalization algorithms, including lowess. The exploratory tools offered

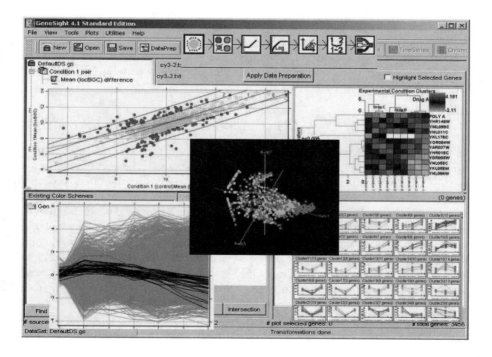

FIGURE 14.12 *(A color version follows page 204)* GeneDirector's data mining and statistical analysis module GeneSight for knowledge discovery from the microarray data.

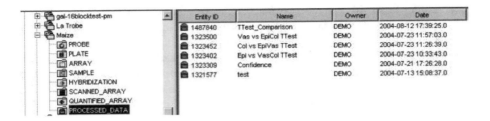

FIGURE 14.13 GeneDirector showing the processed data stored in the database for review and further analysis.

by GeneSight include scatter plot, histogram plot, box plot, principal component analysis (PCA), time series analysis, hierarchical clustering, k-means clustering, self-organizing map (SOM), and chromosome mapping. The statistical confirmatory tools offered include the Significance Analyzer (t-test, ANOVA, Kruskal-Wallis, and Mann-Whitney), and cluster annotation enrichment analysis. GeneSight also offers text-based query building to find genes of interest in the data set. The Integrated Annotation Compiler (IAC) tool of GeneSight allows the users to download annotations and biological pathways information from public domain databases, namely, NCBI LocusLink and GenBank, KEGG, and BioCarta. All these exploratory and confirmatory tools unmask the biological patterns hidden inside the complex and high-dimensional microarray data sets.

The data analysis results of GeneSight are saved as processed data into the GeneDirector database. These processed data can be reviewed with GeneSight for further analysis (Figure 14.13).

CONCLUSIONS

In this chapter, we have described the need for a comprehensive and robust data management solution to store and manage all the data associated with array-based experiments. We emphasized the need for such a system to minimize potential sources of errors as well as provide a mechanism for easy archival and retrieval of data. With more and more publications requiring comprehensive data submission as part of their publication requirements, it has become a necessity for practically all scientists conducting microarray experiments to have a true data management solution. In this chapter, we used BioDiscovery's GeneDirector as a primary example of a system for enterprise deployment designed to address these needs. However, the concepts behind this explicit software implementation are applicable to any tool or project and should be used as a guide.

As the use of microarray technology grows, enterprise data management systems must transform into knowledge management tools, allowing scientists to mine not just data but experimental results as well and combine results obtained through multiple experiments. This is a future direction that we are currently pursuing and will report on in future publications.

REFERENCES

1. Schena, M., Shalon, D., Davis, R.W., and Brown, P.O. Quantitative monitoring of gene expression patterns with a complementary DNA microarray. *Science* 270(5235): 467–470, 1995.
2. Lockhart, D.J., Dong, H., Byrne, M.C., Follettie, M.T., Gallo, M.V., Chee, M.S., Mittmann, M., Wang, C., Kobayashi, M., Horton, H., and Brown, E.L. Expression monitoring by hybridization to high-density oligonucleotide arrays. *Nat Biotechnol.* 14(13): 1675–1680, 1996.
3. http://www.camda.duke.edu.
4. Stivers, D.N., Wang, J., Rosner, G.L., Coombes, K.R., Organ-Specific Differences in Gene Expression and UniGene Annotations Describing Source Material, CAMDA Abstract, 2002. http://www.camda.duke.edu/camda02/papers/days/papers/stivers/paper.
5. Saal, L.H., Troein, C., Vallon-Christersson, J., Gruvberger, S., Borg, Å., and Peterson, C. BioArray software environment: a platform for comprehensive management and analysis of microarray data. *Genome Biol.* 3(8): software0003.1–0003.6, 2002. http://base.thep.lu.se.
6. http://www.stratagene.com/products/showProduct.aspx?pid=538.
7. http://www.affymetrix.com.
8. http://www.biodiscovery.com/genedirector.asp.
9. http://www.microarray.org/sfgf/jsp/home.jsp.
10. http://www.chem.agilent.com.
11. http://www4.amershambiosciences.com.
12. Halgren, R.G., Fielden, M.R., Fong, C.J., Zacharewski, T.R. Assessment of clone identity and sequence fidelity for 1189 IMAGE cDNA clones. *Nucl Acids Res.* 29(2): 582–588, January 15, 2001.
13. http://www.mged.org/Workgroups/MIAME/miame.html.
14. http://www.chem.agilent.com/Scripts/PDS.asp?lPage=51.
15. Petrov, A. and Shams, S. *Microarray image processing and quality control. J. VLSI Signal Process.* 38: 211–226, 2004.
16. http://www.bioconductor.org.
17. Chu, T.M., Deng, S., Wolfinger, R., Paules, R.S., and Hamadeh, H.K. Cross-site comparison of gene expression data reveals high similarity. *Environ Health Perspect.* 112(4): 449–455, 2004. http://www.partek.com.

15 Quality Control of Microarray Data

Huinian Xiao, Albert Leung, and Lynn Yieh

CONTENTS

INTRODUCTION

EVOLUTION OF MICROARRAYS INTO A HIGH-THROUGHPUT TECHNOLOGY

Today's pharmaceutical industry experiences a continuous increase in research and development costs, although the number of new molecular entities reaching the market increases at a significantly lower rate. These trends have prompted pharmaceutical companies to use experimental methods that have the potential to increase the quality and number of novel drug candidates within a shorter time frame. In this capacity, microarray technology has become a favored tool to screen potential drug targets and diagnose diseases.

The utility and prediction power of microarrays in forwarding drug discovery research have proven to be diverse. Gene expression profiles are capable of not only diagnosing certain diseases [1–3], but also determining whether a patient will respond to a certain drug treatment [1–3]. Expression profiles have also been used successfully to identify some classes of toxic response [4,5]. In early drug discovery research, microarrays have been used in the identification of drug targets [6]. Expression profiles, which lead to the elucidation of a cellular pathway, also have the potential to identify multiple drug targets of interest within that pathway [7].

One of the strengths of microarray technology is the ability to perform massive parallel profiling of gene expression from a single sample. A global view of gene expression can provide information not only by identifying families or pathways of genes that are affected, but also by specifying those classes of genes not affected [7]. Hypotheses about genes with unknown function can also be derived by comparing their expression with that of genes with known functions [8]. Thus, microarrays have proven to be adept at explaining how genes act in concert, resulting in a certain biological condition.

Microarrays can detect transcripts present in very low levels; this sensitivity is a great benefit of the technology. However, high sensitivity can lead to problems with subsequent data analysis if a large experimental error interferes with the detection of differential gene expression due to the biological condition being studied. For this reason, microarray research is largely focused on developing algorithms that address quality control (QC) and normalization of data. Many of these methods are statistical and therefore rely on replicated experiments, both technical (hybridization) and biological, to establish characteristics of a population. As a result of this and other such factors, larger numbers of samples are processed by microarrays; this has evolved into a high-throughput technology to accommodate the increasing size of data sets.

Processing larger numbers of samples has not proven to be a problem; instead, the bottleneck lies downstream in the time required for the analysis of expression data. Analysis involves an initial process that checks the quality of data being generated. The success of subsequent stages of analysis is dependent on the data quality. For example, during normalization, a data set may be assumed to fit a certain distribution, which may not hold true if outliers that should have been removed during the QC step are included. High-quality data results in identification of gene expression changes with high confidence.

QUALITY CONTROL OF MICROARRAY DATA

Three main concepts have emerged in the QC of high-throughput microarray data used for drug discovery research. These are (1) high-throughput processing of data, (2) objective and reproducible analysis of quality, and (3) standardization of methods to compare across platforms. Together, these criteria support an ideal in which large quantities of data can be processed rapidly, yet consistently. Consistency of how data is processed and the resulting quality is especially important in extracting information with high confidence from microarrays.

The QC of microarray data is an extremely important process, as further steps in downstream analysis rely on the assumption that effects seen in the data are due to biological changes and not to variations in the technique. Additionally, as the wealth of expression data made available to the public grows, the question of data quality also becomes an issue. To compare data from different laboratories, different platforms, or different years, the ability to analyze quality in a standardized manner is critical. Bias can be introduced into the QC process through an individual's subjectivity, and these minute differences in data handling potentially introduce data artifacts and lead to false conclusions.

As microarrays become a standard experimental method, the requirements about how the data are handled and can be shared become more defined. For example, the Minimum Information about a Microarray Experiment (MIAME) organization is focused on how to define and describe microarray experiments in a strict language that allows any scientist not only to understand the biological experiment, but also to store the information in a database [9]. For public databases, such as Stanford [10] and OmniBus [11], the ability to ascertain what data are in the database is crucial. Likewise, standards for data quality will emerge from drug discovery intent on extracting expression profiles that can be used in assessing compound efficacy or patient stratification in clinical trials [12].

IMAGING ISSUES IN QUALITY CONTROL OF MICROARRAY DATA

There are two basic types of microarrays available: oligonucleotide and cDNA. Affymetrix commercially produces and currently dominates the oligonucleotide array market. Some algorithms and programs have been developed in order to analyze Affymetrix microarray data [13–15], but image analysis has largely been addressed by software included with Affymetrix instrumentation [16]. cDNA arrays, although commercially produced today, are more diverse, as they have also been produced by many organizations in-house as the cheaper alternative to Affymetrix arrays. Because of widespread availability and abundance of data, most academic laboratories have focused on developing algorithms for cDNA array quality. In this chapter, we focus primarily on QC metrics that have been developed specifically for cDNA microarray data.

Obtaining microarray data is a multistep process that is full of opportunities to introduce experimental variation. The workflow begins with two parallel processes: the manufacturing of microarrays and the generation of biological samples, which come together in hybridization. Subsequent scanning and quantitation (imaging) converts an analog signal to a digital signal. Variability and error can be derived at five phases of data acquisition: microarray manufacturing, preparation of samples, hybridization, scanning, and imaging, all of which affect the quality of data. Each phase of the experiment can introduce data artifacts, affecting gene expression profiling and complicating the identification of gene expression changes. With cDNA arrays, variability can arise in the preparation of DNA for spotting. During the spotting process, pin-specific variation is introduced. Errors introduced by sample preparation depend on the protocol and efficiency of reactions, with critical steps being RNA extraction, amplification, and labeling. Scanner settings can be a source of experimental variation. In imaging, software is used for quantification, but human input is often still required for steps such as checking grid alignment and removing poor-quality spots. A daunting challenge is how to assess and filter raw expression data to retain only high-quality data for further analysis.

As a typical microarray array consists of thousands of targets, any process requiring manual inspection would be tedious, slow, and, because of human intervention, potentially subjective. Although it is not easy to avoid variability, many errors arising from limitations of the technology can be categorized as systematic. For systematic errors, it is possible to monitor and generate warnings when aberrant characteristics arise during the experiment. Poor-quality data can then be filtered out and only high-quality data imported to a database for subsequent mining of expression patterns. Errors are often associated with a local region of an array, and, therefore, QC can be implemented at several levels: the pixel, spot, region, or entire array. By automating the identification of systematic variability and limiting human intervention, the QC process can be accelerated, standardized, and made reproducible.

In this chapter, we review some of the more popular methods of QC for cDNA arrays described in the literature. The algorithms use different statistical approaches for identifying abnormal readings in data and frequently focus on a single characteristic, such as spot morphology, that contributes to technical error. Commercial packages incorporate several algorithms to address multiple aspects of QC, and we review some of the widely used programs. Although commercial software is widely used, packages developed by academic groups have their own advantages. Finally, we present an automated QC method that we have implemented in-house to accommodate high-throughput microarray experimentation.

ALGORITHMS DEVELOPED FOR MICROARRAY QUALITY CONTROL

DETECTION OF OUTLIER SPOTS

Outliers are observations that appear to be inconsistent with most of the data population. Identification of outlier spots is an important component of QC for cDNA arrays, linked inherently to image analysis [17–20]. Disparate amounts of DNA deposited in a spot and uneven sample labeling can create inconsistent signal intensity. Different settings can affect the sensitivity of a scanner. Misaligned grids during image quantitation can lead to poor-quality data. Spot morphology and intensity measurements are subject to variability from factors in array manufacturing and in the hybridization process, such as labeling, washing, and scanning. Variation between chips is also possible, although replicate arrays are commonly used to detect spot outliers. Ideally, multiple algorithms would be employed to monitor and identify variability that arises from different aspects of microarray experimentation, which would be adjusted constantly to maintain optimal processing conditions. Ultimately, identifying any factor that results in poor-quality data is helpful in that further data analysis is not skewed.

Detection of Aberrant Spot Morphology

A Bayesian network approach [18] has been proposed, which enables incorporation of *a priori* knowledge and explicit representation of the impact of the features on spot quality. In this method, spot quality is treated as a classification problem (good or bad), and, as with many algorithms, a training set is necessary. Selection of a prior distribution is also controversial, and in the publication, the authors have assumed that the prior is uniformly distributed. Estimation of an appropriate prior is much more involved and would require a training set from which features with high impact on spot quality are first identified, and then the dependencies between the features and spot quality are solved. The software package, BlueFuse [21], uses a Bayesian approach to automatically generate a confidence estimate for spot quality.

Brown et al. [17] proposed a pixel-by-pixel analysis of individual spots to estimate intensity and system error in a 2-dye cDNA array system. Using "best-fit values" of background, computed based on spot intensity, eliminated the problem of a large number of negative intensities calculated when using local background. The normalized standard deviation of the ratio measurement, the spot ratio variability (SRV), was proposed as a simple measure of the irregularity of a spot and correlated with the quality of expression ratios. Additionally, the SRV was used for significance estimates of expression ratios, which improved data quality of a test set of replicate arrays.

Wang et al. [20] first reported the approach of defining quality scores for spots according to their size, signal-to-noise ratio, background level and uniformity, and saturation status, which is commonly used now. A composite score gives an overall assessment of spot quality. Spots with higher scores give less variable measurements and vice versa. According to their study, constructing quality measures is laboratory and system dependent, because variations can be laboratory and system dependent. So a weighted-mean approach that automatically calculates the optimal quality measure for a given microarray experiment is proposed [22]. A similar approach was reported by Chen et al. [23] and is used by Imagene™ [24] and Koadarray [25].

Finding Outlier Spots Statistically

If replicate hybridizations are available, they can be used to identify discrepancies in spot intensities in the data. The most common approach to detect outliers is by calculating the mean and standard deviation (SD) among replicates and eliminating spots greater than 2 SD from the mean. In one report [26], filtering out outlier pairs that deviated greatly from the mean increased the global correlation between replicates dramatically, suggesting that this simple method is effective in identifying spot outliers. Another common approach is based on calculating the coefficient of variance (CV) [26]. With more than two replicates, this can be employed as an iterative process. When a spot exceeds a threshold CV, the spot contributing the most to variance is filtered out, the CV is recalculated with the remaining data points, and the new CV compared again to the threshold. Both methods are easily implemented;

however, outliers affect the result itself. To improve reliability, one strategy "borrows strength" from genes expressed at a similar level by constructing a window encompassing genes whose mean intensities are closest to the gene being tested [27–29]. Another strategy is to use the median and median absolute deviation (MAD) from the median, because the median is better resistant to outliers compared to the mean [28].

DETECTION OF OUTLIER CHIPS

Systematic changes in experimental conditions across multiple chips can seriously affect data quality and lead to false biological conclusions. The detection of outlying samples before sample classification is essential. Robust principle component analysis (PCA) has been used to detect outliers in microarray experiments [19]. A major advantage of the method is that it does not rely on explicit modeling of the microarray process as they are based solely on the distribution of measurements of arrays within the project. The sensitivity of the method improves with increasing study sizes. Because of its multivariate nature, this method is particularly suitable for large-scale microarray experiments in a high-throughput environment.

Another method based on Akaike's Information Criterion (AIC) was developed to detect outliers. AIC was developed to select an optimal model from a set of models. Kadota et al. [30] have applied this method in identifying outlier arrays and have demonstrated an improvement in subsequent sample classification. An advantage of using this criterion is that selection of a significance level is not necessary, which helps objective decision-making in the identification of outlier arrays.

AUTOMATION OF QUALITY CONTROL

Software packages that provide methods for performing QC have focused on developing more complete systems to identify data abnormalities at several levels and require a systematic approach to the identification of variations in data. Many packages are platform specific, though some will accommodate several array types. These packages begin to approach full automation of QC; however, in order to accommodate differences in data handling between customers, they tend to include steps in which a human decision is required. On the other hand, human intervention can lead to variation because of the subjectivity of an individual. Automation of microarray QC is, therefore, a trade-off. Although enabling high-throughput, objective, reproducible decisions in data quality, automation requires knowledge to set thresholds for QC parameters and limits decision-making based on human knowledge.

When choosing or developing an automated QC system that fits the needs of a microarray system, there are several factors that should be considered. Technical error can be observed at spot, region, whole array, or interarray levels. Experimental problems, such as sample degradation, labeling failure, and wash contamination can be identified with specific QC features. Many parameters that indicate quality also require knowledge; acceptance criteria can be chosen by analyzing the technology and instrument or by using training data. Batch-enabled software, the ability to process results in open format (e.g., text), scriptable algorithms, and software development kits (SDK) are also important features that will help users to customize and automate the QC process. In addition to the description of QC packages that follow, a breakdown of automated QC into components, with features of individual packages, is given in Table 15.1.

COMMERCIALLY AVAILABLE PACKAGES

Several software packages are available commercially, and here, we review two of the most widely used packages. Most vendors focus their efforts on the accuracy and effectiveness of feature extraction and data analysis algorithms rather than on full automation for high-throughput analysis of arrays. The release of SDKs helps individual users to customize or automate processes according to their needs.

BioDiscovery is a provider of software solutions for microarray research. Its platform, the independent image quantitation product ImaGene [24], is widely used and provides automatic spot quality flagging [31]. In addition to empty and negative spots, its quality measures include background contamination, signal contamination, abnormal shape regularity, and significant offset from expected position, among others. Poor quality spots are flagged by comparing a spot's intensity with the intensity distribution over the chip; how to use this information in quality control is up to the user. Another BioDiscovery product, GeneSight™ [32], takes the mean or median of replicates that are either on the same slide or on multiple slides after background correction. Outlier spots are identified by defining a cutoff for deviation from the mean and removed. The batch-processing feature enables automatic image processing in a high-throughput environment. Although ImaGene and GeneSight do not completely automate microarray QC processing, it is composed of modules that allow the end user to complete the automation.

Agilent [33] is a major array and scanner supplier. The company markets sophisticated feature extraction software, which generates intensity data from arrays scanned by Agilent scanners. An algorithm identifies and removes data for outlier pixels within a spot as well as in the local background, using the standard deviation or interquartile range. These outlier pixels are omitted before spot statistics (mean, median, and standard deviation) are generated. With replication of spots, a box plot analysis can be used to set an interquartile range cutoff criterion for omission of outlier spots. Background correction is performed using local nearest-neighbor and negative control probe methods. The feature extraction software outputs parameter settings and intensity measurements in multiple formats, which allows results to be loaded into other software for further analysis.

NONCOMMERCIAL QUALITY CONTROL PACKAGES

Microarrays are used primarily as a research tool in academic or research institutes. They are unlikely to be performed on a scale that necessitates a high-throughput, fully automated QC process. Most of the automated processes are algorithm based, focus on reproducibility, and allow users the flexibility to make decisions based on QC indicators.

ExpressYourself [34] is a Web-based platform for processing microarray data. It has been developed in C and Perl, and supports GenePix [35] and ScanAnalyze [36]

TABLE 15.1
Components and Features of Automated Quality Control Packages

Vendor/Author	Software	Detection of Outlier Chips				Detection of Outlier Spots		LIMS Processing		
		Detection of Zero, Negative and Saturation Intensity, High Background	Detection of Streaking, Scratch and Wash Problem	Detection of Hybridization and/or Labeling Problem	Detection of Probe Failure and Sample Degradation	Detection of Aberrant Spot Morphology	Detection of Outlier Spots With Statistical Algorithm	Batch Enabled	SDK or API	Platform Independent
BioDiscovery	Imagene/ GeneSight	Yes	No	No	No	Yes	Yes	Yes	No	Yes
Agilent	Feature Extraction	Yes	No	No	No	Yes	Yes, within chip	Yes	No	Agilent scanner
Luscombe et al.	Express Yourself	Yes	No	Yes	No	No	No	No	No	No
Petri et al.	Array-A-Lizer	Yes	Yes	No	No	No	No	No	No	No
Hijum et al.	MicroPrep	Yes	Yes	No	No	No	No	No	No	No
Affymetrix	GCOS	Yes	No	No	No	No	No	Yes	Yes	No
GE Healthcare	CodeLink	Yes	No	No	No	Yes	Yes	Yes	No	No

file formats. The software performs individual spot and regional filtering, as well as background correction utilizing the nearest-neighbor method. Three measures were established to provide an objective indication of quality. The first is the percentage of good-quality array elements as measured by the proportion of array elements and regions the filtering process has removed. The second is intra-array hybridization quality as measured by the difference in signals between duplicates on the same slide. The third is replicate array hybridization quality as measured by calculating the difference in signal between equivalent spots across multiple slides. ExpressY-ourself has a sound but simple algorithmic approach that could be implemented to produce an automated system.

Array-A-Lizer [37] is a stand-alone package freely available to all (via Web download). The graphical user interface has been built in Borland Delphi and the statistical algorithms in the R language. It supports GenePixPro [38] and Spotfinder [39] file formats and runs under a Microsoft Windows platform. Signal distribution is analyzed to provide quality control measurements. The diagnostic report creates MA plots (where M is log intensity ratio and A is mean log intensity) and red/green scatter plots to identify intensity-dependent biases such as those introduced by scanners. The MA plot has been used by several groups to identify biases and artifacts [40–42]. Included in the diagnostic report is a histogram of intensities, which illustrates the negative and saturated spots. Spatial reports plot the foreground and background intensity, and the position of negative spots shows the gradient, high background, and wash problems. This tool is mainly a visualization tool and lacks objective QC classifiers; therefore, it will not be easy to automate. However, it provides a quick snapshot of the hybridization results. Stanford utilizes a similar visualization tool, which provides simple visualization of biases evident in hybridization and also provides an ANOVA tool to assess variability introduced during microarray printing [43].

MicroPrep [44] is a stand-alone software suite consisting of three modules. PrePreP and PostPreP were developed in Delphi. PreP was coded in visual C++ studio, and the package runs on a Microsoft Windows platform. PrePreP processes output files generated from feature extraction software such as Array-Pro [45], GenePix, and ImaGene. Empty and "bad" spots identified by the imaging software are omitted from further analysis. PreP uses a spotpin-based LOWESS [46,42] to minimize systematic errors such as dye effect, differential amounts of target printed on a chip, and scanner nonlinearity. Visualizations such as MA plot (log intensity ratio vs. mean log intensity), RG plot (log Red vs. log Green), boxplots, and image views allow exploration of the data. PostPreP identifies outlier spots and outlier chips, using the standard deviation. A ratio of the signal intensity standard deviation to the standard deviation minus the spot or slide in question is compared to an empirically derived threshold; when the ratio exceeds the threshold, the spot or slide is labeled an outlier. Although MicroPrep does not have sophisticated image-level QC algorithms, it effectively uses the data generated by image-processing software to identify factors indicative of errors such as bleeding or scratching. Its spot outlier and array outlier QC methods can be easily implemented in an automated fashion.

OLIGONUCLEOTIDE ARRAYS

Although the focus of this review is on cDNA arrays, it is worth mentioning Affymetrix and CodeLink oligonucleotide array alternatives, as they are widely used. Both platforms are complete systems with the software as an integral part of the technology. Affymetrix's GeneChip Operating System (GCOS) [16] manages and analyzes GeneChip arrays. Affymetrix technology is unique (data from 11 perfect and mismatched oligonucleotide pairs are converted into a single intensity measurement); therefore, GCOS handles early stages in data processing and utilizes algorithms that are robust to outliers and returns p values as an indication of data quality.

GE Healthcare's CodeLink™ expression analysis [47] software is used for primary data extraction and acquisition of gene expression values from CodeLink bioarray images. Although CodeLink arrays are oligonucleotide based, the target is spotted onto the array so that QC methods are similar to those used for cDNA arrays. QC printed information for each spot is provided with every array and can be reviewed before hybridization to ensure that a gene of interest is represented on the array. The expression analysis software flags spots with irregularities such as saturation, contamination, and low intensities. The signal intensity variation between replicate arrays is returned in a CV report.

AUTOMATIC MONITORING OF EXPRESSION DATA IMPROVES THROUGHPUT OF DATA PROCESSING AND DATA QUALITY

Similar to some of the organizations whose methods have been presented here, we found a customized solution to address microarray QC to be our best option. At maximum throughput, well over 100 cDNA arrays are processed in a single day by our microarray core facility. Taking advantage of the fact that our complete database consists of over 34,000 samples analyzed by microarrays, we used the data to identify trends and characteristics attributable to failures at specific stages of microarray experimentation. Before implementation of our system, named AutoQC [48], image processing and QC formed the bottleneck for the production of gene expression data. By developing and implementing AutoQC, we were able to improve the speed of image processing dramatically, make our QC process less subjective, and improve the overall quality of expression data in our database.

The system we have designed consists of two components: The QC data collector and QC data analyzer. The QC data collector refers to the experimental design and chip layouts used to collect essential control information about key elements of the process of generating gene expression data. These key elements include RNA preparation, probe labeling, hybridization, washing, and scanning. The QC data analyzer refers to a collection of algorithms used to analyze the information collected by the QC data collector and produce a management report, which summarizes quality control information.

QC DATA COLLECTOR

Because cDNA arrays are printed in our core genomic laboratory, a custom layout design was implemented to collect the essential information for quality analysis.

Two strategies, the use of spiked control genes and replication, are implemented as essential elements in the array layout design.

Two sets of control genes, internal spiked genes and external spiked genes, obtained from Stratagene [49], are spotted on the array. The corresponding internal control RNA is spiked into a reaction before labeling, and external control RNAs, already labeled, are spiked into the hybridization mixture. These controls can be used to determine the quality of the RNA samples, as well as the sensitivity, signal linearity, and consistency of the array. For example, if the internal control signal in the array has a normal dynamic range, but the experimentally derived RNA on the same array does not, it suggests that the sample RNA was degraded.

Adding control genes is helpful but not sufficient to eliminate or evaluate all sources of experimental error; therefore, replication is also important. Intra-array duplication consists of spots within the same array, spotted with the same pin but well spaced to give a better indication of variability across the slide. Interarray replication can be achieved by hybridizing two arrays for each labeled RNA probe sample. Replication provides the ability to use statistical tests to make a decision whether a given intensity is significant or behaves as an outlier.

Besides the special array design, multiple quality measurements based on geometric properties of the spots, provided by the imaging analysis software ImaGene [24], are collected. These measurements are signal standard deviation, background standard deviation, ignored area, signal area, ignored median, and positionoff. Some algorithms were developed based on both the geometric measurements and intensities of spots. They are used not only for assessing the quality of spots, but also for monitoring the quality of laboratory procedures in the array production.

QC Data Analyzer

A quality control package performing data analysis was developed, using the Perl programming language with S-Plus [50] normalization and statistical functions embedded within. A conceptualized workflow is described in Figure 15.1. There are two key components; QC1 is an array-level quality check, and QC2 an experiment-

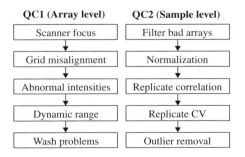

FIGURE 15.1 AutoQC consists of two modules. QC1 checks for failures at the array level. QC2 normalizes data and checks for outlier intensities based on sample replication or spot morphology. Failure at any step of QC results in failure of the array or spot.

level quality check. Data files ready to undergo a quality check wait in a queue. The workflow automatically starts with reading an intensity data file, in which one experiment is a unit. First, partial focus errors and grid misalignments are checked for. Both these errors are correctable by repeating the image scanning; data from repeated scans can then be added to the end of the queue.

There are five categories for the QC1 check in the QC data analyzer: spots with unexpected intensity (including saturated spots, negative spots, and zero-intensity spots), RNA dynamic range, dynamic range of spiked controls, abnormal high background measurements, and patch (wash) problems. A QC1 status ruling is given following these checks. After discarding arrays rejected by QC1, QC2 checks for data consistency and detects outlier spots. The entire process flow ends with a QC2 status ruling, in which experiments with a CV higher than an empirically defined value are discarded.

Analytic data is stored in a relational database, and microarray data curators can access a summary of results through a Web interface. This interactive setup allows curators to act as administrators and interject knowledge and experience into the QC process. Through the Web interface, experiments can be resubmitted for QC analysis and AutoQC decisions can be overridden. Additionally, visualization from QCInspector [51] and ImaGene [24] can be launched from a link in the detailed review.

RESULTS

The AutoQC system has been running at our organization for over 2 years and has been used to evaluate more than 32,000 arrays. Algorithms designed to assess data quality have performed well and have increased the efficiency of our data processing. Replicate hybridizations are less variable as measured by a twofold reduction in the average CV. Whereas one curator could visually screen a maximum of 35 arrays in a day, with the aid of AutoQC, a curator can process up to 200 chips in a day. Thus, image QC is no longer the limiting step in the processing of microarray samples. The following results give specific examples of how the QC system has improved data quality by automatically identifying target failures, hybridization failures, wash problems, and outlier intensities.

QC1: Detection of Intraarray Failures

Our facility database hosts a great amount of cDNA microarray data; therefore, the RNA dynamic range of an array population (represented as an intensity profile) is readily calculated. On the assumption that most genes do not change in expression level across arrays representing thousands of genes, a comparison of the intensity profile of a single array with the profile of the array population is a good indicator of the array quality. For comparison between arrays, we use the concordance correlation coefficient (CCC) [52]. Our study suggests that CCC is more effective than Pearson's correlation coefficient [53] (PCC) in evaluating the similarity between each pair of data spots, because CCC considers how well replicates fit a line with a slope of 1 whereas PCC only evaluates line fitting.

Data from two mouse arrays were used to illustrate the behavior of a target failure. Percentiles of log-transformed intensities were used to describe the dynamic

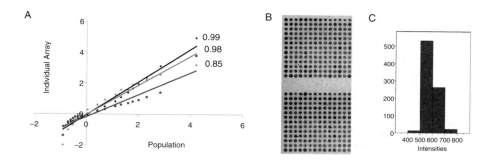

FIGURE 15.2 *(A color version follows page 204)* The concordance correlation coefficient (CCC) detects many target-labeling failures. (A) CCC of array with a good quality (dark blue), poor array (pink), and the array shown in B (light blue). (B) A representative area of the array with low overall hybridization to probes. (C) A histogram of intensities for the array shown in B.

range of the chip and the population. The CCC of the chosen percentiles was calculated to measure the degree of similarity (Figure 15.2A). The CCC of a good array (0.99, dark blue) compared to the population showed a goodness of fit for a line as well as for a regression slope approaching 1; the CCC of a poor array (0.85, pink) displayed a poorer fit. Visual inspection of the array images (data not shown) showed that with the exception of some spiked-in controls, there was little variation in the intensities. Another criterion was required to detect evenly low hybridization over the entire array, often indicative of a degraded RNA sample. The CCC for this array was high (0.98, light blue); however, minimal hybridization was evident with visual inspection (Figure 15.2B). The flat intensity profile of this array (Figure 15.2C) was detectable by comparison of the 75th percentile with the mean and median. If the 75th percentile was similar to the mean or median (ratio between 0.98 and 1.1), the array was designated a failure.

This algorithm was tested on 2404 randomly chosen arrays covering several species. Microarray curators used visual inspection to label 8% (187) of the arrays failures. The algorithm identified 183 failures (98%), missed 4 failures, and called 28 arrays failures that were not detectable by eye. This corresponded to a sensitivity and specificity of 97.9 and 98.9%, respectively. The small percentage of failures that were not agreed upon by AutoQC and curators illustrate disagreement between human and the automated systems, which is why image curators were given the flexibility to override the results of automatic detection.

Degradation of RNA samples can be detected using BioAnalyzer [54] in the early stages after RNA extraction, but a good sample is essential to obtain a trusted expression data profile because errors happen during the hybridization process. Data accumulated in the past few years have supported our hypothesis that labeling or hybridization failures can be detected by combining information about the RNA dynamic range with spiked control genes. Labeling errors are distinguishable from hybridization errors using controls spiked in before and after the target-labeling step. If both control types perform poorly, then the problem is likely in the hybridization process. However, if controls spiked in after the labeling step hybridize well, but

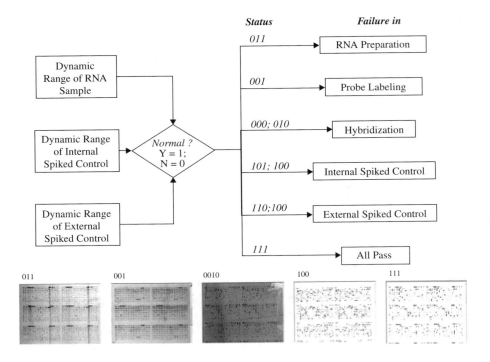

FIGURE 15.3 AutoQC monitors the sample process. While passing through AutoQC, image analysis is monitored computationally. Bit scores of 0 (fail) or 1 (pass) are assigned at each stage. Any score containing a zero indicates failure, whereas a score of 111 means a pass. The status indicates which stage of AutoQC has detected failure without having to manually curate images. Panels below illustrate images that have been assigned different status codes.

not those spiked in before labeling, the failure is in the labeling step. Distinguishing between these types of failures contributes to efficiency in that only necessary steps are repeated to correct for an array failure (Figure 15.3). By assigning bit scores at each step of the QC1 checks, manual visualization of the array is also unnecessary.

Figure 15.4, panels A to F, displays a variety of contaminated images caused by wash problems, which are detected by our algorithm. Most contaminant areas display higher intensity measurements (See Figure 15.4, A to C), but some show lower intensities (Figure 15.4, D and F). Either of these cases increases the variation of background measurements, suggesting that comparison of background measurements over the chip are a reasonable indication of streaking, scratching, or uneven washing of an array.

To identify hybridization errors shown in Figure 15.4, two estimates of background variability were defined and calculated for defined areas of equal size on the array, the CV and the distribution measure (DM). In addition to variation, the DM contains information about distribution and excludes the three top and bottom background values from the calculation. The two parameters together provide information about the type of contaminant. For example, high CV and low DM suggested a small

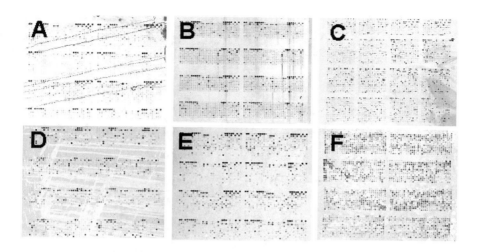

FIGURE 15.4 QC1 detects hybridization errors. A number of problems after hybridization are detected by our algorithms: (A) flagging, (B) streaking, (C) contaminants, (D) scratches, (E) background gradients, and (F) other wash problems.

region of aberrant high intensity. If both were high, it implied a large region of abnormal intensities. Interestingly, we found it unnecessary to discard these arrays, as the following spot outlier algorithm, which was able to identify these regions, removed abnormal intensity values (Figure 15.5).

QC2: Detection of Outliers

Flagging of bad spots on an array is time-consuming, tedious, and subjective; variations between individual curators were striking. We noted trends among four spot curators analyzing two chips. Of the total number of spots rejected by one or more of the curators, less than 15% were agreed upon by all four. Only 56% of the rejected spots were agreed upon by more than one curator. Further examination showed that manual curation tended to focus on spot morphology and disregarded any assessment based on the consistency of replicated spots.

QC2 detects outliers using replicate measurements and spot morphology. When present in at least triplicate, spots are accepted when similar in intensity, regardless of spot morphology. In this case, a Z score is defined, similar to the Z score rule described previously; however, the median is used instead of the mean. The associated p value is adjusted, using a Bonferroni correction, for the large number of genes tested, which results in discarding outliers with at least 95% confidence. For one or two spots, image parameters extracted by ImaGene [24] are used to describe the spot geometry. Spots with poor morphology were labeled outliers. Although the approach to the identification of outlier spot intensities was quite simple, it proved to be very effective. Figure 15.6 shows a scatter plot of intensities before and after outlier removal.

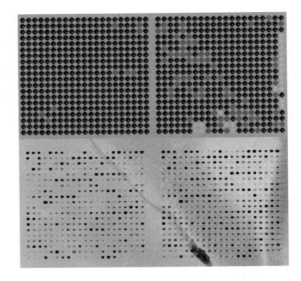

FIGURE 15.5 *(A color version follows page 204)* The outlier algorithm identifies spots affected by hybridization. The top panel is an applet display of regions in the bottom panel. Red-labeled outliers were based on high replicates. Yellow spots were outliers by both variance and morphology.

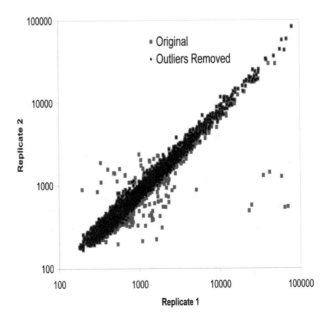

FIGURE 15.6 *(A color version follows page 204)* QC2 identifies and removes outlier intensities from the data set. Replicate arrays were compared with each other in an intensity plot on a log scale. Data are shown before (pink) and after (blue) outlier removal.

The novel AutoQC system that we have created and implemented is well suited to a high-throughput microarray environment. By enabling automatic workflow, the system provides full traceability throughout every processing step in microarray experimentation. Potential errors, which could occur during each procedure, are monitored. Only questionable quality values are reported to QC curators and unambiguous QC failures are processed automatically, reducing the subjectivity introduced by human judgment. The laboratory bottleneck is removed by bypassing the tedious visual check and manual quantitation phases. As expected, implementation of analytic algorithms keeps the QC process more consistent and image quantitation results reproducible.

CONCLUSIONS

Quality control is an important step in the processing of microarray data. For mining data with high confidence, the initial quality of data is crucial. There are several platforms for obtaining microarray data; each platform, including the technology, is slightly different and may require handling of QC in a special manner. Each laboratory is likely to have specialized techniques with errors specific to its own sample handling methods. Nonetheless, there are certain steps in microarray experimentation that are consistent, and variations of QC methods can be applied to many different platforms.

Here, we describe some of the common areas in which systematic error arises. These include stages in sample preparation, as well as in hybridization of microarrays. Many of these errors can be identified using algorithms that require no biological input from scientists, some of which are described. Users can choose to implement preferred algorithms with assumptions that best fit a platform. For groups that run enough microarrays to establish a population, parameters for the algorithms can be determined empirically from the population, thus providing customization of QC in their laboratories.

For many groups hybridizing small numbers of arrays, high throughput is not a necessity. These groups may prefer to monitor results interactively, especially if QC acceptance criteria are dependent on the biological nature of the experiment. Many academic groups fall into this category, and the QC systems from academic groups that are described here are sufficient for small numbers of arrays. Although QC may not be reproducible when human intervention is required, manual curation has the advantage of being flexible when making QC decisions.

Users of large numbers of microarrays (for example, pharmaceutical research and development) would likely prefer an automated QC system that is faster, as well as efficient and reproducible. In our organization, automation became a necessity, as QC was at the stage in which it limited the throughput of microarray experimentation. In addition to standardizing QC through automation, the arrays were customized to provide the most information possible to achieve reproducible and reliable results. In the pharmaceutical industry, standardization and reproducibility of data analysis are extremely important for the future when FDA accepts transcriptional profiling data.

When selecting an appropriate method for QC of microarray data, one must consider the importance of flexibility vs. speed and reproducibility. The priority will help determine whether a partially or fully automated QC process is necessary. Many of the software packages return QC parameters; it is up to the user how to use the information. Therefore, algorithms specific to a platform and developed around these parameters are important; replicating measurements and building a database population are extremely helpful in associating an error type with a QC parameter. These types of explorations and decisions can help customize a QC solution for a group. Although every group may have a system that uses specific algorithms or parameters, it is a necessity for every group to have a QC process to produce high-quality microarray data.

REFERENCES

1. Jain, K.K. Applications of biochips: from diagnostics to personalized medicine. *Curr Opin Drug Discov Dev* 7: 285–289, 2004.
2. Greer, B.T., Khan, J. Diagnostic classification of cancer using DNA microarrays and artificial intelligence. *Ann N Y Acad Sci* 1020: 49–66, 2004.
3. Luo, J., Isaacs, W.B., Trent, J.M., Duggan, D.J. Looking beyond morphology: cancer gene expression profiling using DNA microarrays. *Cancer Invest* 21: 937–949, 2003.
4. Lord, P.G. Progress in applying genomics in drug development. *Toxicol Lett* 149: 371–375, 2004.
5. Suter, L., Babiss, L.E., Wheeldon, E.B. Toxicogenomics in predictive toxicology in drug development. *Chem Biol* 11: 161–171, 2004.
6. Butte, A. The use and analysis of microarray data. *Nat Rev Drug Discov* 1: 951–960, 2002.
7. van Someren, E.P., Wessels, L.F., Backer, E., Reinders, M.J. Genetic network modeling. *Pharmacogenomics* 3: 507–525, 2002.
8. Vilo, J., Kivinen, K. Regulatory sequence analysis: application to the interpretation of gene expression. *Eur Neuropsychopharmacol* 11: 399–411, 2001.
9. MIAME. http://www.mged.org/Workgroups/MIAME/miame.html.
10. Stanford. Stanford Microarray Database (SMD) http://genome-www5.stanford.edu/.
11. NCBI. Gene Expression Omnibus (GEO) http://www.ncbi.nlm.nih.gov/geo/.
12. Hackett, J.L., Lesko, L.J. Microarray data — the U.S. FDA, industry and academia. *Nat Biotechnol* 21: 742–743, 2003.
13. Bolstad, B.M., Irizarry, R.A., Astrand, M., Speed, T.P. A comparison of normalization methods for high density oligonucleotide array data based on variance and bias. *Bioinformatics* 19: 185–193, 2003.
14. Irizarry, R.A., Bolstad, B.M., Collin, F., Cope, L.M., Hobbs, B., Speed, T.P. Summaries of Affymetrix GeneChip probe level data. *Nucl Acids Res* 31: e15, 2003.
15. Li, C., Wong, W.H. Model-based analysis of oligonucleotide arrays: expression index computation and outlier detection. *Proc Natl Acad Sci U S A* 98: 31–36, 2001.
16. Affymetrix. GeneChip Operating System http://www.affymetrix.com/products/software/specific/gcos.affx.
17. Brown, C.S., Goodwin, P.C., Sorger, P.K. Image metrics in the statistical analysis of DNA microarray data. *Proc Natl Acad Sci U S A* 98: 8944–8949, 2001.
18. Hautaniemi, S., Edgren, H., Vesanen, P. et al. A novel strategy for microarray quality control using Bayesian networks. *Bioinformatics* 19: 2031–2038, 2003.
19. Model, F., Konig, T., Piepenbrock, C., Adorjan, P. Statistical process control for large scale microarray experiments. *Bioinformatics* 18(Suppl. 1): S155–163, 2002.

20. Wang, X., Ghosh, S., Guo, S.W. Quantitative quality control in microarray image processing and data acquisition. *Nucl Acids Res* 29: E75–5, 2001.

21. BlueGenome. BlueFuse http://www.cambridgebluegenome.com/products/bluefuse-formicroarrays/overview/index.htm.

22. Wang, X., Hessner, M.J., Wu, Y., Pati, N., Ghosh, S. Quantitative quality control in microarray experiments and the application in data filtering, normalization and false positive rate prediction. *Bioinformatics* 19: 1341–1347, 2003.

23. Chen, Y., Kamat, V., Dougherty, E.R., Bittner, M.L., Meltzer, P.S., Trent, J.M. Ratio statistics of gene expression levels and applications to microarray data analysis. *Bioinformatics* 18: 1207–1215, 2002.

24. BioDiscovery. ImaGene http://www.biodiscovery.com/.

25. KoadaTechnology. Koadarray http://www.koada.com/index.htm.

26. Yang, I.V., Chen, E., Hasseman, J.P. et al. Within the fold: assessing differential expression measures and reproducibility in microarray assays. *Genome Biol* 3: research0062.1–0062.12, 2002.

27. Tseng, G.C., Oh, M.K., Rohlin, L., Liao, J.C., Wong, W.H. Issues in cDNA microarray analysis: quality filtering, channel normalization, models of variations and assessment of gene effects. *Nucl Acids Res* 29: 2549–2557, 2001.

28. Amaratunga, D., Cabrera, J. *Exploration and Analysis of DNA Microarray and Protein Array Data*. Wiley series in probability and statistics. Hoboken, NJ: John Wiley & Sons, 2004, p. 246.

29. Long, A.D., Mangalam, H.J., Chan, B.Y., Tolleri, L., Hatfield, G.W., Baldi, P. Improved statistical inference from DNA microarray data using analysis of variance and a Bayesian statistical framework: analysis of global gene expression in *Escherichia coli* K12. *J Biol Chem* 276: 19937–19944, 2001.

30. Kadota, K., Nishimura, S., Bono, H. et al. Detection of genes with tissue-specific expression patterns using Akaike's information criterion procedure. *Physiol Genomics* 12: 251–259, 2003.

31. Kuklin, A.P.A. and Shams, S. Quality control in microarray image analysis. *G.I.T. Imaging and Microscopy* 2001, p. 1.

32. BioDiscovery. GeneSight http://www.biodiscovery.com/.

33. Agilent. Agilent Feature Extraction Software http://www.chem.agilent.com/Scripts/PDS.asp?lPage=2547.

34. Luscombe, N.M., Royce, T.E., Bertone, P. et al. Express yourself: a modular platform for processing and visualizing microarray data. *Nucl Acids Res* 31: 3477–3482, 2003.

35. GenePix. http://www.axon.com/gn_GenePixSoftware.html.

36. ScanAnalyze. http://rana.lbl.gov/EisenSoftware.htm.

37. Petri, A., Fleckner, J., Matthiessen, M.W. Array-A-Lizer: a serial DNA microarray quality analyzer. *BMC Bioinformatics* 5: 12, 2004.

38. GenePixPro. http://www.axon.com/gn_GenePixSoftware.html.

39. Spotfinder. http://www.tigr.org/software/tm4/spotfinder.html.

40. Yang, H.Y., Dudoit, S., Luu, P., Speed, T.P. Normalization for cDNA microarray data. *Proc SPIE* 4266: 141–152, 2001.

41. Dudoit, S., Yang, Y.H., Callow, M.J., Speed, T.P. Statistical methods for identifying genes with differential expression in replicated cDNA microarray experiments. *Statistica Sinica* 12: 111–139, 2002.

42. Yang, Y.H., Dudoit, S., Luu, P. et al. Normalization for cDNA microarray data: a robust composite method addressing single and multiple slide systematic variation. *Nucl Acids Res* 30: e15, 2002.

43. Gollub, J., Ball, C.A., Binkley, G. et al. The Stanford Microarray Database: data access and quality assessment tools. *Nucl Acids Res* 31: 94–96, 2003.

44. van Hijum, S.A., Garcia de la Nava, J., Trelles, O., Kok, J., Kuipers, O.P. MicroPreP: a cDNA microarray data pre-processing framework. *Appl Bioinformatics* 2: 241–244, 2003.

45. Array-Pro. http://www.mediacy.com/arraypro.htm.

46. Quackenbush, J. Microarray data normalization and transformation. *Nat Genet* 32(Suppl.): 496–501, 2002.

47. Amersham. CodeLink Expression Analysis http://www1.amershambiosciences.com/aptrix/upp01077.nsf/Content/codelink_software.

48. Xiao, H., Liu, X., Leung, A., Carmen, A.A., Bittner, A., Sun, L., Yieh, L., Yao, X., Smith, S., and Wan, J. Methods and algorithms for performing quality control during gene expression profiling on DNA microarray technology. Patent pending, 2004.

49. Stratagene. Stratagene Microarray http://www.stratagene.com/products/showCategory.aspx?catId=18.

50. Insightful. S-Plus http://www.insightful.com/products/splus/default.asp.

51. BioDiscovery. QCInspector http://www.biodiscovery.com/othersoftware.asp.

52. Lin, L.I. A concordance correlation coefficient to evaluate reproducibility. *Biometrics* 45: 255–268, 1989.

53. Pearson, K. Notes on the history of correlation. *Biometrika* 13: 25–45, 1920.

54. Agilent. Agilent Bioanalyzer http://www.chem.agilent.com/Scripts/PDS.asp?lPage=51.

16 Microarray Data Normalization and Transformation

Huinian Xiao, Lynn Yieh, and Heng Dai

CONTENTS

With the sequencing of the human genome [1,2], newer generation microarrays are more capable than ever of presenting a global view of gene expression. Tens of thousands of data points can be obtained from a single assayed sample. With such a large payoff potential, experimental design and subsequent analysis methods are immensely important so that reliable results can be obtained from the data.

Microarray data analysis entails comparing expression data from several samples to obtain information about changes in gene expression among the studied biological states. Data normalization is the process by which the data from several samples in a data set are adjusted or transformed so that they are comparable to each other. Single-color microarray experiments result in intensity measurements, which are proportional to the amount of message RNA in a sample. By comparing intensity measurements among samples, measurable information about changes in gene expression can be derived; therefore, normalizing data is an essential step in the analysis of microarray data. Two-dye systems produce ratio data, but still require normalization to extract reliable results. In this type of system, normalization to correct for differences in the dyes used is most common. In the absence of proper normalization procedures, confounding factors introduced during the experimental process can mask biological changes in gene expression. This increases the probability of selecting false positives and increases the work required to verify results in a secondary assay.

Normalization aids in correcting for systematic variations due to experimental procedures. Technical variation in the data can stem from multiple procedures in microarray processing. RNA isolation, dye incorporation during labeling, variations in spotting, and hybridizations occurring on separate days are just a few factors that can contribute to error in microarray experimentation [3,4]. Regardless of the care in which experiments are performed, some variability will arise simply because of the high sensitivity that can be achieved with microarrays. Therefore, good design of experiments is essential to obtain high-quality data.

Several factors contribute to the design of a good microarray experiment. Replication of experiments, both technical and biological, has emerged as an important principle. Many normalization methods are based on statistical procedures, and general data assumptions can be gleaned from data with sufficient replication. Other methods, such as dye swapping in cDNA arrays, are accepted as a means to identify variations introduced by differential incorporation of label. Removal of poor quality data, both individual spots and whole samples, is also essential and is addressed elsewhere in this book (see Chapter 15).

Conventions of normalizing microarray data have changed dramatically since their first use. Early methods utilized variations of linear scaling. Affymetrix software initially offered two types of linear normalization. The favored method was a global scaling procedure, which assumed that the total fluorescence obtained from each sample should be equivalent. Data from each sample was multiplied by a constant such that the total fluorescence on a chip was achieved. A second method utilized by Affymetrix was to use a housekeeping gene in order to obtain a scaling factor, with the assumption that RNA levels from the housekeeping gene are constant between samples. Stanford Cluster used a variation of the linear scaling method, in which data could be scaled so that the mean or median of each array was equivalent [5].

Data normalization has evolved to incorporate other statistical methods, long used in other applications. For example, several groups have noted that there is a relationship between the variance and the expression levels of genes [3,6]. Because of this, local normalization methods such as splines and quantile-quantile are more commonly used with microarray data today. Likewise, data transformation methods, such as a simple log transformation, reduce the dependence of variance on intensity and are commonly used. Statistical methods are frequently used to identify differential genes expression. Many of these assume a normal distribution with the data;

therefore, data transformations that cause microarray data to resemble a normal distribution are also popular.

In this chapter, we review some of the more popular methods of data normalization and transformation that play a role in the identification of true changes in gene expression between biological conditions. Although these methods are inherently linked to distinguishing between experimental and biological variance, methods that are applied to the determination of data quality (technical variance) are reviewed elsewhere.

OVERVIEW

There are many sources of systematic variation in microarray, which affect expression intensities [3,4,7]. These include unequal quantities of starting RNA, differences in labeling or detection efficiencies between the fluorescent dyes used, and systematic biases in the measured expression levels. Normalization is the process used to remove these variations so that meaningful biological comparisons between microarray samples can be made.

Normalization can be carried out at different levels: within a single array (between replicate spots), between a pair of replicate arrays, and among multiple arrays (over samples to be compared). Normalization between print tips can correct inconsistencies among the spotting tip used to make an array, variability in the array surface, and differences in hybridization conditions across the array. Normalization among multiple arrays makes the comparison between samples and sample groups possible. These methods encompass two distinct types of normalization, between replicates and among different samples; however, the goal is the same, namely, to remove technical variation while retaining the variance from the biology being studied.

LINEAR NORMALIZATION

The linear method of normalization is the simplest type of microarray normalization and was commonly used in early applications. This approach scales all data points on an array by the array mean, median, or quintiles. The principle behind linear normalization is that the intensity distributions of arrays have the same central tendency and a linear relationship passing through the origin. Under these assumptions, a numerical constant, called the normalizing factor, can be used to correct the intensity of each spot on any array without taking into account its intensity level. In some applications, the third quartile is used to obtain a scaling factor (data from each array are scaled such that the third quartile is equivalent) based on the assumption that most genes were not expressed [6]. This method is most effective for normalizing between replicate spots or hybridizations, but the assumptions of linear relationships and similar intensity distributions are not always true for different samples in a data set. It has been observed that genes with low expression intensities require a scaling factor different than those with high expression intensities [3,6]; in these situations, a linear scaling factor is inappropriate. Intensity-dependent normalization is used more widely in the latest applications, in which the normalizing factor is a function of intensity levels instead of a constant value.

INTENSITY-DEPENDENT NORMALIZATION

REFERENCE AND BASELINE ARRAY

Conducting intensity-dependent normalization requires a good reference or baseline array that contains an invariant gene set with expression levels spanning the entire dynamic range of expression observed in the experiments. In addition, normalization for these genes should be representative of the normalization relationship for all the genes.

A small set of negative control plant genes and highly expressed housekeeping genes were the early choices for invariant gene sets in many applications [8]. However, it was found that their expression levels exhibit natural variability across samples and they did not fit the requirement of spanning the whole expression range of the intensity measurements (unpublished observations). More recently, labeled synthetic or cross-species DNA sequences spiked into the probe at various known concentrations are used as control genes for normalization purposes. These controls are more effective, but a small set of genes may not represent the normalization relationship for all the genes in the microarray. Currently, the most common method is to use the entire set of genes on the microarray as the reference in the normalization process. This approach assumes that only a very small percentage of the genes will be differentially expressed across the arrays being normalized and is true in most situations.

Microarray sample pool (MSP) [6] has been proposed; this uses a robust local regression with a set of appropriate controls to aid in intensity-dependent normalization. MSP, a novel sample ensemble, includes all genes on the microarray, and is analogous to genomic DNA without the noncoding sequences. It was titrated over the intensity range of a typical microarray experiment to account for all levels of intensity-dependent bias and was compared to other methods of normalization. Using all genes or MSP works well under specific conditions, but not for every situation. A composite method of normalization was designed (using all genes and MSP together) in which the strengths of the two methods could complement each other. The composite method also provides flexibility, where the techniques that best fit the data can be combined.

Li and Wong have proposed a method to identify genes that are not differentially expressed to use in normalization [9]. The premise is that a gene with invariant expression among samples would also have a similar ranked expression within each sample. They applied an iterative process to the perfect match intensities of a set of Affymetrix samples that calculates the absolute rank difference. Empirically derived cutoffs of the absolute rank difference were used to identify nondifferentially expressed genes at a range of expression levels. This set of genes can then be used to derive a normalization factor. Their data demonstrate that the procedure works well for a pair of arrays.

SMOOTHING FUNCTIONS

There are a few popular methods for intensity-dependent normalization: spline normalization, lowess normalization, and quantile normalization. Both spline and lowess belong to the category of smooth function normalization, which assumes that the majority of genes in a sample are invariant in their expression levels. Spline normalization uses a smooth but flexible function, like a cubic, with a small number of degrees of freedom. The less the degrees of freedom, the smoother is the fit. On the other hand, the smoothness of the lowess curve is controlled by a bandwidth parameter, called span; the larger it is, the smoother the fit. The advantage of these two methods is that neither of them is affected by a small percentage of outliers and is, therefore, widely used in microarray applications.

QUANTILE NORMALIZATION

Quantile normalization is useful for normalization across a series of conditions in which a small but indeterminate number of genes may be differentially expressed, and the distribution of spot intensities does not vary too much. The objective of quantile normalization is to make the intensity distributions of the gene expression as similar as possible across the array set. In other words, all quantiles are adjusted to be equal over the normalized arrays.

QSPLINE NORMALIZATION

Qspline, proposed by Workman et al.[10] is a combination of the spline and quantile normalization methods. Qspline uses quantiles from the full range of array signals to fit smoothing B-splines. The data presented showed similar performance between the qspline, lowess, and invariant gene set methods in all aspects examined. Qspline can normalize over a set of arrays, whereas the lowess method works on a pairwise basis. Therefore, qspline is computationally much less expensive, although generating similar results. The lowess method also uses random sampling such that the normalization results are not stable over data fittings.

Z-NORMALIZATION

Z-normalization is a modified lowess method developed by Wang et al. [11], which incorporates their spot quality model. A dependence of ratio distribution on the quality control score in a two-dye system was observed, which resulted in a novel method of normalizing the microarray data. The method starts with a quality-dependent Z-score. The local standard deviation of the log ratio is calculated using the fraction of neighboring data spots used for lowess smoothing. Their study suggested that Z-normalization is potentially more sensitive in detection at the high-quality end and in generating far less false positives in the low-quality regions.

CHOOSING A NORMALIZATION METHOD

It is difficult to make a conclusion about which normalization methods are the most accurate or adequate for all types of microarray comparison. The general thinking is that the technical replicates can be normalized using smooth function normalization methods (spline or lowess) and biological replicates can be normalized by the quantile method. The quantile method is also appropriate for cross-sample normalization. The prevailing belief, however, is that the best choice of normalization

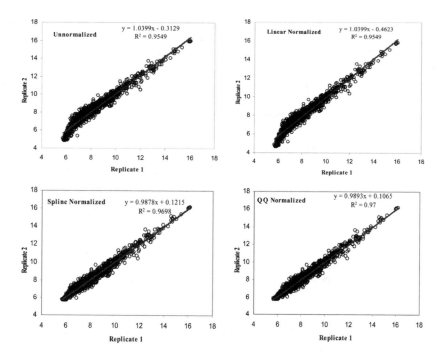

FIGURE 16.1 Comparison of different normalization methods. Spline and Quantile-Quantile normalization provide the best results, especially for low intensities.

methods is application dependent [6]. In a specific application, the better the data assumptions fit the normalization method and the better the reference set used (invariant gene set), the more accurate the results obtained. Figure 16.1 shows a comparison of linear, spline, and quantile-quantile normalization between technical replicates. In this situation, both spline and quantile-quantile normalization provide better performance than linear normalization.

NORMALIZATION FOR OLIGONUCLEOTIDE-BASED ARRAYS

Choosing a normalization method is mainly application and data dependent but not array-platform dependent. All methods discussed earlier are widely used experiments with cDNA arrays; they also work well for oligonucleotide arrays (Affymetrix and CodeLink® arrays, for example). The main difference is that for Affymetrix arrays, normalization at the probe level can be done before the normalization at the probe set level, because there are typically 16 to 20 (currently as few as 11) probe pairs, each interrogating a different region of a gene's sequence. Speed's group proposed three methods of performing normalization at the probe level without using a baseline array, which extended the idea of lowess and quantile [12] methods. There is a trade-off between the risk of choosing a poor baseline and of performing time-consuming calculations.

PERFORMANCE OF NORMALIZATION

Judging the success of normalization is not a trivial job. Amaratunga and Cabrera [13] have suggested using Spearman's rank correlation coefficient, which is a measure of monotone association between two variables, as a guide to whether the normalization is essential. The concordance correlation coefficient can be used as an indicator of how well the normalization procedure worked, as it is an index that quantifies the degree of agreement between two sets of numbers.

DATA TRANSFORMATION

Many statistical procedures assume a normal distribution (or at least symmetrical distribution) of samples. Significant violation of this assumption often leads to a major distortion and increases the chance of a type I or type II error, or a false discovery of (or failure to discover) differentially expressed genes in microarray analysis. These procedures include the *t*-test and ANOVA, which have been commonly applied to microarray expression analysis [14,15].

Normal distributions are clearly not suitable for the measured raw intensities in microarray experiments. A simple test for normality can be done either by visually inspecting the data distribution or calculating skewness and kurtosis. There are also more precise tests available such as Kolmogorov-Smirnov test. When we look at raw intensities of a CodeLink human whole-genome chip experiment with four replicates, we clearly see a deviation from the normal distribution. The curve is "peaked and fat tailed" (also known as *leptokurtic*), and "positively tailed" (positively skewed) (Figure 16.2A). This is also confirmed by statistical measurements. Whereas for a perfect normal distribution, skewness = 0 and kurtosis = 3, for the raw-intensity distribution, skewness = 5.5 and kurtosis = 33. More importantly, when we plot sample deviation of replicates against sample intensity, it is far from constant (Figure 16.2B). And this also violates a very important requirement for accurate *t*-test, making it unsuitable for these raw intensities.

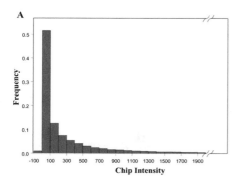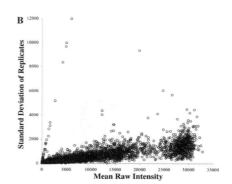

FIGURE 16.2 Nonnormal distribution of chip raw intensities. A. Raw-intensity histogram shows a positively skewed distribution with a long "tail" of high intensities. B. Standard deviations of raw intensities show an upward trend for high intensities.

If we can apply a mathematical transformation to these intensities, and when the transformed data approximately observes a normal distribution across all expression levels, we can still use existing statistical analysis techniques without significant modification. Several transformations have been proposed for microarray data analysis under different situations and have been applied with various degrees of success.

LOGARITHMIC TRANSFORMATION

$$f(y) = \log_a(y)$$

For a positively skewed sample distribution, logarithmic transformations provide "compression" for high intensities; this may partially correct the positive skew. The log base can be 2, e, or 10, and it does not affect the sample distribution. When we investigate the log-transformed raw intensities (Figure 16.3A), the positive data skew is greatly reduced. The variance is also minimized, and it approaches a constant for highly expressed genes (Figure 16.3B). This makes it practical to apply traditional statistical analysis tools. Based on these observations, Chen et al. have recommended using logarithmic transformation for microarray data [16]. Under this transformation, the traditional differential expression ratio between experiments can be simply viewed as the difference between transformed data, as $\ln(y_1/y_2) = \ln(y_1) - \ln(y_2)$.

Although logarithmic transformation has been successful for highly expressed genes, it has problems with weaker-expressed genes. In microarray experiments, intensities are often background-corrected by subtracting an estimate of mean background intensity $\hat{\alpha}$ from measured intensity. This often leads to a negative intensity when expression is low and background is high. For example, 0.87% of the probes on CodeLink chip used in Figure 16.2 have negative values, and they are not eligible for logarithmic transformation. This results in potentially discarding weakly expressing genes with significant changes in the raw data.

Additionally, assumption of a constant variance of log-transformed intensities clearly breaks down when the expression is low (Figure 16.3B). These

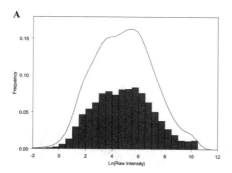
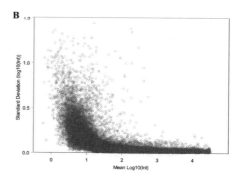

FIGURE 16.3 Distribution of log-transformed chip intensities. A. Log-transformed intensity histogram shows a distribution much closer to normal than raw intensity. B. Standard deviations of log-transformed intensities approach a constant for high intensities.

problems result in discarding all data close to the background, which is clearly suboptimal.

POWER TRANSFORMATION

$$f(y) = y^c (c > 0)$$

A second transformation often used in reducing positive data skew is square root transformation or, more generally, power transformation with a constant $c > 0$. Practically, cubic root transformations are often used for negative values. Although these transformations have also been proposed for certain situations [14], they are usually not recommended because they do not stabilize variance, as logarithmic transformations do, for high intensities. For low intensities, they are not continuous (for intensity values under 1, the transformed values get larger as the raw intensities get smaller), and this is not desirable.

VARIANCE STABILIZING TRANSFORMATION (VST)

In 2001, Rocke and Durbin introduced a two-component model of intensity measurement errors. In this model, measurement error follows a distribution based on the model

$$y = \alpha + \mu e^{\eta} + \varepsilon,$$

where y is the raw expression measurement, α the mean background measurement, μ the true expression level, and η and ε are error terms. η and ε are normally distributed with mean of zero and variances of σ_{η}^2 and σ_{ε}^2, respectively [17].

Under this model, we notice that for highly expressed genes ($y \gg 0$), $y \approx \mu e^{\eta}$ or $\ln(y) \approx \ln(\mu) + \eta$. This suggests that logarithmic-transformed data will have a constant variance, consistent with the earlier observation of a constant coefficient variation for highly expressed genes. On the other extreme of weakly expressed genes ($\mu \rightarrow 0$), $y \approx \alpha + \varepsilon$. This indicates that y will have a normal distribution around background intensity α with a constant standard deviation σ_{ε}; it also means log-transformed intensity $\ln(y)$ will have an asymptotic variance that approaches infinity [17]. This explains why the logarithmic transformation breaks down for weakly expressed genes. For measurements between the two extremes, the variance of y equals $\mu^2 S_{\eta}^2 + \sigma_{\varepsilon}^2$, where $S_{\eta}^2 = e^{\sigma_{\eta}^2}(e^{\sigma_{\eta}^2} - 1)$.

GENERALIZED AND MODIFIED LOGARITHMIC TRANSFORMATIONS

An ideal transformation $f(y)$ will stabilize variance across full intensities range; i.e., $AV(f(y)) \approx$ Const. In addition, it is desired that variance of $df = f(y_1) - f(y_2)$ will be approximately constant. Based on the error model, Durbin et al. and Huber et al. independently proposed two different versions of generalized logarithm

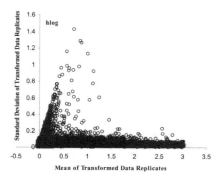 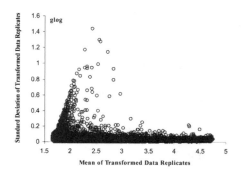

FIGURE 16.4 Standard deviation of glog and hlog transformed intensities. Both glog and hlog successfully stabilize variance except for intensities very close to background.

(glog) transformations:

$$g(y) = \ln(y - \alpha + \sqrt{(y - \alpha)^2 + \sigma_\varepsilon^2 / S_\eta^2}) \quad {}^{18}$$

$$h(y) = \log(a + by + \sqrt{(a + by)^2 + 1}) \quad {}^{19}$$

$$a = -\alpha \frac{S_\eta}{\sigma_\varepsilon} \quad b = \frac{S_\eta}{\sigma_\varepsilon}$$

They have shown that for both transformations, the variance is approximately constant at S_η^2 [18]. As seen in Figure 16.4, both glog and hlog transformations successfully stabilize variance except for intensities very close to the background.

The parameters σ_ε and σ_η can be estimated from expression intensities of well-defined strong-expressing and weak-expressing genes in replicate experiments. In most commercial microarray platforms, σ_ε can be estimated from standard deviation of negative controls (it can be further validated by $y < a + 2\sigma_\varepsilon$). σ_η can be estimated from the standard deviation of the logarithmic-transformed positive controls in replicate measurements.

Although glog transformation satisfies the stable variance requirement, it may still be convenient to have ratio information in some situations. Roche et al. evaluated two modifications of logarithmic transformations, started logarithm and log-linear hybrid, to retain the advantages of logarithmic transformation [20].

Started logarithm takes a general form of $g(y) = \log(y - \alpha + c)$, with $c > 0$. The addition of a positive constant c will reduce the problems of discarding all genes expressed below background. It also will limit the variance to α^2 / c^2 when $\mu \rightarrow 0$, instead of infinite. Roche et al. have shown that the optimal conditional to achieve minimal maximum deviation from constant variance will take a form of $c = \sigma_\varepsilon / \sqrt[4]{2S_\eta}$ [20].

Log-linear hybrid transformation takes a general form of

$$g(y) = \begin{cases} (y - \alpha) / k + \ln(k) - 1, & y - \alpha \leq k \\ \log(y - \alpha), & y - \alpha > k \end{cases}$$

k is a cutoff constant. Rocke et al. have shown that the optimal conditional to achieve minimum deviation from constant variance can be achieved with $k = \sqrt{2}\sigma_\varepsilon / S_\eta$ [20].

In theory, glog should have the best performance, although simulation and experimental data show that the three different transformations (glog, started, and log-linear hybrid) perform almost equally well [18]. If ratio or fold change is important for interpretation or visualization, such as pathway analysis, a log-linear hybrid approach may be the best compromise.

CONCLUSION

Microarray data can generate powerful results leading to the elucidation of molecular pathways that contribute to biological states such as disease progression or drug effects. Central to the ability to detect meaningful biological changes in gene expression are the normalization and transformation of raw expression data. Here, we review some of the popular methods for rendering data so that samples are comparable within an experiment. Knowledge of variations in data handling is essential, as nuances of each array system can determine which method is best for normalizing and transforming data. Data from genes expressed at high levels have different properties in comparison with data from low-expressing genes, which creates an additional challenge in normalizing array data. The most recent methods for addressing these special issues are hybrids, and it appears that data normalization and transformation will continue to evolve and improve by combining the best characteristics of existing methods.

REFERENCES

1. Venter, J.C. et al. The sequence of the human genome. *Science* **291**, 1304–1351, 2001.
2. Lander, E.S. et al. Initial sequencing and analysis of the human genome. *Nature* **409**, 860–921, 2001.
3. Quackenbush, J. Microarray data normalization and transformation. *Nat Genet* **32(Suppl.)**, 496–501, 2002.
4. Churchill, G.A. Fundamentals of experimental design for cDNA microarrays. *Nat Genet* **32(Suppl.)**, 490–495, 2002.
5. Eisen, M.B., Spellman, P.T., Brown, P.O., and Botstein, D. Cluster analysis and display of genome-wide expression patterns. *Proc Natl Acad Sci U S A* **95**, 14863–14868, 1998.
6. Yang, Y.H. et al. Normalization for cDNA microarray data: a robust composite method addressing single and multiple slide systematic variation. *Nucl Acids Res* **30**, e15, 2002.
7. Schuchhardt, J. et al. Normalization strategies for cDNA microarrays. *Nucl Acids Res* **28**, E47, 2000.

8. Luo, L. et al. Gene expression profiles of laser-captured adjacent neuronal subtypes. *Nat Med* **5**, 117–122, 1999.

9. Li, C. and Wong, W. H. Model-based analysis of oligonucleotide arrays: expression index computation and outlier detection. *Proc Natl Acad Sci U S A* **98**, 31–36, 2001.

10. Workman, C. et al. A new nonlinear normalization method for reducing variability in DNA microarray experiments. *Genome Biol* **3**, research0048, 2002.

11. Wang, X., Hessner, M.J., Wu, Y., Pati, N., and Ghosh, S. Quantitative quality control in microarray experiments and the application in data filtering, normalization and false positive rate prediction. *Bioinformatics* **19**, 1341–1347, 2003.

12. Bolstad, B.M., Irizarry, R.A., Astrand, M., and Speed, T.P. A comparison of normalization methods for high density oligonucleotide array data based on variance and bias. *Bioinformatics* **19**, 185–193, 2003.

13. Amaratunga, D. and Cabrera, J. *Exploration and Analysis of DNA Microarray and Protein Array Data.* John Wiley & Sons, Hoboken, NJ, 2004.

14. Tusher, V.G., Tibshirani, R., and Chu, G. Significance analysis of microarrays applied to the ionizing radiation response. *Proc Natl Acad Sci U S A* **98**, 5116–5121, 2001.

15. Pan, W. A comparative review of statistical methods for discovering differentially expressed genes in replicated microarray experiments. *Bioinformatics* **18**, 546–554, 2002.

16. Chen, Y., Dougherty, E.R., and Bittner, M.L. Ratio-based decisions and the quantitative analysis of cDNA microarray images. *J Biomed Opt* **2**, 364–374, 1997.

17. Rocke, D.M. and Durbin, B. A model for measurement error for gene expression arrays. *J Comput Biol* **8**, 557–569, 2001.

18. Durbin, B.P., Hardin, J.S., Hawkins, D.M., and Rocke, D.M. A variance-stabilizing transformation for gene-expression microarray data. *Bioinformatics* **18(Suppl. 1)**, S105–110, 2002.

19. Huber, W., von Heydebreck, A., Sultmann, H., Poustka, A., and Vingron, M. Variance stabilization applied to microarray data calibration and to the quantification of differential expression. *Bioinformatics* **18(Suppl. 1)**, S96–104, 2002.

20. Rocke, D.M. and Durbin, B. Approximate variance-stabilizing transformations for gene-expression microarray data. *Bioinformatics* **19,** 966–972, 2003.

17 Amplification Strategies and DNA Biochips

Barbara Ruggeri, Laura Soverchia, Massimo Ubaldi, Roberto Ciccocioppo, and Gary Hardiman

CONTENTS

INTRODUCTION

Gene expression analysis has progressed rapidly over the past decade, from the historical analyses of individual genes to the sophisticated genome surveys routinely performed today. DNA biochips have facilitated this progression and helped accelerate target validation and drug discovery efforts by the pharmaceutical industry. The increasing use and acceptance of biochips to study genetic and cellular processes is clearly demonstrated by the large number of publications in recent years and the emergence of several robust commercial microarray platforms [1–3]. Affymetrix (Santa Clara, CA) pioneered this field and has been the market leader for many years, applying photolithographic technologies derived from the semiconductor industry to the fabrication of high-density 25-mer oligonucleotide microarrays. Alternative commercial platforms equally as robust have emerged, including array platforms from Agilent, Amersham, Illumina, and Applied Biosystems (see Chapter 1 of this book for further discussion of the technology platforms).

Microarrays are primarily used for gene expression studies, where transcriptional profiling is carried out on RNA extracted from cells or tissues under different physiological conditions. Microarrays have been utilized by the pharmaceutical industry for both *in vitro* pharmacology and toxicology analyses. Chip technology allows questions to be asked at a qualitatively different scale than has been possible in the past. Additional biochip applications in routine use include DNA sequencing, genotyping, comparative genomic hybridization studies, and chromatin immunoprecipitation assays (ChIP/chip) [3–7].

To date, the greatest application of microarray technology has been in the fields of cancer research and genetics. This technology has now matured to the point where it is being widely used in many other biomedical fields. One of the more exciting applications is in the area of neuroscience, where we are seeing an expansion of our understanding of the brain's functions and the ability to study at a molecular level the complex disorders that affect the central nervous system. When one considers the intricacy of the brain, it is not surprising that difficulties emerge with expression analyses of this complex organ [8]. The greatest difficulty is the challenge of extracting sufficient amounts of high-quality RNA. The mammalian brain is characterized by a unique cellular and anatomical complexity. Heterogeneous cells (e.g., neurons and glia cells) contribute to form complex cytoarchitectures that can differentiate the various brain areas or, within the same nucleus, can delineate the structure of different subregions. Furthermore, these cells are organized in complex synaptic networks that have distinct neurophysiological, neurochemical, morphological, and molecular characteristics. Given the intricate organization of the brain, it is extremely difficult to obtain homogeneous samples to use for RNA extraction and subsequent array experiments [9–11].

LASER CAPTURE MICRODISSECTION (LCM)

A major reason for the variability in brain tissue sample preparation and RNA extraction is the low amounts of starting material that are typically obtained from small nuclei or single cells. In order to properly interrogate neuronal functions, the investigator must isolate and collect sufficient RNA quantities from a single cell, which requires the use of Laser Capture Microdissection (LCM) [12,13]. This technology was developed over a decade ago at The National Institutes of Health (NIH) and subsequently commercialized by Arcturus Bioscience, Inc. (Mountain View, CA). LCM initially found use amongst cancer researchers who utilized the technology to segregate benign and malignant cells [14]. The cells are visualized through a thermoplastic film, which is attached to the bottom of an optically clear microfuge-tube cap. A laser pulse is directed onto the target cells through the film. This causes the film to melt and allows it to flow onto the targeted area where it cools and binds with the underlying cells. The film including the adhered cells or clusters is then lifted, and the captured cells can be used for mRNA-expression profiling studies. A second method of laser capture has also been commercialized by P.A.L.M. (P.A.L.M. Microlaser Technologies GmbH, Bernried, Germany) that utilizes a cutting laser to perform microsurgery and free individual cells from surrounding tissue. LCM has been applied to the study of the central nervous system (CNS) [15]. As it permits the selection and capture of cells, cell aggregates, and discrete morphological structures deriving from thin tissue sections, LCM is now well established as a tool for enriching cells from tissue sections, thus overcoming the issue of tissue heterogeneity.

MICROARRAY TARGET LABELING

Two dominant target-sample preparation methodologies have emerged, a one-color approach utilized by the majority of the commercial platforms where the sample of interest is assayed on a single biochip, and a two-color competitive hybridization utilized solely by Agilent where an additional sample, typically a control, is also included [15,16]. The Agilent platform can also be utilized as a single-color platform, although historically these arrays have been hybridized with two samples, one labeled with cyanine 3 and the other with cyanine 5.

In the early studies, fluorescent targets were prepared by reverse transcription of RNA with direct incorporation of Cy3- or Cy5-labelled deoxyribonucleotides (dNTPs) into the synthesized first-strand cDNA. A popular alternative method utilizes a chemically reactive nucleotide analog, an amino allyl-dUTP, which is incorporated into the cDNA and then subsequently labeled with monoreactive Cy3- or Cy5 dyes. These methods require relatively large amounts of cellular RNA, which makes them inadequate for use with samples, where RNA is limiting, or a tissue such as the brain where a study of specific cell types is often required. The amount of RNA in a single cell is in the range of 0.1 to 1 pg, which is difficult to manipulate experimentally and also two orders of magnitude less than the minimum amount required for direct or amino-allyl-based labeling schemes. Direct labeling demands 10 to 20 μg of total RNA or 0.5 to 2 μg polyA+ mRNA, the amounts of RNA corresponding to approximately 10^7 cells or several milligrams of tissue [17]. RNA samples extracted from cells harvested by needle biopsy, cell sorting, or laser capture microdissection thus require an amplification step [18]. Two different amplification approaches have been employed, one based on enhancement of the fluorescent signal and the other on global enrichment of the mRNA.

Signal Amplification

Signal amplification methodologies such as dendrimer technology [19] and tyramide signal amplification (TSA) [20] boost the fluorescence signal emitted per mRNA molecule. Dendrimers are essentially branched polymers that permit attachment of various molecules including fluorescent moieties to their branched ends. A multitude of fluorescent markers are attached to the dendrimers in addition to a capture oligonucleotide DNA sequence complementary to primer used to initiate cDNA synthesis. This ensures that all the cDNA molecules have a binding site for detection with the dendrimers. The overall dendrimer signal amplification process is a two-step process. Initially, the extracted mRNAs are converted into cDNA targets, which are hybridized to the microarray. The complementary oligonucleotide-branched dendrimers are then subsequently hybridized to the array and interact with the bound target cDNAs via base pairing interaction between the oligonucleotide capture sequence, and the branched dendrimers capture sequence oligonucleotide. Approximately 360 molecules of dye per labeled DNA are obtained when the dendrimers are hybridized to the modified target sequences [19]. The labeling is independent of the composition of the DNA sequence. This system is therefore advantageous in that a constant amount of fluorescent dye is incorporated per DNA molecule. Utilizing this labeling system, the amount of DNA hybridized can be determined on a molar basis, permitting the calculation of a stoichiometric unit representing the ratio between a defined amount of a target cDNA, hybridized to a defined amount of the corresponding homologous array probe [21].

Tyramide signal amplification differs from branched dendrimers in that it utilizes an enzymatic reaction. This methodology was originally developed to increase the sensitivity of immunohistochemical detection schemes. In a similar manner to the dendrimer approach, a cDNA target is first synthesized from the extracted RNA. During the reverse transcription process, a biotin dNTP is incorporated into the synthesized cDNA. The cDNA target is hybridized to the microarray and subsequently detected with streptavidin-horseradish peroxidase. This enzymatic reaction causes the deposition of numerous cyanine 5 fluorophores on the array, thereby amplifying the signal intensity. The use of different haptens coupled to dNTPS, detected with different antibody or protein conjugates, permits two-color labeling approaches using this scheme.

GLOBAL mRNA AMPLIFICATION

Experimental fidelity is dependent on both even and uniform amplification of the entire population of mRNA species with minimal bias and the preservation of the relative transcript levels. Global mRNA amplification methods therefore increase the number of transcript equivalents [22] with the objective of generating sufficient nucleic acid target to permit standard labeling techniques. In order to obtain adequate RNA from small tissue samples such as biopsy material or single cell isolations, global amplification techniques based either on isothermal linear RNA polymerase amplification [23], exponential PCR [24], or single primer isothermal amplification have been employed [25].

T7 RNA POLYMERASE-BASED AMPLIFICATION

The T7 RNA Polymerase amplification method described by Eberwine and coworkers [23] has found wide application in the microarray field. This method employs a synthetic oligo (dT) primer fused to a phage T7 RNA polymerase promoter to prime synthesis of first strand cDNA by reverse transcription of the polyA+ RNA component of total RNA. Second-strand cDNA is synthesized with RNase H by degrading the polyA+ RNA strand, followed by second-strand synthesis step with *E. coli* DNA polymerase I. Amplified antisense RNA (aRNA) is synthesized via *in vitro* transcription of the double-stranded cDNA (ds cDNA) template using T7 RNA polymerase. This approach enables amplification of the starting RNA material by up to 200-fold. A second round of amplification can also be carried out by annealing random hexamers to the newly synthesized cRNA and performing first-strand cDNA synthesis, permitting further amplification. Linear RNA amplification is preferred over exponential amplification as RNA Polymerase activity is less influenced by template sequence or concentration than Taq DNA polymerase. Additionally, comparison of RNA-based amplification with a nonamplified control revealed stronger correlation and less bias than with PCR-based amplification [25]. The resultant cRNA products are nevertheless biased toward the 3' end of their cognate transcripts because of the initial priming occurring at the polyA+ tail. This generally does not pose a problem as microarray probes are designed from sequence at the 3′ end of mRNAs [26–29].

T7 linear amplification protocols currently in use differ, based on a few criteria, namely whether a template-switching mechanism is used in the synthesis of second-strand cDNA, the enzymes utilized, the method used to purify the double-stranded cDNA prior to *in vitro* transcription step. The level of bias, introduced into gene expression profiling experiments by RNA amplification, has been shown to be relatively low. Zhao et al. [18] utilized a virtual array approach with repeatedly amplified samples to examine and minimize experimental variation. They observed that slight differences in T7 linear amplification protocols did not greatly affect the correlation of amplified samples with unamplified samples. Expression profiles obtained with aRNA were seen to closely resemble expression profiles of the original sample. The degree of reproducibility in microarray experiments with amplified material presented a greater concern. Samples amplified on the same day were seen to have a much greater correlation than samples amplified on different days [18]. Additionally, the amount of total RNA input was also seen to affect the amplification process. Within the range of 0.3 to 3 μg total RNA, decreasing the input RNA did not adversely affect either the fidelity or reproducibility of amplification. However, when reducing the input starting total RNA to less than 300 ng, the yields of aRNA were less than 3 μg and not sufficient to permit even one hybridization experiment. The fold of amplification was observed to be greater with smaller starting quantities of template RNA, but the absolute yield of aRNA was typically smaller.

Gold et al. [30] utilized Affymetrix GeneChips (U95Av2) with samples extracted from normal human tracheobronchial epithelial cells (NHTBE) and human pulmonary mucoepidermoid carcinoma cells (NCI-H292) to examine the effect of two rounds of amplification, a double IVT (dIVT) requiring 200 ng of total RNA compared to a single IVT requiring 5 μg of total RNA. In both cell lines, approximately 10% more genes were detected with IVT than with dIVT.

PCR-Based Amplification

Linear isothermal RNA procedures require multiple steps and are dependent on nucleic acid extractions, reverse transcription, and purification steps, all of which may introduce bias. Additionally, these steps are both labor intensive and time-consuming, and do not preserve accurately the abundance information from smaller starting amounts of sample. PCR-based amplification is more straightforward but is prone to bias the abundance relationships. PCR-based approaches have been applied to microarray studies. One approach utilized terminal deoxynucleotide transferase to append a homomeric tail to the 3′ end of the first strand cDNA, followed by PCR with a homomeric primer (dN) and one incorporated during the reverse transcription step at the 5′ end of the cDNA. Nonspecificity arising from the use of homomeric primers for PCR poses problems with this approach. Another approach, termed three-prime end amplification (TPEA), facilitated global amplification of the 3′ end of all mRNAs present in a sample [31,32]. PCR amplification was performed with a primer, incorporated into the first strand, during reverse transcription, and a second primer was used to initiate second-strand synthesis. The second-strand primer contained partially degenerate sequence at the 3' end, which promoted annealing every 1 kb, resulting in uniformly sized amplicons. This was advantageous, as all mRNA species were amplified, regardless of the initial size of the transcript. Iscove et al. [33] were the first group to demonstrate that the exponential approach, when performed carefully, actually preserved abundance relationships through amplification as high as 3×10^{11} fold. Their approach involved reverse transcription of a first strand cDNA primed by oligo (dT) and addition of an oligo (dA) tail with terminal transferase, followed by exponential amplification with an oligo(dT) containing primer. This global RT-PCR protocol was reported to be applicable to 10 pg of starting RNA, or as the authors noted, a single cell.

SINGLE PRIMER ISOTHERMAL AMPLIFICATION TECHNIQUE

The single primer isothermal amplification technique, SPIA™, has been developed and commercialized by NuGEN™ Technologies and is discussed in depth elsewhere in this book (see Chapter 18, Dafforn et al.). It allows linear amplification of DNA by employing chimeric oligonucleotide primers to generate cDNA that serves as a substrate for amplification, producing multiple copies of first-strand cDNA representing all mRNAs in the sample [34]. This amplification scheme generates microgram amounts of amplified cDNA from 5 to 100 ng of total RNA in a single amplification round and is performed in less than 4 h. The combination of this amplification approach with a simple fragmentation and labeling method generates amplified cDNA, ready for hybridization to Affymetrix® GeneChip® arrays or spotted arrays.

CONCLUSION

In conclusion, we are facing a robust expansion of gene array studies in many fields and particularly neuroscience, but in order to take full advantage of this technology, it will require considerable attention to all experimental variables that may influence the resulting data. When working with samples that are limiting, amplification of nucleic acid targets is a necessity. Experimental fidelity is dependent on both equal and consistent amplification of the entire population of mRNA transcripts. The ability to accurately preserve relative transcript levels is the most important issue with any amplification methodology.

REFERENCES

1. Hardiman, G. Microarray platforms — comparisons and contrasts. *Pharmacogenomics* 5(5): 487–502, 2004.
2. Stafford, P. and Liu, P. Microarray technology comparison, statistical analysis, and experimental design. *Microarray Methods and Applications — Nuts and Bolts*. DNA Press, Eagleville, PA, 2003, pp. 273–324.
3. Wick, I. and Hardiman, G. Biochip platforms as functional genomics tools for drug discovery. *Curr. Opin. Drug Discov. Dev.* 8(3): 347–354, 2005.

4. Waring, J.F., Ciurlionis, R., Jolly, R.A., Heindel, M., and Ulrich, R.G. Microarray analysis of hepatotoxins *in vitro* reveals a correlation between gene expression profiles and mechanisms of toxicity. *Toxicol. Lett.* 120: 359–368, 2001.

5. Hamadeh, H.K., Amin, R.P., Paules, R.S., and Afshari, C.A. An overview of toxicogenomics. *Curr. Issues Mol. Biol* 4(2): 45–56, 2002.

6. Johnson, J.A. Drug target pharmacogenomics: an overview. *Am. J. Pharmacogenomics.* 1(4): 271–281, 2001.

7. Kruglyak, L. and Nickerson, D.A. Variation is the spice of life. *Nat. Genet.* 27: 234–236, 2001.

8. Soverchia, L., Ubaldi, M., Leonardi-Essmann, F., Ciccocioppo, R., and Hardiman, G. Microarrays — the challenge of preparing brain tissue samples. *Addict. Biol.* 10(1): 5–13, March 2005.

9. Coombs, N.J., Gough, A.C., and Primrose, J.N. Optimization of DNA and RNA extraction from archival formalin-fixed tissue. *Nucl. Acids Res.* 27(16): e12, 1999.

10. Van Deerlin, V.M., Gill, L.H., and Nelson, P.T. Optimizing gene expression analysis in archival brain tissue. *Neurochem. Res.* 27: 993–1003, 2002.

11. Preece, P., Virley, D.J., Costandi, M., Coombes, R., Moss, S.J., Mudge, A.W., Jazin, E., and Cairns, N.J. An optimistic view for quantifying mRNA in post-mortem human brain. *Brain Res Mol Brain Res* 116: 7–16, 2003.

12. Bonner, R.F., Emmert-Buck, M., Cole, K., Pohida, T., Chuaqui, R., Goldstein, S., Liotta, L.A. Laser capture microdissection: molecular analysis of tissue. *Science* 278: 1481–1483, 1997.

13. Simone, N.L., Bonner, R.F., Gillespie, J.W., Emmert-Buck, M.R., Liotta, L.A. Laser-capture microdissection: opening the microscopic frontier to molecular analysis. *Trends Genet.* 14: 272–276, 1998.

14. Emmert-Buck, M.R., Bonner, R.F., Smith, P.D., Chuaqui, R.F., Zhuang, Z, Goldstein, S.R., Weiss, R.A., Liotta, L.A. Laser capture microdissection. *Science* 274: 998–1001, 1996.

15. Luo, L., Salunga, R.C., Guo, H., Bittner, A., Joy, K.C., Galindo, J.E., Xiao, H., Rogers, K.E., Wan, J.S., Jackson, M.R., Erlander, M.G. Gene expression profiles of laser-captured adjacent neuronal subtypes. *Nat. Med.* 5: 117–122, 1999.

16. Chee, M., Yang, R., Hubbell, E., Berno, A., Huang, X.C., Stern, D., Winkler, J., Lockhart, D.J., Morris, M.S., Fodor, S.P.A. Accessing genetic information with high-density DNA arrays. *Science* 274: 610–614, 1996.

17. Brown, P.O. and Botstein, D. Exploring the new world of the genome with DNA microarrays. *Nat. Genet.* 21: 33–37, 1999.

18. Zhao, H., Hastie, T., Whitfield, M.L., Borresen-Dale, A.L., and Jeffrey, S.S. Optimization and evaluation of T7 based RNA linear amplification protocols for cDNA microarray analysis. *BMC Genomics* 3: 31, 2002.

19. Stears, R.L., Getts, R.C., and Gullans, S.R. A novel, sensitive detection system for high-density microarrays using dendrimer technology. *Physiol. Genomics* 3: 93–99, 2000.

20. Karsten, S.L., Van Deerlin, V.M., Sabatti, C., Gill, L.H., and Geschwind, D.H. An evaluation of tyramide signal amplification and archived fixed and frozen tissue in microarray gene expression analysis. *Nucl. Acids Res.* 30: E4, 2002.

21. Rouse, R.J., Espinoza, C.R., Niedner, R.H., and Hardiman, G. Development of a microarray assay that measures hybridization stoichiometry in moles. *Biotechniques* 36(3): 464–470, 2004.

22. Nygaard, V., Loland, A., Holden, M., Langaas, M., Rue, H., Liu, F., Myklebost, O., Fodstad, O., Hovig, E., and Smith-Sorensen, B. Effects of mRNA amplification on

gene expression ratios in cDNA experiments estimated by analysis of variance. *BMC Genomics* 4: 11, 2003.

23. Van Gelder, R.N., von Zastrow, M.E., Yool, A., Dement, W.C., Barchas, J.D., and Eberwine, J.H. Amplified RNA synthesized from limited quantities of heterogeneous cDNA. *Proc Natl Acad Sci USA* 87: 1663–1667, 1990.

24. Lukyanov, K., Diatchenko, L., Chenchik, A., Nanisetti, A., Siebert, P., Usman, N., Matz, M., and Lukyanov, S. Construction of cDNA libraries from small amounts of total RNA using the suppression PCR effect. *Biochem Biophys Res Commun* 230: 285–288, 1997.

25. Puskás, L.G., Zvara, A., Hackler, L., Jr., and Van Hummelen, P. RNA amplification results in reproducible microarray data with slight ratio bias. *Biotechniques* 32: 1330–1334, 2002.

26. Mori, M., Mimori, K., Yoshikawa, Y., Shibuta, K., Utsunomiya, T., Sadanaga, N., Tanaka, F., Matsuyama, A., Inoue, H., and Sugimachi, K. Analysis of the gene-expression profile regarding the progression of human gastric carcinoma. *Surgery* 131: S39–47, 2002.

27. Sotiriou, C., Khanna, C., Jazaeri, A.A., Petersen, D., and Liu, E.T. Core biopsies can be used to distinguish differences in expression profiling by cDNA microarrays. *J. Mol. Diagn.* 1: 30–36, 2002.

28. Scheidl, S.J., Nilsson, S., Kalen, M., Hellstrom, M., Takemoto, M., Hakansson, J., and Lindahl, P. mRNA expression profiling of laser microbeam microdissected cells from slender embryonic structures. *Am. J. Pathol.* 160: 801–813. 2002.

29. Wang, E., Miller, L.D., Ohnmacht, G.A., Liu, E.T., and Marincola, F.M. High-fidelity mRNA amplification for gene profiling. *Nat. Biotechnol.* 18: 457–459, 2000.

30. Gold, D., Coombes, K., Medhane, D., Ramaswamy, A., Ju, Z., Strong, L., Koo, J.S., and Kapoor, M. A comparative analysis of data generated using two different target preparation methods for hybridization to high-density oligonucleotide microarrays. *BMC Genomics* 5: 2, 2004.

31. Dixon, A.K., Richardson, P.J., Lee, K., Carter, N.P., and Freeman, T.C. Expression profiling of single cells using 3 prime end amplification (TPEA) PCR. *Nucl. Acids Res.* 26: 4426–4431, 1998.

32. Freeman, T.C., Lee, K., and Richardson, P.J. Analysis of gene expression in single cells. *Curr Opin Biotechnol.* 10: 579–582, 1999.

33. Iscove, N.N., Barbara, M., Gu, M., Gibson, M., Modi, C., Winegarden, N. Representation is faithfully preserved in global cDNA amplified exponentially from sub-picogram quantities of mRNA. *Nat. Biotechnol.* 20(9): 940–943, September 2002.

34. Kurn, N., Chen, P., Heath, J.D., Kopf-Sill, A., Stephens, K.M., and Wang, S. Novel isothermal, linear nucleic acid amplification systems for highly multiplexed applications. *Clin Chem.* 51(10): 1973–1981, 2005.

18 Ribo-SPIA™, a Rapid Isothermal RNA Amplification Method for Gene Expression Analysis

Alan Dafforn, Pengchin Chen, Glenn Y Deng, Michael Herrler, Dawn M. Iglehart, Sriveda Koritala, Susan M. Lato, Susheela Pillarisetty, Reshma Purohit, Leah Turner, Martin Wang, Shenglong Wang, and Nurith Kurn

CONTENTS

INTRODUCTION

Gene expression profiling provides a snapshot of complex, regulated gene expression processes that link the genotype of an organism with a corresponding phenotype. The new analytical tools developed in recent years for the determination of gene expression profiles have greatly advanced the understanding of the regulation of normal and pathogenic cell development and function. Large-scale gene expression profiling currently relies on a variety of methods, including microarrays for parallel determination and quantification of thousands of gene transcripts [1] and any of various quantification methods for individual gene transcripts. A major limitation for large-scale gene expression profiling is the large quantity of RNA required for analysis when using either microarrays or the quantification of large numbers of specific gene transcripts by methods such as quantitative PCR. Thus, most expression profiles have been obtained for samples derived from large numbers of cells, tissues, or organs, representing mixtures of large numbers of cell types. They represent average expression profiles of all cells in the mixture and clearly do not reflect the expression profile of any of the specific cell types present. In contrast, gene expression profiling in defined cell populations enables the elucidation of gene expression patterns that are specific for defined genotype and cell function in normal and pathological states. However, this approach typically requires the analysis of sample from relatively small numbers of cells. Recent advances in sample acquisition [2], the isolation of pure cell populations [3], and methods for RNA amplification are being implemented toward establishing expression profiles and patterns at the level of homogeneous cell populations.

Various RNA amplification schemes have been developed for this purpose. Insofar as expression profiles are expected to provide an accurate representation of all the mRNA species in a given sample, it is clear that the amplification method must provide high representation fidelity. Thus, linear amplification of all transcripts in a sample is essential. One of the most common methods is that first described by Eberwine and colleagues [4,5]. In this method, T7 RNA polymerase transcription is used to generate cRNA for microarray-based analysis. Additional rounds of transcription are used to allow analysis of small samples. Although capable of reasonable representation of transcripts in the unamplified material [6,7], this approach is rather lengthy, tedious, and requires highly skilled operators. A serious limitation of this method is the requirement for different numbers of amplification rounds for preparation of hybridization targets from samples of various sizes, leading to complications in interpretation of the data obtained. Each additional round of amplification also reduces the template fidelity.

PCR-based methods for global amplification have also been described [8,9]. These are also somewhat cumbersome, and fidelity of transcript representation may be reduced because of the exponential nature of PCR. Thus neither approach offers a rapid, simple procedure for global gene expression analysis suitable for all samples, including samples with RNA input in the low-nanogram range.

We have developed a novel, isothermal, linear RNA amplification, Ribo-SPIA™, which generates microgram amounts of amplified cDNA from 5 to 100 ng of total RNA in a single amplification round and is performed in less than 4 h [10]. This scheme provides a single method for target preparation from all samples. The combination of this amplification technology with a simple fragmentation and labeling method generates amplified cDNA, ready for hybridization to Affymetrix® GeneChip® arrays or spotted arrays from as little as 5 ng of starting total RNA in one day. This report summarizes applications of this new technology and demonstrates its accuracy and fidelity of representation.

RiBO-SPIA™: A NOVEL METHOD FOR GLOBAL ISOTHERMAL LINEAR AMPLIFICATION OF mRNA

Ribo-SPIA™ allows global mRNA amplification while preserving representation of relative expression levels. This simple, rapid process reproducibly achieves amplification from as little as 5 ng of total RNA, generating single-stranded DNA products that are complementary to mRNA (homologous to the first strand cDNA). Ribo-SPIA amplification products are ideally suited for quantifying expression levels by any commercially available nucleic acid measurement technique such as real time PCR or various microarrays.

Ribo-SPIA incorporates a single primer isothermal amplification technique, SPIA™, developed by NuGEN™ Technologies, for the linear amplification of DNA. The method employs unique features of chimeric oligonucleotide primers to generate cDNA that serves as a substrate for amplification, producing multiple copies of first strand cDNA representing all mRNAs in the sample. The Ribo-SPIA reaction for amplification of mRNA is schematically described in Figure 18.1. A chimeric primer comprising a 3' DNA portion and a 5' RNA portion is employed for synthesis of first, strand cDNA. The chimeric primer hybridizes to all mRNA targets in the sample, at the beginning of the polyA tail. Primer extension by reverse transcriptase is initiated to produce first strand cDNA. The heteroduplex produced at the end of replication of all mRNA targets by reverse transcriptase contains a unique RNA tail at the 5' end of the newly synthesized cDNA strand, which was incorporated by the chimeric primer.

Second-strand cDNA is synthesized in the next step. The first-strand DNA is replicated by DNA polymerase and the RNA tail incorporated by the chimeric

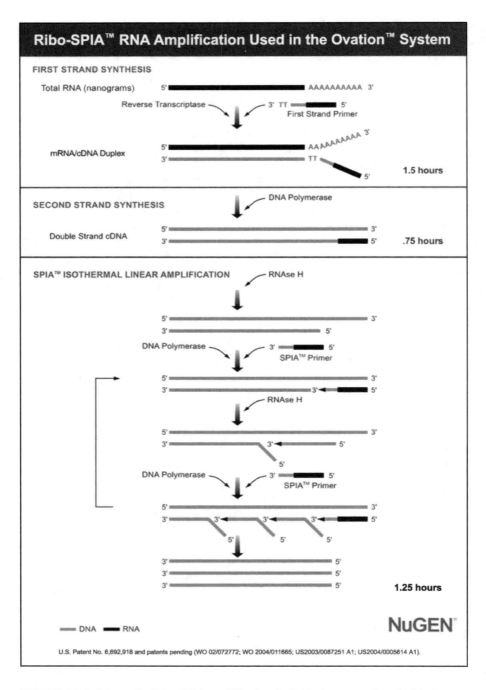

FIGURE 18.1 Scheme for Ribo-SPIA amplification. Individual steps are described in the text.

primer is simultaneously reverse transcribed. The double-stranded cDNA produced by this process comprises a DNA/RNA heteroduplex at one end, containing a unique sequence that was introduced by the 5′ end of the chimeric primer. The RNA portion of the heteroduplex is cleavable by RNase H, which specifically cleaves an RNA strand in a DNA/RNA heteroduplex. Cleavage of the RNA strand in the heteroduplex portion of the double-stranded cDNA leads to formation of a unique partial duplex of first- and second-strand cDNA. The single-stranded portion of the partial duplex, at the 3′ end of the second-strand cDNA, is the complement of the unique sequence introduced by the 5′-RNA portion of the chimeric primer.

This sequence is employed as the priming site for SPIA amplification, using a second chimeric SPIA primer that is complementary to the unique single-stranded sequence. After binding to the priming site, the primer is extended isothermally by a strand-displacing polymerase. Once extension begins, RNase H can again digest the new DNA/RNA heteroduplex and regenerate the priming site. Binding and extension of new molecules of primer leads to the continuous, isothermal generation of multiple copies of cDNA complementary to the original mRNA. Modified bases can be incorporated if needed to allow labeling reactions.

Ribo-SPIA PROTOCOL FOR RAPID ROBUST AMPLIFICATION OF TOTAL mRNA IN SMALL TOTAL RNA SAMPLES

The simple NuGEN protocol for linear, isothermal mRNA amplification from total RNA can be carried out by minimally trained laboratory personnel in less than 4 hours (Figure 18.2). The continuous nature of the linear amplification reaction makes it possible to generate microgram amounts of cDNA from as little as 5 ng of total RNA input. This product is suitable for both array-based expression profiling and expression analysis of specific genes. Products of the linear amplification exhibit a size distribution with most material below 2 kb for optimal performance on arrays designed for other 3′-biased IVT methods. Specific gene transcripts may be quantified directly by various quantitative PCR (qPCR) methods such as TaqMan®, SYBR® Green detection, or others.

Alternatively, array-based expression profiling can be carried out, following labeling of the amplification products to produce targets suitable for the desired array platform. NuGEN is currently commercializing the Ovation family of products designed for target preparation for quantitative analysis. The Aminoallyl Ovation kit incorporates aminoallyl-dUTP during DNA synthesis, to allow subsequent reaction with dyes such as Cy3/Cy5. The amplified product can be used for gene expression profiling on spotted arrays employing two-dye detection schemes, such as the Agilent oligonucleotide arrays, home brew-spotted arrays, cDNA arrays, and the like. The Biotin Ovation kit provides reagents that allow the fragmentation and biotin labeling of the cDNA product, generating targets suitable for hybridization to Affymetrix GeneChip arrays or GE's CodeLink arrays for gene expression profiling. Finally, the Ovation RNA Amplification System kit is especially tailored for preamplification

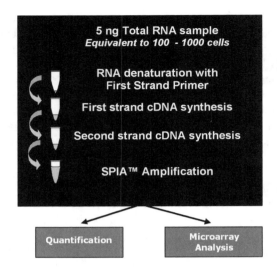

FIGURE 18.2 Flow diagram of the steps in the Ribo-SPIA procedure. The entire process is run in a single tube, with an addition between each of the incubations shown. Steps following the amplification are specific to the desired application. More detail is provided in Table 18.1 for the case of target generation for the Affymetrix GeneChip.

for qPCR applications. Table 18.1 summarizes in more detail the Biotin Ovation protocol, illustrating its relative simplicity and suitability for automation. The entire process, from total RNA (5 to 100 ng) to either fragmented, biotin-labeled product ready for hybridization to GeneChip arrays or Cy3/Cy5-labeled product ready for hybridization to any of various spotted arrays, can be completed in a single 8-hour day.

TABLE 18.1
Protocol for Ribo-SPIA Amplification and Biotin Labeling

1. Mix 1–100 ng of total RNA in 5 μl of water with 2 μl of primer, denature 5 min at 65°C
2. Add 13 μl of first-strand reaction mix, incubate 1 h at 48°C and 15 min at 70°C
3. Add 20 μl second-strand reaction mix, incubate 30 min at 37°C and 15 min at 75°C
4. Add 120 μl SPIA amplification mix (primer, enzymes, dNTPs, and buffer), incubate for 60 min at 50°C and 5 min at 95°C
5. Purify amplified cDNA on one Nucleospin column (BD Biosciences); elute in 30 μl total volume of water. Amplification efficiency may be determined by qPCR if desired either before or after purification
6. Add 5 μl each of fragmentation reagents 1 and 2, incubate 30 min at 50°C
7. Add 5 and 2.5 μl of labeling reagents 1 and 2, incubate 30 min at 50°C
8. Purify by gel filtration using a DyeEx spin kit (Qiagen)

ACCURACY, REPRODUCIBILITY, AND LINEARITY OF Ribo-SPIA ASSESSED BY qPCR

Essential requirements of any global mRNA amplification for gene expression analysis are high reproducibility, linearity, and accuracy for maintaining fidelity of representation of all transcripts in the sample. Ribo-SPIA reproducibility for global mRNA amplification from 20 ng total RNA UHR (Universal Human Reference RNA, Stratagene), is shown in Figure 18.3. Primers and probe sets were designed for quantification of GAPDH cDNA at two positions relative to the 3' poly-A end of the mRNA, GAPDH 3' (330 nucleotides) and GAPDH 5'(1000 nucleotides), and real time PCR (TaqMan) was employed for the quantification of Ribo-SPIA amplified cDNA. The data presented was obtained from 88 amplification reactions carried out by multiple users on multiple days and using various Ovation kits, and demonstrates the high degree of reproducibility of the method.

The linearity and accuracy of the Ribo-SPIA global mRNA amplification is demonstrated by the comparison of amplification efficiency of various housekeeping gene transcripts in the sample. real time PCR, with SYBR Green, was employed for the quantification of a set of 19 housekeeping gene transcripts in nonamplified and amplified cDNA generated from 20 ng HeLa total RNA, representing about a 1000-fold abundance range. The amplification efficiency is denoted by the delta Ct (cutoff threshold) values for nonamplified cDNA (produced by the Ribo-SPIA enzyme system) and the amplified cDNA, as shown in Figure 18.4. The delta Ct, representing the fold amplification for the various transcripts is very consistent, thus indicating equal amplification of all mRNA species in the sample.

Differential expression of 68 genes in total RNA from either Universal Human Reference RNA (UHR) sample or from human skeletal muscle was compared in

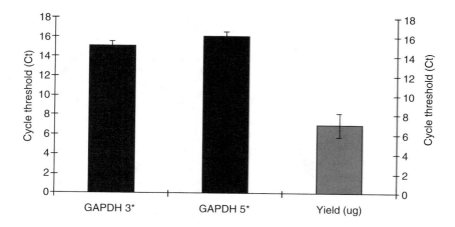

FIGURE 18.3 Reproducibility of the Ribo-SPIA process. The amount of amplified product from two loci in the GAPDH gene was determined by TaqMan for 88 separate Ribo-SPIA amplifications over an extended period. Data is summarized as observed cutoff threshold (Ct) without any normalization. Yield data (μg) is included for the same data set.

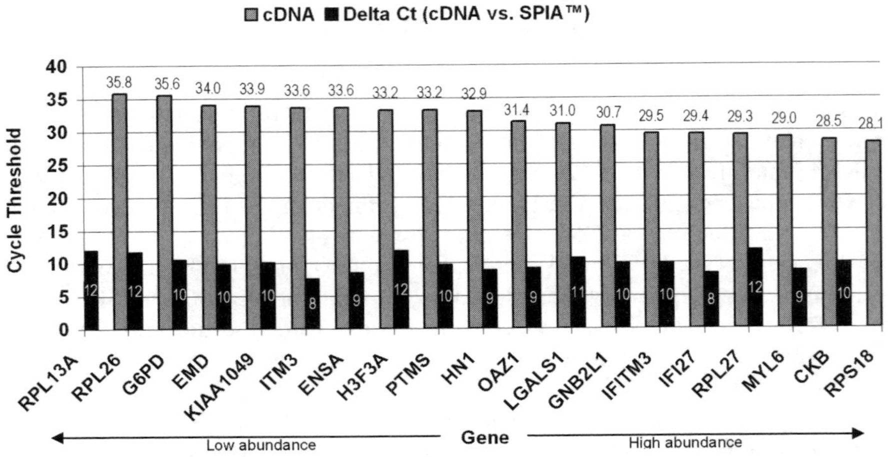

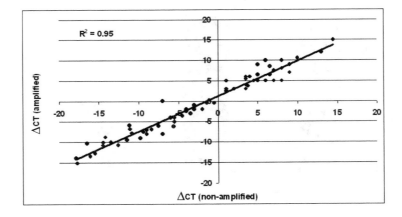

□ cDNA ■ Delta Ct (cDNA vs. SPIA™)

FIGURE 18.4 Accuracy of Ribo-SPIA™ amplification of different genes. Transcripts from 19 genes were quantified before and after amplification using real time PCR. The cutoff threshold (Ct) before amplification for cDNA from each transcript and the extent of amplification, delta Ct, the change in Ct after amplification are shown. The amount of amplification remains consistent over a broad range of inputs.

nonamplified or amplified cDNA. Figure 18.5 shows an excellent linear relationship for $\log_2$ of relative expression levels between the two samples before or after amplification (expressed as difference in threshold cycle Ct) over a range of 30 Ct, or 9 orders of magnitude, with a correlation coefficient R^2 of 0.95. Thus Ribo-SPIA provides a reliable representation of changes in transcript abundance in nonamplified

FIGURE 18.5 Differential expression of 68 genes before or after amplification. $\log_2$ of relative expression in UHR compared to human skeletal muscle (expressed as difference in Ct) is compared after (vertical axis) or before amplification.

mRNA. Preamplification by Ribo-SPIA before qPCR analysis, as illustrated here, is especially useful in multiplexed gene expression analysis in general, as well as microfluidic applications. In cases where a small sample must be divided extensively, some aliquots may not contain even a single molecule of the template mRNA. Linear preamplification of the sample is then indispensable to provide statistically meaningful amounts of template in each aliquot for transcript quantification. The amplification product generated by the Ribo-SPIA procedure can be directly quantified by any quantitative PCR method, as it is amplified cDNA, and provides sufficient material for quantification of hundreds of transcripts per sample.

EXPRESSION ANALYSIS USING THE Ribo-SPIA AMPLIFICATION AND GeneChip ARRAYS

FRAGMENTATION AND LABELING OF RIBO-SPIA PRODUCT FOR MICROARRAYS

Labels for detection of Ribo-SPIA product on microarrays may be introduced using any of the conventional techniques for labeling of DNA. Platforms such as Affymetrix® GeneChip® arrays and GE CodeLink arrays also require fragmentation of the labeled nucleic acid before hybridization. The Ovation™ Biotin kit provides a rapid, simple procedure, and reagents for fragmentation and biotin labeling of Ribo-SPIA cDNA products. The process involves only two 30 min incubations, followed by purification to remove unincorporated label (Table 18.1) and thus is readily amenable to automation. Figure 18.6 shows a typical analysis of the size distribution of products on an Agilent BioAnalyzer, illustrating the fragmented material with a peak at about 75 bases and most material below 200 bases in length. For comparison,

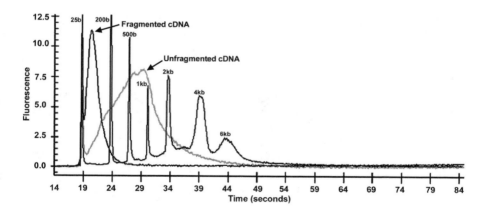

FIGURE 18.6 Bioanalyzer analysis of Ribo-SPIA product before and after labeling and fragmentation. Sharp peaks are internal standards at 25, 200, 500, and 1000 bases; 2kb, 4kb, and 6 kb. Unfragmented Ribo-SPIA product gives a very broad peak with maximum intensity typically in the 500–1000 base region. Fragmented material gives a narrow distribution with maximum intensity around 75 bases.

unfragmented material gives a broad peak, mostly below 2 kb but extending to about 6 kb. Biotin incorporation is easily demonstrated by a gel-shift assay. Product aliquots are incubated with excess streptavidin and resolved by gel electrophoresis using a nondenaturing gel. Gel electrophoresis analysis of replicate fragmented products demonstrated that most of the fragmented and labeled products were shifted to larger size by reaction with streptavidin. The extent of labeling may be quantified by image analysis. The gel-shift analysis is also useful to determine qualitatively the reproducibility of fragmentation, but does not provide an accurate picture of fragment size because it is performed under nondenaturing conditions.

GeneChip Hybridization Results

Ribo-SPIA performance on microarrays was tested most extensively on Affymetrix® GeneChip arrays because this system has come to occupy a central role in expression analysis. Fragmented and labeled targets, produced by amplification from 20 ng of UHR total RNA as described above, performed as well on the HG-U133A GeneChip arrays as cRNA produced from 10 μg total RNA input by the T7-based standard Affymetrix protocol [11] even though only 2.5 μg cDNA was used for hybridization compared to 10 μg for cRNA. Reproducibility of gene expression analysis on the U133A GeneChip arrays was excellent for replicate-independent target preparations. Figure 18.7 compares signals from two independently generated Ribo-SPIA products made by amplification of 20 ng total UHR, giving a signal correlation R^2 of 0.992.

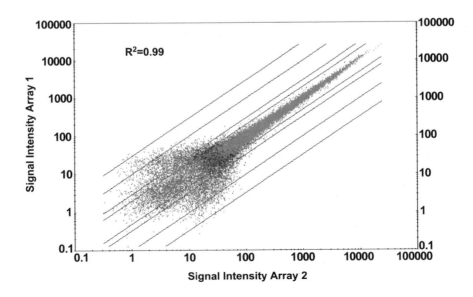

FIGURE 18.7 Signal correlation on Affymetrix HG-U133A GeneChip between two biological replicates amplified by Ribo-SPIA.

TABLE 18.2
Signal Correlation and Call Concordance between Independent Ribo-SPIA Reactions as a Function of Total RNA Input

a. Signal Correlations (R^2)

	1 ng	5 ng	20 ng	100 ng
1 ng	0.98			
5 ng	0.98	0.99		
20 ng	0.96	0.98	0.99	
100 ng	0.91	0.94	0.96	0.98

b. Call Concordance (%)

	1 ng	5 ng	20 ng	100 ng
1 ng	87.3			
5 ng	87.4	89.8		
20 ng	85.9	88.7	90.9	
100 ng	85.8	88.6	89.5	89.3

Signal correlation coefficients (R^2) were 0.98 to 0.99 over the entire range of 5 to 100 ng for comparison of duplicates at the same input and as high as 0.94 for correlation between 5 and 100 ng total RNA input (Table 18.2). Call concordance over the same range is also highly reproducible, ranging from 89 to 91%. Although outside of the range currently recommended for the Ovation™ Biotin system, excellent reproducibility was achieved with as little as 1 ng total RNA input (signal correlation coefficient of 0.98 and call concordance of 87% for replicate-independent amplification products).

Gene expression profiling with the Ovation™ system was further shown to be highly accurate and comparable to that achieved with targets generated by the standard T7-based protocol. The standard protocol represents GeneChip array-based gene expression analysis of samples after relatively little amplification, as these are carried out with high-input total RNA (10 μg total RNA per target preparation reaction). Three replicate cRNA targets were prepared using the standard protocol and hybridized to U133A GeneChip arrays, all by an independent laboratory, yielding an average Present Call = 51.5 ± 1.7%. Hybridization of 13 Ribo-SPIA-derived cDNA targets yielded an average Present Call = 53.6 ± 2.7%. Most other quality-control metrics were in the range defined as acceptable by Affymetrix®, including 3′/5′ ratios less than 3 for GAPDH, except for an average value of about 20 for the 3′/5′ ratio for beta-Actin. Increases in 3′/5′ ratios have also been noted in T7-based amplification of small (10–30 ng) samples and are attributed to decreased average length of the product with increasing amplification [6].

The high efficiency and accuracy of the Ribo-SPIA single-cycle linear-amplification method is also demonstrated by the high degree of call concordance on the U133A GeneChip arrays between targets prepared by Ribo-SPIA amplification of mRNA from 20 ng total RNA as compared to targets prepared by the T7-based standard protocol. Call concordance of greater than 85% was demonstrated between targets prepared by the two methods. Similar call concordances for Ovation™ amplification targets and T7-labeled cRNA targets were obtained for RNA samples from various other tissues and cell lines including human liver, placenta, skeletal muscle, and spleen.

DIFFERENTIAL GENE EXPRESSION USING THE RIBO-SPIA METHOD AND GENECHIP ARRAYS

More extensive data were obtained in an experiment comparing Ribo-SPIA to the standard T7 Affymetrix® procedure, using a series of mixtures of placenta and spleen total RNA as input. The samples employed for this experiment were generated by mixing the two total RNA samples to generate a series of samples with 100% placenta, 90% placenta-and-10% spleen total RNA, 50% of each, 10% placenta-and-90% spleen, and 100% spleen total RNA. Targets for expression profiling on U133A GeneChip arrays were generated by the Ovation™ Biotin system (20 ng total RNA input) and Affymetrix® standard T7-based protocol (10 μg total RNA input), in triplicates. A total of 15 GeneChip arrays were hybridized with the corresponding 15 target preparations generated with each of the two methods (3 replicates for each of 5 mixed total RNA samples).

Signal correlation coefficients, R^2, were calculated using all possible pairs of arrays for each sample or mixture. Average signal correlation coefficients, representing the reproducibility of independent target preparations by either procedure, were similar for targets obtained by the Ovation™ Biotin system and T7 cRNA targets, with average $R^2 = 0.981$ for the Ovation™ system and average $R^2 = 0.988$ for the T7-based standard Affymetrix® protocol, in spite of the 500-fold difference in RNA input (20 ng vs. 10 μg). Excellent call concordance between the Ribo-SPIA and T7 products prepared from this large number of targets was also demonstrated, as shown in Table 18.3.

The accuracy of Ribo-SPIA amplification of mRNA for differential gene expression determination with the GeneChip high-density array was confirmed by correlation of relative expression levels. Highly reproducible differential gene expression determination, $R^2 = 0.94$, using the Ovation™ Biotin system was demonstrated by correlation of Log_2 of signal ratios obtained with independently prepared replicate samples, as shown in Figure 18.8 (UHR vs. liver total RNA). Targets prepared with the Ovation™ Biotin system and targets generated with the Affymetrix® method were also well correlated, with $R^2 = 0.83$, representing the correlation

---

TABLE 18.3
Call Concordance (%) between cRNA and Ovation™ cDNA

Sample (Total RNA Composition)	Call Concordance (Total)
100% Placenta	89.7% (17387)
P90% & S10%	89.1% (17290)
P50% & S50%	88.7% (17346)
P10% & S90%	88.2% (17571)
100% Spleen	88.7% (17756)

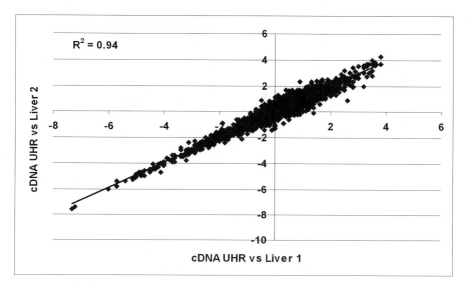

FIGURE 18.8 Correlation of differential expression measurement between two replicate Ribo-SPIA amplifications. Log_2 of signal ratios for each transcript in UHR vs. human liver total RNA are plotted for replicate reactions.

of gene expression profiling in amplified and nonamplified samples, as shown in Figure 18.9 (UHR vs. liver total RNA).

RIBO-SPIA LINEARITY

Linearity of mRNA amplification by the Ribo-SPIA technology and target preparation by the Ovation™ Biotin system was assessed by amplification from mixtures of placenta and spleen total RNA samples, as described above, and analysis of gene expression levels using the U133A GeneChip arrays. Amplification linearity was assessed from the signal correlation as a function of dilution of the placenta total RNA into spleen total RNA in a sample. Placenta-specific transcripts were identified by comparing array results obtained from triplicate independently prepared targets of placenta and spleen total RNA, selecting those that were in full concordance on all triplicate preparations by the Ovation™ Biotin system (20 ng input) and T7 standard protocol (10 μg input), and gave signal > 200 for the placenta-only sample. Excellent linear correlation of signals as a function of % placenta total RNA in the sample (mixed with spleen total RNA) was obtained with an average $R^2 = 0.922$ gained for 127 placenta-specific transcripts. Similar linear correlation was obtained for targets obtained by the T7 standard protocol, with $R^2 = 0.885$. Examples of the linear correlation of signals of three placenta-specific transcripts as a function of dilution (% placenta RNA in a sample) are shown in Figure 18.10.

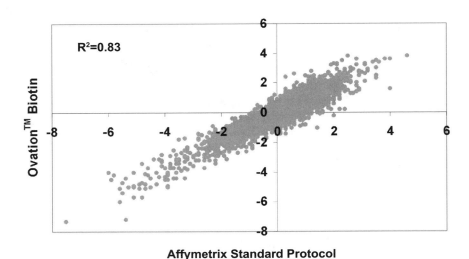

Affymetrix Standard Protocol

FIGURE 18.9 Correlation of differential expression between Ovation™ Ribo-SPIA amplification and the Affymetrix® method. Tissues and plots are as described earlier.

PERFORMANCE OF Ribo-SPIA-AMPLIFIED PRODUCT ON TWO-COLOR SPOTTED ARRAYS

Ribo-SPIA product can easily be labeled with multiple dyes by incorporating aminoallyl-dUTP during amplification. The Aminoallyl side chains can then be reacted with derivatives of Cy3 and Cy5 or other dyes using standard methods. The Ovation™ Aminoallyl kit includes all the reagents required to produce Aminoallyl product suitable for reaction with any dyes appropriate for the experiment. As with biotin, this kit enables the amplification of mRNA in 5 to 100 ng total RNA to give microgram quantities of labeled product ready for hybridization within a single day. Performance was assessed, using the Agilent Human 1A (v2) Oligo Microarray Kit. Replicate Ribo-SPIA amplifications of UHR total RNA were labeled with either Cy3 or Cy5, then hybridized and scanned. Signal correlation between slides was excellent, with $R^2 = 0.96$ to 0.98 for comparison of duplicates.

CONCLUSIONS

Gene expression profiling often requires more input mRNA than is available from the sample, especially when analyzing very small samples representing homogeneous cell populations. The ability to linearly amplify all mRNA species in a sample enables transcriptome analysis from small samples and enhances the sensitivity of analysis of very low abundance transcripts. Achieving the ability to determine expression profiles of homogeneous well-defined cell populations greatly increases the

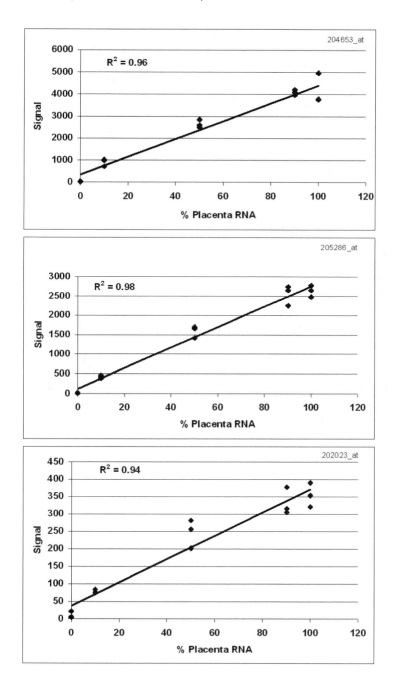

FIGURE 18.10 Examples of linear correlation between signal measured on GeneChips for placenta-specific transcripts and the percent of placenta RNA in the mixed placenta-spleen input.

reliability of our understanding of cellular processes. In order to achieve accurate gene expression profiling from small samples, it is also imperative to employ a global amplification methodology that is linear, nonselective, and capable of providing sufficient efficiency in a single round of amplification.

The present report describes a rapid, highly efficient linear RNA amplification methodology, Ribo-SPIA, which, when applied to mRNA, satisfies these requirements. Unlike current transcription-based mRNA amplification methods, Ribo-SPIA generates single-stranded cDNA products suitable for direct analysis using any of the currently available analysis platforms, in a single round of amplification. Global mRNA amplification is dependent on the generation of unique, partially double-stranded cDNA, which serves as a substrate for single-primer isothermal amplification using a DNA/RNA chimeric primer, and employs a DNA polymerase with strand-displacement activity and a RNase H enzyme. The simple, highly efficient and robust procedure enables amplification of mRNA from as little as 5 ng total RNA, yielding microgram amounts of cDNA in less than 4 h. The linearity and accuracy of the Ribo-SPIA amplification method was demonstrated by quantification of amplification products, using both quantitative PCR and hybridization to GeneChip arrays. The results demonstrated high correlations between amplified products and nonamplified samples, an indication of high amplification fidelity.

The simple procedure and speed of the Ribo-SPIA amplification technology, combined with a straightforward labeling and fragmentation procedure, renders this technology suitable for high-throughput gene expression profiling of all samples, large and small. Options are available, tailored for Affymetrix® GeneChip arrays, for spotted arrays, and for preamplification before quantitative PCR. Because of its operational simplicity, the technology should be readily amenable to automation. Further advances of this new technology are expected to enable single-cell expression-profiling analysis on the one hand, and the development of procedures and reagents for the amplification of the total-sequence content of the transcriptome.

MATERIALS AND METHODS

MATERIALS

Universal Human Reference total RNA and HeLa total RNA were obtained from Stratagene (La Jolla, CA), human spleen and placenta total RNA were from Ambion (Austin, TX), and human skeletal muscle total RNA was from Clontech (Palo Alto, CA). RNA quality was confirmed using an Agilent 2100 Bioanalyzer (Agilent, Palo Alto, CA). PCR primers were from QIAGEN (Valencia, CA) or Integrated DNA Technologies (Coralville, IA).

RIBO-SPIA

Ribo-SPIA amplifications were performed with the Ovation™ Biotin RNA Amplification and Labeling System for GeneChip experiments or the Ovation™ Aminoallyl RNA Amplification and Labeling System for all other experiments (NuGEN Technologies, Inc., San Carlos, CA), as described by the manufacturer [12].

AMPLIFICATION AND LABELING BY *IN VITRO* TRANSCRIPTION

RNA samples were labeled, using the Enzo® BioArray™ High Yield RNA Transcript Labeling Kit. All labeling reactions using *in vitro* transcription were carried out by Expression Analysis (Durham, NC), an Affymetrix® authorized GeneChip service provider.

QUANTITATIVE PCR (qPCR)

Quantification of amplified and nonamplified HeLa, UHR, and skeletal muscle mRNA for the study of amplification fidelity, and the fidelity of differential expression determination before and after Ribo-SPIA amplification, used detection by SYBR green. PCR primer pairs were designed using Primer Express® software (ABI, Foster City, CA). RNA samples were reverse-transcribed into cDNA, then amplified using Ribo-SPIA™ reagents. Aliquots of the reaction mixtures, following the second-strand cDNA synthesis step and after amplification, were diluted into TE buffer, and transcripts present were quantified using the QuantiTect® SYBR® Green PCR Kit (QIAGEN®), following manufacturer's instructions. Real-time PCR reactions were monitored using an MJ Opticon® (MJ Research, Waltham, MA). Data on reproducibility of amplification, as determined by the quantification of GAPDH cDNA (Figure 18.3), were obtained with TaqMan primers and probes (ABI, Foster City, CA), monitored on an ABI Prism™ 7700 Sequence Detector, as described by the manufacturer.

AFFYMETRIX® GENECHIP ANALYSIS

Samples were labeled either with the Ovation™ Biotin kit or the Affymetrix® standard protocol and hybridized to HG-U133A GeneChip® arrays (Affymetrix, Santa Clara, CA). Hybridization, staining with streptavidin-phycoerythrin having antibody amplification, and scanning were carried out according to the manufacturer's protocols, with the following exceptions: (1) only 2.5 μg per chip of cDNA target prepared by the Ovation Biotin system was hybridized to the GeneChip array, compared to 10 μg cRNA; (2) target denaturation before hybridization was for 2 min at 99°C, and (3) hybridization was carried out for 20 h. cRNA samples, labeled by the Affymetrix® protocol, were prepared, hybridized, and scanned by Expression Analysis, following their standard protocol. Array data was analyzed using MAS5 software (Affymetrix).

ANALYSIS OF OVATION™ AMINOALLYL KIT PRODUCT ON AGILENT MICROARRAYS

Product aliquots from amplification of 20 ng UHR were reacted with either Cy3 or Cy5 NHS esters (Amersham, Piscataway, NJ) following manufacturer's instructions. Extent of labeling was determined by UV-visible spectroscopy. Samples (0.75 μg) were hybridized to Agilent Human 1A arrays as recommended by Agilent, except that slides were washed after hybridization with 6x SSPE, 0.005% sarcosine, then with 0.06x SSPE. Arrays were scanned with a GenePix 4000B scanner (Axon Instruments, Hayward, CA). Reproducibility was evaluated from log–log plots of unfiltered signal.

ANALYSIS OF RIBO-SPIA PRODUCT BY BIOANALYZER

Aliquots of product before or after fragmentation and labeling (about 100 ng in 1 μl water) were analyzed using Agilent RNA 6000 Nano Chips and the RNA Nanosmear program on the 2100 Bioanalyzer. This option gave the most consistent results for the single-stranded cDNA product. Accordingly, the RNA 6000 Ladder (Ambion, Austin, TX) was used as internal standard.

DETERMINATION OF BIOTIN INCORPORATION BY GEL SHIFT ANALYSIS

Two sample aliquots were prepared, each containing about 50 ng of fragmented, labeled cDNA in 3 μl of TE. Streptavidin solution (3 μl of 2.5 mg/ml Sigma streptavidin (Cat # S4762 in water) was added to one and allowed to incubate for at least two minutes. Glycerol-bromophenol blue loading dye (3 μl) was added to each aliquot, and solutions were loaded on a 4 to 20% TBE PAGE gel (Novex pre-cast, Invitrogen, Carlsbad, CA) with 1x TBE running buffer. The gels were stained with SYBR Green II RNA gel stain (Molecular Probes, Eugene, OR) and imaged, following manufacturer's instructions.

ACKNOWLEDGMENTS

We are pleased to acknowledge Liz Robertson's assistance in preparation of the figures.

REFERENCES

1. Schulze, A., Downward, J. Navigating gene expression using microarrays — a technology review. *Nat Cell Biol* 3: E190–E195, 2001.
2. Dolter, K.E., Braman, J.C. Small-sample total RNA purification: laser capture microdissection and cultured cell applications. *Biotechniques* 30: 1358–1361, 2001.
3. Emmert-Buck, M.R., Bonner, R.F., Smith, P.D., Chuaqui, R.F., Zhuang, Z., Goldstein, S.R., Weiss, R.A., Liotta, L.A. Laser capture microdissection. *Science* 27: 998–1001, 1996.
4. Van Gelder, R.N., von Zastrow, M.E., Yool, A., Dement, W.C., Barchas, J.D., Eberwine, J.H. Amplified RNA synthesized from limited quantities of heterogeneous cDNA. *Proc. Natl. Acad. Sci. USA* 87: 1663–1667, 1990.
5. Eberwine, J., Yeh, H., Miyashiro, K., Cao, Y., Nair, S., Finnell, R., Zettel, M., Coleman, P. Analysis of gene expression in single live neurons. *Proc. Natl. Acad. Sci. USA* 89: 3010–3014, 1992.
6. Luzzi, V., Mahadevappa, M., Raja, R., Warrington, J.A., Watson, M.A. Accurate and reproducible gene expression profiles from laser capture microdissection, transcript amplification, and high density oligonucleotide microarray analysis. *J. Mol. Diagn.* 5: 9–14, 2003.
7. Wilson, C.L., Pepper, S.D., Hey, Y., Miller, C.J. Amplification protocols introduce systematic but reproducible errors into gene expression studies. *BioTechniques* 36: 498–506, 2004.

8. Zhumabayeva, B., Chenchik, A., Siebert, P.D., Herrler, M. Disease profiling arrays: reverse format cDNA arrays complementary to microarrays. In Scheper, T., *Adv. Biochem Eng Biotechnol* 86: 191–213, 2004.

9. Seth, D., Gorrell, M.D., McGuinness, P.H., Leo, M.A., Lieber, C.S., McCaughan, G.W., Haber, P.S. SMART amplification maintains representation of relative gene expression: quantitative validation by real time PCR and application to studies of alcoholic liver disease in primates. J. *Biochem. Biophys. Methods* 55: 53–66, 2003.

10. Dafforn, A., Chen, P., Deng, G.Y., Herrler, M., Iglehart, D.M., Koritala, S., Lato, S.M., Pillarisetty, S.V., Purohit, R., Wang, M.J., Wang, S., Kurn, N. Linear mRNA amplification from as little as 5 ng total RNA for global gene expression analysis. *Biotechniques*, 37: 854–857, 2004.

11. *GeneChip® Expression Analysis Technical Manual*. Affymetrix, Santa Clara, CA. 2001.

12. http://www.nugeninc.com (accessed August 2004).

19 Genomics, Transcriptomics, and Proteomics: Novel Detection Technologies and Drug Discovery

Phillip Stafford

CONTENTS

BACKGROUND AND INTRODUCTION

What drives the development of novel technological innovations in the biomedical field? What is the catalyst that makes researchers notice the lack of some ability or capacity? What is the fundamental element that fosters the inventive spirit? For example, the expression microarray answered a pressing need; observing expression of mRNA was difficult and time-consuming and was low density. Researchers wanted to know expression levels over the entire transcriptome for several different biological phenotypes but who first asked this question? Is the quest for new block-buster drugs responsible for this tidal wave of biological data, or are we as scientists simply asking increasingly detailed questions that require much broader views of the cell?

We are enmeshed in the process of data accumulation precisely because we have adopted high-throughput measurement technologies. What was luxury has become essential. We routinely design experiments that measure miniscule amounts of biological molecules, sometimes obtained from a mere handful of cells that were laser-captured, and often measured at the very limit of detection. We often find ourselves tipping the scales away from thoughtful and elegant experimental design toward rather blunt methods of data accrual and high-throughput analysis because our hypothesis is open ended. How do we turn all of this biological data into insight and foster inspiration and discovery? How do we leverage new technology and become truly translational?

First and foremost, we should emphasize that disease has been, and likely will continue to be, the single major impetus behind most of the biological advances in the 21st century. Although research into environment, ecology, population biology, and other high-level aspects of biology will continue to provide us with a view into the natural world at a macroscopic level, much of the funded research from the National Institutes of Health and other government agencies continues to use human disease as the biological paradigm. Pharmaceutical companies have invested enormous resources into finding the next blockbuster drug for the most prevalent and dangerous human diseases. The growth of systems biology has certainly aided this search, via the development of novel algorithms and data representations that allow difficult or intractable biological experiments to be conducted *in silico* (i.e., PhysioLab from Entelos, www.entelos.com). It is the engineering advances in miniaturization, however, that have played a pivotal role in the advancement of genetic research. Much of our current understanding of oncogenesis was made possible using the high-density expression microarray. As new technologies utilize microminiaturization and high-density molecular probes, they will continue to ensure that glimpses into the complex molecular interrelationships within the cell will prove fruitful at a very practical level. Disease is a dynamic and often host-specific biological state, and will likely guarantee that personalized medicine is the next great medical development. Microarrays and other high-density/high-throughput devices will play an increasing role in helping to convert today's generic treatment regimes into more personalized methods that accommodate individual genetic factors. This will help reduce the risk of off-target effects, increase the chance of killing unwanted cells, and provide a more effective combination of drugs that work synergistically.

Microarray technology, miniaturization of fluidics pumps, high-intensity quantum dots, novel fluorophores, and new detection technologies have been driven by a need to measure cellular functions at many levels or dimensions. Researchers have found that measuring a cell really means measuring a dead cell frozen in time. That cell represents a phenotype of the cell at a particular point in its life. This has some drawbacks including measuring the effects of killing the cell, and measuring molecules that supposedly represent the cell immediately before its death; molecules are dynamic, always changing, shifting, denaturing. Measuring a living cell means watching a cell accommodate perturbations. Unfortunately, measurements taken within living cells are often prone to nonspecific responses that result from introduction of molecular probes that are abiotic or even toxic, illustrating a biological corollary of Heisenberg's Uncertainty Principle — one can thoroughly visualize a cell, but what you see may not represent the cell in its native state. Many of these problems can be accommodated through proper experimental design. Let us refer to a case where a researcher wants to measure a cell by extracting its DNA, RNA, and proteins. In this case, each molecular species is a dimension, and here each dimension can be thought of as an "-ome" (Figure 19.1). Biologists have always had the desire to look at more than one dimension of the cell at a time to determine whether each aspect correlates well with other aspects; whether the gene sequence causes a direct and predictable mRNA quantity; whether mRNA causes a predicted quantity of a single species of protein, whether proteins are quantitatively modified, etc. As we are discovering, the cell is much more complex than we imagined. These different biological molecules represent the historic traces of cellular process. The traces are often the remnants of complex molecular interactions that in essence drive an energetically unfavorable process within the cell's life. The traces also record the cell's attempt to return to an equilibrium state after perturbation. The scientist

FIGURE 19.1 Biological life can be described in terms of complexity. At the genome level, a single molecule of DNA (error rate: 10^{10}) defines the code necessary for life. RNA, a more dynamic copy of DNA, adds complexity, functionality, and error (error rate: 10^{-4} to 10^{-5}). Proteins (error rate: 10^{-4}) are even more functional and dynamic than RNA and have an even higher error rate, but the process of translation absorbs mistakes through many inherent accommodations. As complexity grows, so does the amount of information contained within the molecular profile. Protein–protein networks combine to drive signaling, metabolic pathways, and functional specialization in cells. Tissues are combinations of specialized cells, and organs are complex connections of specialized tissues. Organisms are functional groups of optimized organs working for mutual survival, and populations are groups of individual organisms interacting and often modifying the local environment and other organisms to form functional groups.

measures the endpoint of a cell, a state that cannot return to a normal equilibrium because of the interference of a drug, or because of disease (cancer being a prime example). Often, we wish to understand the behavior of a cell in response to an environmental insult, but for years we were unable to visualize more than a few molecular interactions at one time, and these were often transitory and not fully captured by our technology. Researchers can stop the process of transcription (actinomycin D or edeine) or translation (clotrimazole, cyclohexamide) and study cellular responses, but these drugs often have nonspecific or epistatic effects that may confuse the transcriptional picture. The advent of the microarray, attributable to many labs simultaneously in the 1980s (which is itself an interesting story), has led to an enormous leap forward in detecting the presence of an entire "-ome" in a homogenous population of eukaryotic cells. This was followed closely by another "-ome"; the entire *S. cerevisiae* genome was the first fully sequenced eukaryote and now another biomolecule was thoroughly known, if not completely understood. We then had the ability to measure two "-omes" completely and with reasonably high accuracy, resulting in discoveries that could only have been made in the context of the genome and transcriptome. Because the knowledge gained from observations of only two biomolecules has been so synergistic, the push is on to develop high-resolution, high-density microarrays and microdetection technologies that measure genomic polymorphisms (variome), protein abundance (proteome), protein interactions (interactome), modified molecules (e.g., methylation arrays, CpG islands, glycosylation and other posttranslational modification of proteins), and even siRNA and drug interaction arrays. The gain in insight as we added more observations has been much more than additive, and we fully expect that continuing this trend will increase our insight into complex causal relationships even more. Early prototype protein abundance and siRNA arrays have led to rapid progress in the field of systems and integrative biology, where classification and prediction algorithms can now fully accommodate interactions between thousands of biological molecules. These methods are far from perfect, but to simply compare the state of knowledge now with where we were 5 years ago gives us an understanding of the tremendous value of systems biology and high-throughput screening. As biological databases grow, the benefits reaped are exponentially greater than the investment, and we are now in an ideal position to leverage public data as we search for the next blockbuster drug.

FOSTERING BIOLOGICAL KNOWLEDGE: CROSS-DISCIPLINE DATA

The development of high-density, high-resolution microarrays and microfluidics devices has given us a wealth of data but has simultaneously created a need for data storage, data representations, and integration. We most often see data stored in native formats (see www.ncbi.nlm.nih.gov/geo/ as an example of storing expression, CGH and SNP arrays). This is the most intuitive method for posting and retrieving data and NCBI has done a good job in providing the necessary infrastructure. However, we need to consider new data representations, a transformation of our current observational data into a new format that is both unified and consistent, yet fully represents the information contained within each of the "-omes" that we measure.

The cancer bioinformatics grid (caBIG, http://cabig.nci.nih.gov) at the National Cancer Institute is working diligently to create new biological data representations, to create databases that allow sharing and interaction from one class of biological observations to another, and to allow users to integrate their own software with the highly open and utilitarian software being developed by caBIG. Other efforts exist, and many will flourish, but time will filter out those not destined to become a universal standard.

UNIFYING THE BIOLOGICAL (DATA) WORLD

Biological data centers are collections of biological data sets. Some of the most widely used are those that represent a genome (Genebank), polymorphisms (dbSNP), protein structure and function (Swiss-Prot, UniProt, TrEMBL, BIND, PDB), synthetic lethal (gene interaction) data, siRNA libraries (Dharmacon, Ambion, Qiagen, and see www.proteinlounge.com), and many more. Each of these data centers is a repository for native formats of a particular measurement, whether DNA sequence, mRNA abundance, protein mass spectrometry data, protein abundance, etc. Each of these elements is measured as fluorescence level, binary presence/absence, mass spectrometry data, even fluorescence images or tissue microarrays. With this much variation across data types, it is difficult to assign a new numeric scale that fully represents the range of data, or the implications and subtleties that are inherent in each measurement. Quantized or Boolean data (binomial, trinomial, or other discrete representation) can describe several conditions: state change, scalar range, affectation status, cell growth, drug response, etc., and may prove to be the least loss-prone and most cross-platform representation available. It would be highly useful for data integration and analysis, and it is an appropriate representation for manipulations in Boolean space, feature selection and classification methods, principal components, naïve Bayes, maximum likelihood estimators, etc. Thus, as long as the methods for converting data into a discrete range of values, representing as much of the data's original dynamic range as possible are applied, the inherent problem of losing precision is overcome by the flexibility of combining multiple data types. The combination of a federated system of biomedical databases, a standards-driven metadata server, and remote databases throughout the physical world logically connected, coupled with intelligent data transformation will lead us eventually to the integration and application of multidisciplinary biological data.

THE LEGACY OF EXPRESSION ARRAYS

Expression arrays have provided much useful information about the transcriptional state of a cell. This single bit of information about mRNA abundance continues to add to our understanding of cellular response to perturbations, disease progression, gene regulatory networks, and metabolic pathways. It has had a direct impact on prediction of tumor development, disease susceptibility and recurrence, drug response, and other medical consequences of gene expression. Today, we are using expression data to identify the best anticancer drugs for late-stage cancer patients, and to predict whether a tumor may become metastatic (see www.agendia.com

for their X-Print technology). However, to be even more comprehensive, we could expand our knowledge of the disease by using such resources as arrayCGH, protein abundance arrays, tissue microarrays, and SNP arrays. If we knew that a region of chromosome is amplified using arrayCGH, we might also suspect some form of cancer development. We can validate that genes implicated in cancer progression are expressed or mutated through expression and SNP arrays. We can say with a little more certainty that the chromosome amplification event was important in the transformation from healthy to diseased state. Using these array-based technologies and highly parallel informatics methods, we can now look at how tumors respond to drugs and drug combinations (see www.combinatorix.com for examples) in the context of the patients' genome, and we can start to make a direct impact on patients' lives.

All of the aforementioned technologies are considered high density and most are fairly high precision. As we celebrate the 18th birthday of the first commercial microarray and the 16th anniversary of the first microarray publication, we are seeing an increase in the amount and type of data that is required to identify complex disease phenotypes. Many of the early disease discoveries had strong genetic components — one gene, one phenotype. New research into complex multigenic diseases like Alzheimer's clearly shows that quite often a large number of genes interact with varying degrees of synergy to eventually contribute to disease susceptibility and/or progression. Thus, as our technology improves, our ability to successfully tackle highly complex and seemingly intractable disease improves. Additionally, the ongoing miniaturization of pumps, valves, heaters, and mixing chambers has brought sample preparation to a new level of simplicity. These microfluidics chips can process raw tissues acquired onsite and be directly plugged into any observational technology such as an expression, SNP, or protein microarray without the need of sophisticated laboratory environments. These automated systems are being developed for use in defense applications, where rapid identification of unknown biological compounds is important but clinical applications where speed and precision are important are also being targeted. Sample preparation was one of the biggest sources of variation for expression arrays; automation and miniaturization of tissue extraction systems now has the potential for reducing that variability to a minimal level, and moves the preparation away from the laboratory to almost anywhere on the Earth.

SYSTEMS BIOLOGY — DRIVING DRUG DEVELOPMENT THROUGH DATA INTEGRATION

Systems biology is driving the integration of many types of cellular observations into one mathematically defined model with the eventual goal of modeling the entire cell *in silico*. The goal of systems biology is to transform observations that range from transcript number, protein structure and abundance, drug efficacy, genome sequence, etc. into a form where one can create an algorithm that is sufficiently trained to accurately simulate cellular molecular profiles. It seeks to integrate the fundamental aspects of each type of data that makes that measurement unique and important. For example, in some cases, DNA sequence is key to whether an individual will respond to a drug or not, in other cases that same DNA sequence has no effect. Each measurement must be examined in the proper context, and that is the supreme challenge for systems biology. If the aforementioned data is analyzed individually, those results can be used to support or reject an alternate hypothesis on a case-by-case basis, but systems biology has expanded the analysis paradigm to a point where multidisciplinary, multivariate analyses can foster much more complex hypotheses than univariate models. The trouble with multidata-type analysis is the necessity of generating a specific, sophisticated hypothesis that accommodates the corresponding limitations or usefulness of our data. There is always a limitation to the depth and quality of metadata, and that limitation has been a significant bottleneck for many years, regardless of the efforts of groups like MGED (www.mged.org). However, the tremendous benefits and insight that should result from examination of complex disease traits using molecular profiles should more than offset the difficulty in mining the vast metadata repositories and establishing an appropriate and all-encompassing hypothesis.

CONTROLS AND STANDARDS

As the proliferation of new technology continues, little effort has gone into developing molecular standards for each of the measurement technologies. It is incumbent on adopters to identify a resource where known and fixed quantities of biomolecules are compiled. The metric system is well served by identifying standards for weight, length, and temperature and storing or accurately describing those standards so anyone anywhere can reproduce a metric measurement precisely. The National Institute of Standards and Technology (http://www.cstl. nist. gov/biotech/workshops/ ERCC2003/, http://www.cstl.nist.gov/biotech/workshops/ERCC2004/, and www.fda. gov/nctr/science/centers/toxicoinformatics/maqc/) in Gaithersburg, MD, hosts a grass-roots effort that is aimed at creating a series of spike-in controls for at least 100 well-characterized clones for expression arrays. The FDA is clearly interested in establishing a standards-based approach to microarray analysis because so many new and upcoming clinical devices rely on expression profiling. If one creates a series of spike-in experiments using known targets at three different concentrations (at least), one can calibrate any expression experiment in a simple manner, and obtain a precise measurement of the actual copy number by referencing fluorescence value of the spike-in experiments. This essentially means that the expression platform with the highest precision wins the game — accuracy would no longer be an issue, only repeatability. In this respect, at least in expression arrays, a manufacturer has already made reproducibility a high priority and has been making the highest-precision arrays for a number of years. Any detection technology that measures molecule abundance can and should be calibrated with artificial mixtures of biomolecules. Even self-calibration can be useful — Shyamsundar et al. [1] used genomic DNA as a way to accommodate differences in fluorescence in expression arrays. Genomic DNA should be present at a known concentration per gene, while the number of transcripts fluctuates. This ratio is not as precise a measure as a true mRNA spike-in, but clearly emphasizes the point that some internal or external control is needed to ensure precision translates into accuracy.

BIOMARKERS AND THE HISTORY OF GENETICS

For years, scientists have been investigating processes such as heritability, genetic linkage, patterns of linkage disequilibrium, and genetic instability in order to better understand complex heritable traits. Much of the early work focused on single-locus, single-allele human disease genes, as well as diseases of commercially important agricultural species. Phenotypes are marked by either discrete qualitative traits, similar to those identified by Gregor Mendel in the pea plant, or continuous quantitative traits, such as height or weight. In the fields of commercial agriculture, animal husbandry, host–pathogen interaction, and disease resistance, we find that many of the simple one-trait, one-gene models have been thoroughly studied, leaving only the complex multilocus models. Traits with many genes contributing to a single complex phenotype are taxing the resources of traditional geneticists and have led to the progression from basic genetics experiments (like the three-point cross) to whole-genome haplotype mapping.

Early genetic studies would analyze polymorphic markers that spanned the genome. These markers might have included variable nucleotide tandem-repeats (VNTRs or minisatellites, 15 to 100 bp), microsatellites or STRs (dinucleotide repeats or short tandem repeats), or RFLP (restriction fragment-length polymorphisms). These markers were often analyzed using fluorescent or radioactive gels and blots requiring large investments of time to analyze the combinatorial patterns of genetic inheritance within a family. Association studies required far too many samples and too much resolution to fulfill the promise of a true case/control study, so most early analyses examined pedigreed families with well-characterized histories (i.e., the CEPH family via Foundation Jean Dausset, www.cephb.fr/, or the Coriell Institute for Medical Research, locus.umdnj.edu/nigms/ceph/ceph.html). Many familial diseases have single-gene mutations, and most have been identified quite early in the history of human genetics. With the completion of the human and mouse genome, great advances in understanding gene and gene functions have propelled disease study forward. Unfortunately, many dangerous and poorly understood diseases are multigenic in nature, and some are mixtures of genetically defined disease susceptibility, environment, and health, and immunologically enhanced disease onset. The genetic load for these diseases ranges from 70% to 80% on the high end to a low 20%; enough to warrant research into genetic risk factors. Even risk factors themselves are incredibly hard to identify. Although Alzheimer's disease has been studied for years, only now are we accumulating enough data to clarify the distinction between the causative agents of the disease and the well-known risk factors that interact so closely with environment, such as APOE, beta amyloid production, cholesterol synthesis, and others.

These new diseases and disease paradigms require much more precision and depth than is possible using microsatellite markers. Although highly informative, microsatellites are too far apart to cover the resolution necessary for high-density genetic mapping. SNPs (single nucleotide polymorphisms) are abundant (11 million SNPs in humans with minor allele frequencies above 1%). These low-information, high-density markers that have been well-characterized are due in large part to the success of the human genome project. The principal benefit of using SNPs for mapping genetic diversity is the abundance. Out of 3.2 billion nucleotides, roughly one SNP/300 bases is available as a potential marker. This density is adequate to identify relatively small regions of the genome in association studies, and resequencing costs are dropping to less than a penny per genotype. Thus, as genetic studies continue and the dreams of personalized medicine continue to drive our research, we find that technology has supplied a number of highly sensitive, precise and economical devices that allow us to look at the linkage between phenotype and gene in almost any organism we choose, at high density in a short amount of time. It will be shown that data availability is not always the end; in fact a great deal of supercomputing, algorithm development and novel combinatorics may be required to make sense of our growing genetic databases.

SNP DETECTION

Single nucleotide polymorphisms, or SNPs, are DNA sequence variations that occur when a single nucleotide (A, T, C, or G) in the genome is altered relative to an allele that occurs most often in a particular population. For a variation to be considered a SNP, it must occur in at least 1% of the population, but populations are defined based on the scope of the experimenter. Often, SNPs are defined within a certain tiny population, such as a race or a geographically isolated group, and may not be considered SNPs in the larger world population. SNPs, which make up about 90% of all human genetic variation, occur every 100 to 300 bases along the 3-billion-base human genome. Two of every three SNPs involve the replacement of cytosine (C) with thymine (T). SNPs can occur in both coding (exonic) and noncoding (intronic) regions of the genome and may be present in transcription factor binding sites, promoters, enhancers, splice junctions, 3' UTR sequences, or other subtle functional locations. Many SNPs have no effect on protein structure, transcription, or gene replication and are simply markers that cosegregate with a functional polymorphism. The most widely studied SNPs are those that could predispose people to disease or influence their metabolic response to a drug. Several groups have worked to identify novel SNPs, and ultimately create maps of the human genome. Among these groups were the U.S. Human Genome Project (HGP) and a large group of pharmaceutical companies called the SNP Consortium (TSC) (http://snp.cshl.org and www.ncbi.nlm.nih.gov/SNP/). The TSC used a pool of 24 individuals' DNA across several racial groups who were incontrovertibly identifiable in order to identify the variation within and across subpopulations of human beings. The likelihood of duplication among the groups was quite small because of the estimated 3 million potential SNPs that could be identified. The international HapMap project (www.hapmap.org) continues to elicit data from and provide data to the international community from its groups of Japanese, Chinese, European-American, and African populations. The HapMap project seeks to establish the most common haplotypes that exist in populations worldwide. Haplotypes are those patterns of SNPs that are commonly inherited together. Haplotypes add depth to the normally quite sparse amount of information contained in a single SNP, and will allow researchers to spend less money on sequencing each and every SNP and concentrate instead on Tag SNPs — those polymorphisms that are dominant and unique to particular combinations of haplotypes. As data from more individuals continues to grow, our genetic databases grow in size and utility. NCBI's dbSNP database is now (build 126) over 10 million unique human SNPs and continues to grow.

Although SNPs are much less informative on an individual basis than microsatellites, the fact that 10 million validated SNPs exist throughout the human genome means that on average a useful validated SNP exists every 300 bases in the human genome, and likely there will be much more coverage as remaining SNPs are identified and annotated. Although some regions of the genome are SNP rich and others are very sparse, the sheer coverage allows scientists to design experiments that can identify and map small regions of the chromosome that show linkage disequilibrium. The field of pharmacogenomics specializes in those SNPs that are associated with an effect drug metabolism. The cytochrome P450 family of genes was an early target because several well-described P450 SNPs were associated with differential drug metabolism. Pharmaceutical companies could use this information to design clinical trials and eliminate physiological variation and to foster the development of personalized medicine.

SNPs are not ordinarily associated directly with disease, but they can help determine the likelihood that someone will develop a particular disease or may suffer more than usual from the deleterious effects of a disease. One of the genes associated with Alzheimer's, apolipoprotein E or *ApoE*, is a good example of how SNPs affect disease development. This gene contains two SNPs that result in three possible alleles for this gene: E2, E3, and E4. Each allele differs by one DNA base, and the protein product of each gene differs by one amino acid. Each individual inherits one maternal and one paternal copy of *ApoE*. Research has shown that an individual who inherits at least one E4 allele will have a greater chance of getting Alzheimer's during his or her lifetime. Inheriting the E2 allele, on the other hand, is strongly associated with a positive outcome, or a small likelihood of Alzheimer's.

SNPs are not absolute indicators for disease development; rather they show linkage between a marker and a phenotype. Someone who has inherited two E4 alleles may never develop Alzheimer's, but another who has inherited two E2 alleles may. *ApoE* is one gene that has been linked to Alzheimer's, but, as with many diseases, AD is polygenic. Many *cis*- and *trans*-acting effects from a broad panel of several genes contribute differentially to the disease. The polygenic nature of many of our most dangerous disorders is what makes genetics so complicated. The problem, until recently, was how to detect a large number of SNPs in a sizable case-control or pedigree-based experiment. Detection of SNPs was often done by direct sequencing of the individual regions of interest, or by PCR amplification. The need for measuring thousands or even millions of genotypes at a single time from many individuals has led to the development of SNP microarrays and other high-density SNP detection methods. In this chapter, we will investigate those companies and research organizations that have developed and/or are marketing SNP detection technologies, and what each platform offers in terms of flexibility, cost, density, and accuracy. Because of the nature of the technology, we will list some similar technologies, such as the protein array, simply because the detection technology in many cases is similar to SNP detection.

SOME DETECTION TECHNOLOGIES OF INTEREST

Several companies have now or are in the process of developing new technologies of interest to biologists and drug researchers. As such they should be addressed in a chapter about biological detection technologies. We begin this section by introducing a few companies that have contributed intellectual property (patents or concepts), but have not contributed to the technological or engineering development of a device. In the early years of genomics and transcriptional profiling, some companies pursued a goal of protecting some sweeping intellectual property that covered fairly generic ideas. Two such companies, Xenometrix (subsumed by Discovery Partners International, www.xeno.com) and Oxford Gene Technology (OGT, founded by Edwin Southern in 1995, www.ogt.co.uk) are worthy of mention because their business plan was based on licensing technology. OGT licensed the general "technology" or "concept" of expression microarrays [2,3], where spotted features on a slide would report the mRNA content of a cell. Xenometrix [4,5] had obtained a license for the general concept of "gene expression profiling," where eukaryotic cells are exposed to a pharmacological agent and the resulting expression patterns are analyzed for changes caused by the treatment effect. These companies are an interesting study in the history of microarray commercialization because they cleverly obtained broad and highly controversial patents on basic biological processes and research techniques, and have (or had, because Xenometrix is no longer a recognizable entity) no obvious external liabilities or development overhead. Questions arise on whether these global patents should have been awarded at all, as many believe they drive up the cost of basic research and drive down the incentive to improve existing technology. As patent offices have become more technologically savvy and resistant to scientific jargon, it is (gratifyingly) more difficult to obtain these sorts of basic methodology patents. Litigiousness aside, most companies are simply not finding it fruitful in the long run to remain in business only to sue investigators who are in the business to discover knowledge. It remains to be seen whether one of the last and very public business plans devoted to litigation (SCO, Santa Cruz Operation) can successfully sue all users of LINUX; the consequences of that battle may permeate the biotechnology field.

Research into disease development drives much of the research in molecular detection technology. One example is the study of aberrant methylation patterns that are common in all major human cancers. CpG islands are genomic elements that comprise about 2% of the human genome (or approximately 30,000 islands) and about half of these are found within the promoter region of genes. Normally the cytosines in CpG dinucleotides are unmethylated, but cytosines of CpG dinucleotides outside of CpG islands are methylated at the C5 position, a reaction catalyzed by DNA methyltransferase (DNMT). These mutations are frequently observed in the promoter and exon-1 regions of genes. CpG island hypermethylation can cause chromatin structures in the promoter to be altered, preventing normal interaction with the transcriptional machinery [6,7]. If this occurs in genes critical to growth inhibition (tumor suppressors), the resulting silencing of transcription could promote tumor progression and in fact has been seen in most solid tumors studied to date [8–11]. In addition to classic genetic mutations, promoter CpG island hypermethylation has been shown to be a fairly typical mechanism for transcriptional inactivation of classic tumor suppressor genes [6,8] as well as genes important for cell cycle regulation [6,8], and DNA mismatch repair [12].

The Human Epigenome Project (http://www.epigenome.org) aims to collect and partition genomewide DNA methylation patterns of all human genes in most major tissues, similar in many fundamental respects to the GeneAtlas and SymAtlas from the Genomics Institute of the Novartis Research Foundation (http://expression.gnf.org and http://symatlas.gnf.org). Many medical research centers specialize in identifying methylation patterns as part of their larger effort to characterize the molecular profiles that distinguish cancer types (http://www.mdanderson.org/departments/ methylation/). Several human CpG island arrays are available from small research houses but an important advance is the availability of MSOs, or methylation-specific oligos [13]. DMH (differential methylation hybridization) is run on tumor samples vs. normal tissues in order to distinguish potentially silenced (often tumor-suppressor) genes due to hypermethylation and is a unique technology that looks at biomolecular modifications that commonly lead to oncogenesis (Figure 19.2A, Figure 19.2B) [14].

NONSNP TECHNOLOGY: THE PROTEIN CHIP, MICROFLUIDICS, AND ASSORTED ENGINEERING MARVELS

Another rapidly developing technology is the protein microarray. Several manufacturers are scrambling to get a robust and inexpensive quantitative protein array to market in order to measure protein abundance. Many approaches have been attempted, because a protein abundance array is difficult and expensive to create. Several examples include protein–protein interaction, where a purified protein or hapten is immobilized on the surface of an array and proteins and protein complexes are allowed to hybridize to the haptens. Proteins are cross-linked to freeze their interaction state onto the array, and all other proteins are washed off. Mass-spectrometry is used to identify the proteins that bind. Antibodies are also used to bind specific proteins, and the relative abundance of bound protein is measured by fluorescence or other simple methods. Other approaches include HPLC/MS-MS to filter complex mixtures of proteins by size, ionic strength, or molecular weight before using MS to identify specific protein species. Further, as mentioned above with the methylation array, scientists are examining the potential for measuring posttranslational modifications using a technology such as the glycoarray. These arrays utilize a neoglycolipid conjugation technology to create a multivalence surface that allows sufficient and robust interactions to provide a strong signal-to-noise ratio.

Glycominds Ltd. (Lod, Israel, www.glycominds.com, Figure 19.2C) has developed several protocols for detecting carbohydrate moieties from a variety of substrates, including intact cells [15]. The concept of a "glycomics profile" is becoming widely accepted, as many debilitating diseases (such as Crohn's disease) are characterized by variant carbohydrate levels in the blood. Protein arrays face numerous challenges; narrowing the field of protein detection down to a clinically relevant

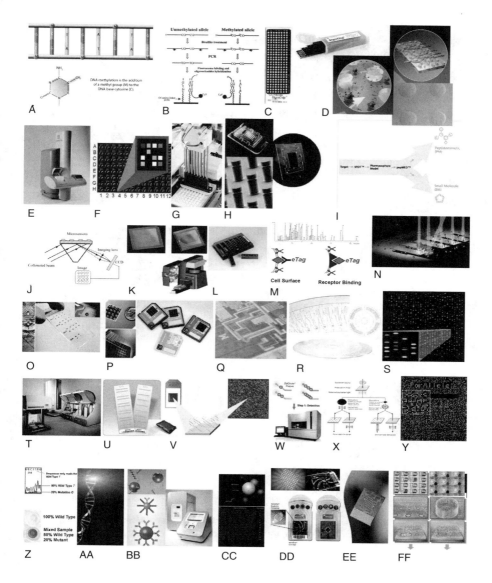

FIGURE 19.2 *(A color version follows page 204)* A: DMH (differential methylation hybridization). B: DNA Methylation. C: Glycominds, Ltd. D: Biocept. E: Ciphergen. F: High-Throughput Genomics. G: Protagen. H: Zyomyx. I: Jerini. J: Genoptics. K: HTS Biosystems. L: Zeptosens. M: Aclara/Virologic. N: Protiveris. O: Advalytix. P: Calipertech. Q: HandyLab. R: Gyros Systems. S: Xeotron. T: Sequenom. U: Illumina. V: Affymetrix. W: Applied Biosystems. X: PerkinElmer. Y: Parallele. Z: Tebu-Bio. AA: Orchid. BB: Nanosphere. CC: Lynx Therapeutics. DD: Nanogen. EE: CMS. FF: Combimatrix. GG: GeneOhm. HH: Nanoplex. II: BioArray Solutions. JJ: Luminex. KK: PamGene. LL: Metrigenix. MM: Solexa. NN: Graffinity. OO: Febit. PP: Genospectra. QQ: Epoch Biosciences. RR: Exiqon. SS: Nimblegen. TT: Perlegen. UU: ArrayIt. VV: Genomic Solutions. WW: Idaho Technology. XX: Asper Biotech. YY: Agilent Technologies.

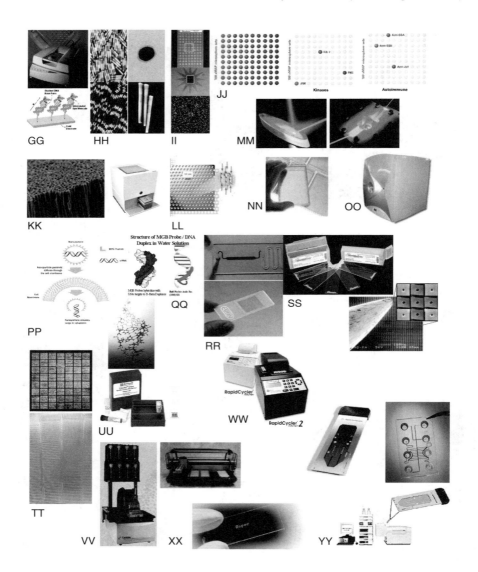

FIGURE 19.2 (Continued)

subset of posttransitionally modified proteins means the developer can create an assay that is extremely stringent and detects only glycosylated proteins.

Biocept (San Diego, CA, www.biocept.com, Figure 19.2D) has created a cell enrichment and extraction concept that extends beyond their original technology, primarily microarrays that utilized hydrogel pads (Figure 19.2D bottom). These original hydrogels were semifluid hemispherical microdroplets, distinct from both fluids and solid substrates. PEG (polyethylene glycol) was used as the polymer to form the matrix, and each microdroplet of hydrogel spanning 300 microns by

30 microns thick contained 10^{10} to 10^{11} molecules of a single probe. Although very promising during development because it allowed more fluid transfer across each probe than normal surface-spotting technologies, Biocept now embraces MEMS technology (MicroElectroMechanical Systems) in order to capture and analyze entire cells. Specific antibodies for low-copy or unique protein are bound on an array through which cells are concentrated, enhancing the capture of cells containing these rare proteins. PreCEED is a product designed to enrich for fetal trophoblasts from endocervical mucus, a clinically useful application to a technology that combines cell acquisition and highly sensitive protein detection. The end analysis is a standard FISH test but the sensitivity and accuracy is greatly enhanced through preselection by the CEE (Cell Enrichment and Extraction) technology [16].

Ciphergen (Fremont, CA, www.ciphergen.com, Figure 19.2E) now has the Protein Chip 4000 system (both Enterprise and Personal editions) for integrating the ProteinChip arrays into a SELDI-TOF-MS system for identifying unknown proteins. The ProteinChip Assay binds entire classes of biomolecules using highly specific functional groups and active moieties to capture proteins from complex mixtures, but relies on the power of mass spectrometry to identify individual proteins [17]. The mass spectrometer relies on the ability of the ProteinChip Assay to lower the complexity of the input, increasing the specificity and resolution of its output. Ciphergen has patented a system known as Surface-Enhanced Laser Desorption/ Ionization (SELDI), designed for the direct analysis of protein–protein interactions and individual protein resolution. The ProteinChip System provides molecular weight-based analytical information about the protein and peptide components of biological samples, whether the sample of interest is a pure protein, a chromatography fraction from a purification application, a bioprocessing sample, or a crude clinical or biological sample. The ProteinChip reader is based on time-of-flight mass spectrometry (TOF-MS). Typically, less than 10 μl of crude sample is necessary and can be analyzed in 5 to 10 min. SELDI-TOF MS technology can be used for real-time analysis of protein expression samples from cell culture systems. Both tagged and untagged proteins can be selectively captured on an array providing qualitative and quantitative data about individual and groups of proteins.

High Throughput Genomics (Tucson, AZ, www.htgenomics.com, Figure 19.2F) has developed and marketed a system of 96-well microplate containing a 16-target Universal Array in each well that can be customized to simultaneously measure 16 DNA, RNA, or protein targets. HTG also markets a Quantitative Nuclease Protection Assay (qNPA) that simultaneously allows a measurement of gene expression as sets of genes, rather than one at a time. aNPA technology allows RNA to be quantitatively measured from samples of fewer than 1,000 cells (0.001ng total RNA) [18]. The assay can detect RNA at 10ng total mRNA without amplification and protein at 1 pg/ml and can distinguish single nucleotide differences.

Protagen (Dortmund, Germany, www.protagen.de, Figure 19.2G) markets protein separation and identification technologies. Large-gel technology (LGT) enables resolution of up to 10,000 proteins ranging in pH from 3 to 12 and a size range of almost three orders of magnitude. Protein identification uses mass spectroscopy, either MALDI (Matrix-Assisted Laser Disorption and Ionization) or ESI (Electrospray Ionization). Currently Protagen owns a library containing a collection of clones of over 10,000 different human recombinant proteins. From this collection, Protagen developed a biochip containing 2,500 different human recombinant proteins known as the UNIchip®. This protein biochip has been successfully used to characterize binding molecules like monoclonal antibodies for their binding specificity. Protagen continues to use these technologies in-house for its continuing research into autoimmune diseases and cancer biomarker detection. The technology relies on total recombinant protein isolated from bacterial clones and affixed to a proprietary substrate in most cases retaining the native three-dimensional structure and conformation.

Zyomyx (Hayward, CA, www.zyomyx.com, Figure 19.2H) creates protein-profiling biochips. The Zyomyx system utilizes microfluidics and microimmunoassay technologies to miniaturize and increase the density of the necessary components for protein detection. The Assay 1200 Workstation is a fully automated system that applies the sample, hybridizes the sample to the biochip, and automatically performs washes. Detection technology relies on standard biomolecule markers and common wavelengths such that a standard microarray scanner such as the Axon GenePix 4000B can be used [19].

Jerini (Berlin, Germany, www.jerini.com, Figure 19.2I, Jerini 2005©) is primarily a drug company specializing in liver cirrhosis, angioedema, tissue burns, and macular degeneration but is also a technology company created in 1994 to commercialize the SPOT™ technology. Currently the company continues to create small custom peptides designed to bind to active sites of proteins. The pepSTAT microarray is a platform that immobilizes Jerini's custom peptide molecules in a matrix such that proper folding and protein–protein interactions can still occur in solution. Jerini has created a 2000-protein biochip utilizing standard fluorescent tags and a proprietary hybridization buffer that limits background and promotes protein–peptide interactions that can be seen at a similar level seen in two-hybrid yeast systems. The most significant advance made by Jerini is the noncontact attachment of biodetector molecules on a low-background substrate that works for almost any biological molecule. Of the three distinct proteomic systems available for purchase today (MS of single or complex mixtures or proteins, abundance measurements using antibodies, and protein–protein interaction using whole or partial protein fragments), it is likely that protein–protein interaction chips will gain a strong foothold in high-density/ high-throughput proteomic analysis, and should prove extremely useful for systems biologists. Companies that provide products that can accommodate these needs will likely find themselves in an ideal market position.

Genoptics (Orsay, France, www.genoptics-spr.com, Figure 19.2J), a French company that began operations in 2001, utilizes an optical approach similar to the Biacore technology (www.biacore.com), where surface plasmon resonance reports the interactions of biomolecules' differential interference of the angle of polarized light at the surface of a biochip on which a biomolecular interaction occurs. Two instruments, the Interactor and the SPRiLab, when combined with Genochips, report the interaction between two biological molecules. When light of a certain wavelength and polarization angle strikes gold, a resonance is set up in the electrons from the metal, causing a large drop in reflected light. When protein–Ab or protein–protein interactions occur near the surface of the biosensor (the gold-plated detector), a quantitative change in the resonance angle occurs. Genoptics has leveraged that technology to measure a large number of protein interactions simultaneously on the surface of a proprietary biochip. Sensors that detect the SPR scan the surface of the chip where many protein interactions can occur, thus effectively generating a detection system that can simultaneously screen thousands of protein–protein, protein–Ab, or protein–peptide interactions.

HTS Biosystems (East Hartford, CT, www.htsbiosystems.com, Figure 19.2K) FlexChip™ Kinetic Assay system essentially utilizes the Biacore (Surface Plasmon Resonance) technology but adds a diffraction grating (grating-coupled SPR, or GCSPR) for enhanced sensitivity and discrimination. The FlexChip has since been acquired by Biacore AB, but the technology remains essentially as developed. HTS continues to use this system for parallel kinetic analysis of hundreds to thousands of binding events simultaneously. Genoptics utilizes fiber optics to channel the light information to the detectors. HTS utilizes direct sensors and grating–coupling to enhance detection of minute interactions while lowering the background. HTS also produces the ChemiFlex microarray device and the PhaseFlex device that measures fluorescence lifetimes.

Zeptosens, a Swiss/German company (Witterswil, Switzerland, www.zeptosens.com, Figure 19.2L), produces the SensiChip™ and ZeptoMARK™ products [20,21]. They utilize a method where a coherent light source is coupled into a thin (150 to 300 nm) film of a highly refractive material (Ta_2O_5 or TiO_2) overlaying glass or polymer by a diffractive grating that is etched in the film. This material forms a so-called waveguide that tends to propagate the coherent light creating a strong evanescent field perpendicular to the direction of light propagation. Essentially this light is much more focused and intense than confocal illumination, and can thus selectively illuminate fluorophores in a very confined area. This allows even weak binding events to be monitored, and is sensitive and nondestructive enough to allow real-time kinetic studies. The detection system is similar to surface plasmon resonance in that no molecular tags are necessary, but the evanescence interference has the potential to exceed the detection limits of SPR and confocal microscopy significantly. Subnanometer and sub-kM interaction strengths are detectable with this system and repeatability is only limited by the purity of the protein samples. Density per chip is relatively low but has the potential to exceed several thousand samples per chip.

MICROFLUIDICS AND ALTERNATE SMALL-SCALE DETECTION DEVICES

Microfluidics systems are becoming increasingly popular due to the fact that only very small amounts of reagents are needed, application, hybridization and washing steps are all possible simultaneously, and because computer-controlled nanopumps and relays have precision comparable to full-scale peristaltic and pneumatic pumps. Companies investing in this area include **Aclara, Advalytix, Calipertech, HandyLab, Protagen, and Xeotron**.

Aclara/Virologic (Mountain View, CA, www.aclara.com, Figure 19.2M) markets the eTag™ system where a system of tagged antibodies elucidates tags from antibodies that coat proteins of interest [22,23]. The system provides a unique method of covering those proteins that the researcher is interested in with tagged antibodies and releasing only specific tags from those antibodies that are proximal to activated molecular scissors, extremely useful for cell-surface receptors, signaling proteins, and ligands. The tags are identified and the researcher immediately knows the specific tags that correspond to the antibodies of interest, and directly knows the proximity of tagged proteins to one another. The system works on a microfluidics system where microtiter plates containing hundreds of wells are used as individual bioreactors. Aclara is using the system to pursue oncology-specific pathway discovery through the discovery of new and complex protein–protein interactions.

Protiveris (Rockville, MD, www.protiveris.com, Figure 19.2N) is a company that utilizes a unique physical cantilever approach to the problem of protein detection, a true nanotechnology utilizing the exertion of force caused by complementation of two biomolecules [24,25]. In 1996 at Oak Ridge National Laboratories scientists identified label-free microcantilevers that physically react by creating force on a nanometer scale in direct response to a target molecule. The system acts much like a bimetallic strip that bends due to differential expansion between two metals. The cantilever is coated on one side with a biomolecule such as an antibody, DNA or RNA molecule, or protein that is expected to interact with another species of molecule. The cantilever will exert force through distance when the opposite side of the system is exposed to a complementary or binding partner to the first biomolecule. Exposing the cantilever to interacting molecules is the job of an advanced microfluidics system, a system that prevents complex mixtures from overwhelming the detector system.

Advalytix (Brunnthal, Germany, www.advalytix.de, Figure 19.2O) is a German company that uses a slightly different approach to microfluidics, specifically the use of nanopumps that utilize surface acoustic waves (SAW) that are generated through the application of radiofrequency electric pulses to the surface of biochips. Through the appropriate use of biochemical reactions, labels and marker molecules, a large variety of molecular interaction events can occur and be detected rapidly on the surface of a small chip. Any type of biochemical reaction that uses standard markers and labels can be performed using SAW technology to move reactants together into the reaction vessel. One enormous problem with fluid dynamics at the micro- or nanoscale is the differential flow of water and other liquids due to the much larger effects of surface tension.

Calipertech (Hopkinton, MA, www.calipertech.com, Figure 19.2P) is a company that has developed a system of microfluidics devices that utilize electro-osmosis: a method whereby electrodes attached to reservoirs at each end of a channel create a current that physically moves fluids of the appropriate type within a defined reservoir. Calipertech offers a unique system whereby continuous biochemical experiments can be performed by designing the flow sequence, rate, and volume. The technology developed at Calipertech is applied to many different technological problems including electrokinetic (charged) and pressure (mixed charged and uncharged) driven flows, microenvironment temperature control, microfluidic flow cytometry, and on-chip sample preconcentration to enhance sensitivity.

HandyLab (Ann Arbor, MI, www.handylab.com, Figure 19.2Q) relies on thermopneumatic pumps to move nanoliter-sized plugs of fluid from compartment to compartment on a small, siliconized chip in order to complete biochemical reactions on a nanoscale. No special changes other than rescaling are necessary to convert full-scale reactions to nanoscale reactions. The chips are available in a number of sizes, and reservoirs support a large number of configurations. These self-contained microfluidic systems are capable of integrating and automating several bioprocessing steps on a single chip. Primary benefits include high speed and sensitivity in point-of-service devices requiring minimal skills to operate, and a high specificity/low noise environment. Reproducibility derives primarily from a "hands-off" approach to processing, ensuring that any biases in sample preparation are known and accommodated internally.

Gyros Systems (Monmouth Junction, NJ, www.gyros.com, Figure 19.2R) utilizes a CD-sized reaction chamber that creates motion force for its microfluidics device via centrifugal force. This Amersham-owned company utilizes CD spindle mechanisms to produce a balanced low-g inexpensive system of chambers and tubes that use differential centripetal forces to move fluid through a designed series of reservoirs and channels. Properly designed Gyros discs encompass entire biochemical detection events and increase the efficiency and reduce variability by reducing human interaction.

OTHER BIOMOLECULE DETECTION METHODS

Xeotron (acquired mid-2004 by Invitrogen, www.xeotron.com, Figure 19.2S) was founded in 2001 by Xiaochuan Zhou and Xiaolin Gao in Houston, Texas, and has combined microfluidics and digital photonics to create a unique microarray technology that utilizes a system similar to TI's DLP micromirror system (found in many home large screen TVs and projectors). This process can create a biochip from customer data in a matter of hours. Essentially this technology is *in situ* parallel combinatorial maskless synthesis of nucleotides in three-dimensional nanochambers [26]. XeoChips can be composed of nucleic acids or peptide sequences, greatly expanding the utility for this technology, as Invitrogen's acquisition has clearly shown. Affymetrix holds that this level of flexibility tends to be niche-oriented but companies that have been offering customization at low cost have reaped the benefits (e.g., Agilent's custom expression arrays). The actual process uses computer generated light patterns to control a projection device, which in turn projects light onto the chip at each elongation cycle to create a specific oligonucleotide (or polypeptide), avoiding the use of individual photomasks. The localized light energy generates deprotection; only these deprotected sites couple with the incoming monomer. The major difference in the XeoChip™ synthesis process is the use of a photogenerated acid (PGA) rather than an acid in the deprotection step to control the parallel synthesis. In a synthesis cycle, upon light activation, acid forms in seconds, removing the DMT group. An incoming phosphoramidite nucleoside monomer is then coupled to the growing oligonucleotide chain. The synthesis cycle is repeated for each additional monomer until an array of thousands of oligonucleotides in a microfluidic chip is formed [27]. Xeotron has synthesized a 150-mer DNA oligonucleotide using the above process. XeoChip™ synthesis is not limited to DNA oligonucleotides. RNA sequences, peptides, peptide analogs, biomolecular conjugates, and a variety of organic molecules can be utilized and are highly flexible. Invitrogen has acquired the technology and a portion of Xeotron's staff.

Other companies have oligosynthesis technology that can be used to generate SNP chips, although the companies have not created a specific SNP product. **Agilent Technologies** has a completely customizable 44K array comprised of 60-mers generated by sequential addition of nucleotides through inkjet spotting [28]. Each spot on each array is essentially custom, thus any manufacturing run could generate a 44K SNP chip for the price of a custom expression array, given the appropriate bioinformatics expertise (not endorsed by Agilent). Rather than focus on the highly competitive and fragmented SNP market, Agilent has continued its track to provide scientists highly precise measurement tools for expression, chromosomal copy number, proteomics and metabolomics, microfluidics, and most laboratory development of chromosomal instability with its oligo CGH arrays, ChIP-on-chip technologies (chromatin immunoprecipitation), and unique high-throughput mass spec technologies targeted at protein identification and metabolomics. Amersham (now part of GE) has the **CodeLink SNP chip**, which detects a panel of 72 human P450 genes. This technology used the same platform and scanning infrastructure as its expression arrays, but was dropped due to lack of interest underscoring the highly dynamic and competitive genomics market. Affymetrix has led the way in SNP arrays, producing in rapid succession the 10K, 100K and soon the 500K human genome SNP chips. Although the density of new SNP chips exceeds previous efforts by almost two orders of magnitude, candidate gene screening and fine mapping still require that almost all available SNPs are detected, almost guaranteeing that mass spectrometry and sequencing will not be replaced by SNP arrays in the near future.

SNP DETECTION: MAJOR COMMERCIAL TECHNOLOGIES

Several companies have risen to the top in SNP arrays. One important distinction between expression array detection technology and SNP detection technology is the relative lack of competition to license the detection methodology. The high-density array has already been developed, and there are so many interesting and accurate detection methodologies, that it is unlikely that any one company will stymie development by imposing a sweeping and litigious licensing program simply to make money based on patents. This becomes quite apparent as we examine the top SNP detection technologies — arrays, mass spectrometry, fiber optics, and sequencing, none of which have a substantial upper hand. Price per genotype and density will likely determine the winner but it seems just as likely that there will still be uses for each of the leading technologies.

Sequenom's MassArray 7K and 20K (San Diego, CA, www.sequenom, Figure 19.2T) focuses on both SNP genotyping and SNP discovery. The MassArray system identifies individual SNPs in a multiplex reaction (up to 16 individuals per reaction chamber) using MALDI-TOF to selectively measure the atomic fingerprint of nucleotides in order to unambiguously determine the SNP sequence [29–35]. This system uses SNP probes supplied from Sequenom, so the system can measure SNPs not only contained in dbSNP but also those the user has discovered. Sequence patterns are produced by running four cleavage reactions compared to patterns of known reference sequences; the variations between the two indicate sequence changes caused by SNPs. As little as 25 ng DNA per patient pool (83 pg/individual patient) can be used, and sensitivity is on the order of allele frequencies as low as 3% (a SNP is defined as prevalence of at least 1%, see www.hapmap.org) and can scan nearly 200,000 individuals per hour.

Further advancing the utility of nanobeads in SNP detection is **Illumina** (San Diego, CA, www.illumina.com, Figure 19.2U). The BeadArray™ technology has enabled Illumina to achieve high-throughput and high-density test sites and to format arrays in a pattern arranged to match the wells of standard microtiter plates. Illumina utilizes fiber-optic bundles to convey fluorescence information throughout a microchip, delivering light to sensors precisely located for a unique application [36]. The Oligator system synthesizes many different short segments of DNA to meet the requirements of large-scale genomics applications. The miniaturization of the BeadArray provides substantial information content per unit area than similar competing arrays. A wide variety of conventional chemistries are available for attaching different molecules such as DNA, RNA, proteins, and other chemicals to beads. By using beads, Illumina is able to take advantage of these alternate chemistries to create a wide variety of sensors, which are assembled into arrays. Additionally the fiber-optic bundles can be manufactured in multiple shapes and sizes and organized in various arrangements to optimize them for different applications. The high density of beads in each array enables multiple copies of each individual bead type, measured simultaneously, allowing comparison of each bead against its own population of identical beads, statistically validating each bead's data. Illumina has been researching human populations to determine variations among races and geographically isolated populations. Through this basic research Illumina continues to gather novel SNP data, and because its technology is unique it offers detection of certain SNPs that other technologies find intractable.

Affymetrix (Santa Clara, CA, www.affymetrix.com, Figure 19.2V) is winning the overall volume race for SNP chips, and offers the SNP 10K and the SNP 100K arrays, soon 500K arrays. These arrays utilize hybridization to perfect and mismatch probes to identify major and minor alleles. Probes are created in order to detect a SNP approximately every 210kb on the human genome for genome-scanning applications.

Applied Biosystems (Foster City, CA, www.appliedbiosystems.com, Figure 19.2W) offers a 7900HT sequence detection system, a high-throughput real-time PCR system that detects and quantifies nucleic acid sequences. qRT-PCR/TaqMan® has long been regarded as the gold standard for quantifying mRNA levels within a cell and ABI has been an innovator in RT-PCR detection systems for years. ABI has been in genotyping development for at least as long, and through many state-of-the-art technologies including SNPPlex, genotyping can be performed at either high-density, high-resolution or at low-density genome-scanning levels. Many laboratories have sequencing systems utilizing ABI technology, and much of that hardware can be reused with new reagents designed for genotyping rather than sequencing. ABI offers over 50K SNPs per reaction set, and can be used for whole genome or fine mapping applications, depending on the reaction set purchased. ABI's SNPPlex genotyping system uses OLA (oligonucleotide ligation assay) where ligation only occurs when there is a perfect match between probe and SNP. Reporter probes bind to biotin and utilize a so-called zipchute probe. Products are run on ABI sequencers where the mobility modifiers are used to control the rate of SNPs that pass past the laser scanners. This system also uses multiple dyes to increase multiplexing.

SNP DETECTION DIVERSITY

SNPs can be detected in a sequence-specific or nonspecific way. Target capture, cleavage or mobility shifts are based on the match or mismatch between allelic or single-stranded DNA, or the conformation of the DNA duplex or heteroduplex in denaturing gels. Specific detection is preferred over nonspecific detection simply because one is able to ensure the actual genotype. Several methods are in wide use today.

Single nucleotide primer extension has been widely used to detect SNPs. It is a method where an amplicon (target DNA) is amplified by PCR and used in a separate reaction where primers with a known SNP sequence promote extension (or not) from the potential polymorphism. The primer binds when the single nucleotide that makes up the SNP matches perfectly, and does not bind when there is a mismatch. Often when the SNP is an unknown sequence but the position is known, four different primers can be used to verify the true sequence of the SNP. Extension products are fluorescently labeled and detected using a variety of methods. The amplification reaction is so robust that it can actually be performed on the surface of microarrays (such as the CodeLink SNP chip, using the SurModics™ surface). Amplification products are detected and compared to a spot where there should be no signal (a mismatch primer) and the ratio of match to mismatch will indicate the presence of a major allele, minor allele, or heterozygote.

Electronic detection is another popular method for detection of sequence-specific DNA targets. These techniques take advantage of the fact that DNA differentially conducts electrical current based on the sequence. Although these methods require substantial knowledge about the target sequence, they can often detect SNPs in a highly heterogeneous mixture. Some technologies utilize probes (CMS/Motorola uses a ferrodoxin probe) and some detect the DNA directly with no intermediate.

Bead-based detection uses Dynal® beads to conjugate a probe and fluorescently labeled dye to a small glass bead which then finds its complementary match (for example, Luminex). Other methods use gold beads (Nanogen) conjugated to specific probes, which can then be used in extremely heterogeneous mixtures, such as whole cells.

MPSS®, or massively parallel signature sequencing, developed at Lynx Technologies [37], uses a large number of identifiable tags and fluorescent beads coupled with FACS (fluorescently activated cell sorting) to quantitatively identify all available mRNAs, but this technology also identifies SNPs because it is a sequence-based technology.

Many methods use variations on those listed above; occasionally the technology involves flow-through arrays that greatly decrease the detection time, and some rely on existing technologies modified to a specific new purpose (Sequenom's mass spectrometer and ABI's sequencers, for example).

PerkinElmer (Wellesley, MA, www.perkinelmer.com, Figure 19.2X) uses FP-TDI (fluorescence polarization template directed dye terminator incorporation) to perform SNO genotyping [38,39]. In single-base extension, the target region is amplified by PCR followed by a single-base sequencing reaction using a primer that anneals one base short of the polymorphic site. Several detection methods can be used: one can label the primer and apply the extension products to gel electrophoresis, or the single-base extension product can be measured by MS. Fluorescence detection of the terminal dideocynucleotide terminator can also be detected on a microarray, such as the CodeLink SNP Chip, but in PerkinElmer's case they chose to detect the fluorescence polarization. The fluorescence polarization technique takes advantage of the fact that small molecules labeled with a fluorophore spin rapidly and large molecules labeled with a fluorophore spin slowly. When a polarizing light is used to excite the fluorophores small molecules emit light in all directions while large molecules emit light in a single plane, producing highly polarized fluorescence. Several of the benefits include low cost because it does not require labeled primer. Reagent costs are around 50 cents per SNP and FP reading instruments are typically $30k. Single-base incorporations are simple to optimize, and FP does not require high intensity and FP is not sensitive to differential incorporation of fluorophores.

Parallele (San Francisco, CA, www.parallelebio.com, Figure 19.2Y) supplies an inverse PCR method for improving the Affymetrix system, and supplies a Molecular Inversion probe detection system, using several different detection platforms, although Affymetrix remains the largest single application [40]. Molecular Inversion Probes are named to reflect the oligonucleotide central to the amplification, and self-ligation reaction undergoes a unimolecular rearrangement from a retractable unamplifiable molecule to one that can be amplified. Enzymes are used to fill in the gaps in an allele-specific manner. The circularized probe can be separated from cross-reacted or unreacted probes through exonuclease digestion. Additionally Parallele offers TrueTags, a library of 21-mer tags for detection of genomic targets from complex biological mixtures. These tags work for PCR, array hybridization, bead-based detection and conjugation reactions.

Tebu-Bio (www.tebu-bio.com) has developed a similar method through a process called Mutector (TrimGen Genetic Technology, Sparks, MD, www.trimgen. com, Figure 19.2Z) that utilizes a primer extension reaction, specifically known as STA, or Shift Termination Assay. A detection primer complementary to the target DNA (just upstream of a potential SNP) is designed to hybridize near the SNP. If the SNP, or target base, is the expected wild type nucleotide, the primer extension reaction will be terminated at the target base position without incorporating any labeled nucleotides. If any mutation is present at the target base, the primer extension will continue incorporating labeled nucleotides, yielding a strong fluorescent or colorimetric reaction. The Mutector technology has a sensitivity of 1 mutated molecule in 100.

Orchid CellMark (Princeton, NJ, www.orchidbiosciences.com, Figure 19.2AA) also makes use of the enzymatic addition of a fluorophore that can be read by Affymetrix's GeneChip™ Scanner. This primer extension method is identical to many of the above methods and differs only in the detection matrix, but Orchid has worked diligently to remain part of the new clique of genomics technologies, in part by contributing heavily to the TSC (The SNP Consortium, snp.cshl.org [41]). Orchid also provides genomic testing in several forms: Orchid Identity Genomics, Orchid Cellmark, Orchid GeneScreen, Orchid Diagnostics and Orchid GeneShield. Each of these business units is built from the genomic detection methods developed in the late 1990s at Orchid.

Nanosphere's (Northbrook, IL, www.nanosphere-inc.com, Figure 19.2BB) Verigene platform uses nanoprobes that are bound to gold particles and bind specifically to specific DNA or protein sequences [42]. This renders the system capable of operating in dense and heterogenous cellular milieux and has spawned several practical disease detection application as well as sensitive and precise detection of unamplified DNA and RNA [43,44]. Nanosphere takes full advantage of the specificity of antibodies with the detectability and robustness of gold nanoparticles and the ease of amplification of biobarodes [45].

Lynx Therapeutics (Hayward, CA, www.lynxgen.com, Figure 19.2CC) uses a unique massively parallel sequencing system (MPSS) to sequence every identifiable cRNA or DNA on beads separated by FACS sorting [37]. The method is very complex but highly reproducible and accurate. Megaclone™ transforms a sample containing millions of DNA molecules into one made up of millions of microbeads, each of which carries approximately 100,000 copies of one of the DNA molecules in the sample. MPSS™ identifies the DNA sequence of the molecules on each bead in a parallel process.

SNP DETECTION: ELECTRONIC DETECTION

It is notable that many e-detection SNP kits are targeting the clinical market with low-density disposable chips. Some SNPs are so prevalent and pharmacologically relevant that insurance carriers will reimburse the cost for the test, to determine correct drug dosages. For example, the P450 gene shows enormous variation within the human population, and this particular gene effectively determines the rate of metabolism of many drugs. Knowledge of the behavior and individual drug turnover would greatly facilitate care-providers' prescriptions and dosages, and could bring the hope of personalized medicine much closer to reality.

Although robotic spotted arrayers are becoming increasingly popular and much more cost-effective, some companies have taken advantage of the fact that DNA behaves in a predictable manner in an electrical field in the presence of certain chemicals or metals. Electronic biomolecular detection relies on the ability to produce an electromagnetic field that provides a method for discriminating biomolecules in an extremely heterogeneous mixture of biological detritus.

Nanogen's (San Diego, CA, www.nanogen.com, Figure 19.2DD) technology uses amplicons and biotinylated probes but makes use of differential resistance and the partial charge inherent in DNA to move, concentrate, and detect single species of DNA (SNPs). One added benefit to utilizing the charge of nucleic acids is the speed at which samples can be moved and concentrated. Using charge to concentrate DNA, a speed increase of 1000 times over passive hybridization techniques can be routinely seen. Nanogen's electronic addressing technology utilizes individually addressable units on a microchip to anchor solutions containing nucleic acid probes in-place. The technique involves placing charged molecules at specific sites on a NanoChip® array. When a biotinylated sample solution is added to the array, the negatively charged sample moves to the selected positively charged sites where the sample is concentrated and bound to the streptavidin in the permeation layer. Currently Nanogen offers a 2 mm² array containing 100 test sites of 80 μm diameter spaced 200 μm apart. CMOS-prepared arrays may extend to 10,000 sites or more.

The minimum sample volume is 60 μl and a maximum loading of 10^9 fragments per site can be accurately detected. These arrays have been shown to detect SNPs, STRs, insertions, deletions, and other mutation analyses. Of note is a new technology from Nanogen and Becton, Dickinson and Company for Strand Displacement Amplification (SDA) for exponential detection of low levels of target. Relying on the ability of the Nanogen system to concentrate DNA at sequence-specific sites, amplification is specific, rapid and quantitative, suggesting point-of-service or field work applications.

CMS (Clinical Microsensors, a division of Motorola Life Sciences, Pasadena, CA previously, now acquired by Osmetech, PLC as of July 26, 2005 — www.osmetech. plc.uk, Figure 19.2EE) utilized a similar technology where bioelectronic signal probes (2-ribose Ferrocene with multiple redox potentials) hybridize to a capture probe that targets a genomic amplicon, a segment of DNA where a unique nucleotide polymorphism can be found. A completed hybridization completes a circuit detectable by monitoring the first harmonic of signal produced by alternating current voltammetry. Motorola Life Sciences was developing a cost-effective and easy to use eSensor™ Cytochrome P450 DNA Detection System that simultaneously detects ten well-characterized mutations belonging to the 2D6, 2C9 and 2C19 genes of the Cytochrome P450 superfamily, but Motorola has decided to deactivate the company pending further investigation. Note that the current iteration of Motorola Life Sciences no longer includes the fluorescence detection SNP chip (CodeLink) and depending on the outcome of the board of directors at Motorola, all life sciences may be suspended or divested completely.

Another unique approach for electronic biological molecule detection is **CombiMatrix'** (Mukilteo, WA, www.combimatrix.com, Figure 19.2FF) application of electronic current for altering or enhancing chemical reactions through application of microcurrents [46]. CombiMatrix' core technology is the Lab-on-a-Chip integrated circuit. Sensitivity is approximately 0.5 pM and a dynamic range of approximately three orders of magnitude for nucleic acids, although this system is flexible enough for smaller and larger molecules including proteins, protein complexes and even entire cells. These integrated circuits contain arrays of microelectrodes that are individually addressable using logic circuitry on the chip. The first generation of these integrated circuits was designed and produced in 1996 under a controlled process developed by CombiMatrix. Each microelectrode is uniquely addressed to generate chemical reagents by means of an electrochemical reaction. These chemical reagents facilitate the *in situ* synthesis of complex molecules such as DNA oligonucleotides. These molecules are synthesized within a proprietary porous reaction layer (PRL) that coats the chip. Once the chip is placed inside a small reaction chamber, thousands of simultaneous reactions are initiated by logically addressing spots on the chip. The chip synthesizes hundreds or thousands of different molecules in parallel within the PRL above each electrode. Because a different product is synthesized at each site, this technology eliminates the need to synthesize different products in individual flasks by conventional methods. Conventional CMOS processes are used to fabricate the semiconductor chips. The first-generation chips had a density of 1,000 microelectrodes/cm^2, the second-generation chips have a density of 18,000 microelectrodes/cm^2, new internal designs exceed 500,000 microelectrodes/cm^2

using conventional CMOS manufacturing technologies. Laboratory-scale experiments have achieved densities as high as 1.7 million electrodes per cm^2.

GeneOhm Sciences (San Diego, CA, www.geneohm.com, Figure 19.2GG) uses the flow of electric current through DNA to detect genetic mutations (single nucleotide polymorphisms, or SNPs). The system has high sensitivity and accuracy, critical for SNP detection. GeneOhm was formed in 2001 with technology developed throughout the previous decade by Professor Jacqueline Barton at CIT and merged with Infectio Diagnostic of Quebec, Canada, in 2004. GeneOhm's technology uses electrochemistry for the detection and readout of genetic variations in DNA or RNA. A known electrical current flows through DNA (or RNA) based on sequence composition. If a DNA sequence shows single or multiple mismatches in comparison to a control sequence, the current is altered regardless of the sequence. GeneOhm is targeting the point-of-care clinical market by distributing single-use disposable detection devices and low-density arrays. The company is leveraging the benefit that direct detection provides. Such a technology that uses the electrical properties of DNA, not hybridization, to detect SNPs, is independent of the gene sequence and could prove to be much more flexible and robust, especially in detection systems that operate in the field under harsh conditions. The system will also provide the ability to work with either RNA or DNA, simplifying sample preparation.

SNP DETECTION: BEAD-BASED

Electronic detection has provided an alternative detection method for biomolecules in complex solutions, but new technologies have continued to develop, some brand new, and some novel applications to older technologies. No longer will molecular interactions occur on a flat substrate. Now, detection systems are taking advantage of three-dimensional space and the access of probe and target, and the enormous changes in fluid dynamics on the submicron scale. Companies are looking at grafting detection molecules onto microbeads, be they magnetic, or coated with reporter molecules that fluoresce, or to particles that have intrinsic identifiers. The largest player in the evolution of bead-based detection systems has been **Dynal** (www.dynal.no) Industries. In 1979, Professor John Ugelstad of Norway created polystyrene spheres of uniform size. These beads were made paramagnetic (magnetic in a strong EM field), and were subsequently induced to acquire very uniform and useful properties. Paramagnetism allows resuspension when the magnetic fields are removed and has become very useful in the isolation and separation of biological materials (target-specific antibodies, nucleic acid probes, peptide probes, etc.). Dynal, in Oslo, Norway, was formed in 1986 as a joint venture between Dyno Industries ASA and A.L. Industries AS, combining knowledge from the chemical, pharmaceutical and biotechnology fields. In July 2001, Dynal Biotech was acquired by the Swedish equity firm Nordic Capital. Today, the supraparamagnetic particles supplied by Dynal Biotech are known as Dynabeads® and the nonmagnetic beads called Dynospheres®, both widely used in bead-based array detection systems.

It is interesting to note that beads are not the only nanoparticle used in biomedical imaging. **Nanoplex Technologies** (Menlo Park, CA, www.nanoplextech. com, Figure 19.2HH) provides nanoparticles that are cylinders coded with differential gold and silver bands, knows as Nanobarcodes®, as mentioned above [47]. These cylindrical nanoparticles are encoded with submicron stripes used for creating unique tags that can be used for tracking particles and other tiny biological particles. Multiplexing has always relied on the number of distinguishable fluorescent tags in order to increase sensitivity, throughput, and mixture complexity [48].

A combination electrical detection and bead-based detection system is represented in the electronic detection system from **BioArray Solutions** (Piscataway, NJ, www.bioarraysolutions.com, Figure 19.2II). They have developed a unique, proprietary technology for the rapid and flexible analysis of DNA (LEAPS, Light-controlled Electrokinetic Assembly of Particles near Surfaces), proteins and cells on semiconductor chips. BioArray Solutions' optically programmable bead array technology enables a universal assay platform for clinical and research-based biomolecule detection systems. BioArray Solutions combines semiconductor physics, extensive bead chemistry (Dynal) and molecular biology to bring selectivity and sensitivity to quantitative DNA, protein and cellular detection. BioArray Solutions' assay can be used for any homogeneous assay that monitors the binding of one molecule to another using color-encoded beads.

As mentioned above, LynxGen is a company with a similar product line and technology. Lynx Technologies uses Dynal microbeads with unique combitag sequences hybridized to a cDNA library that has a unique and arbitrary sequence bound to the beads. The combitag on each bead is created by MPSS (Massively Parallel Signature Sequencing) invented by Sydney Brenner [37]. MPSS is based on Megaclone, Lynx's technology for cloning DNA molecules onto microbeads. It measures expression or sequence (SNPs) by counting the number of individual mRNA molecules or DNA sequences on a per-gene basis. MPSS does not measure transcript abundance in an analog fashion, but uses a digital approach where transcripts are counted using FACS (Fluorescence-Activated Cell Sorting). MPSS counts more than 1 million transcripts per sample, providing quantitative expression or SNP data at single-copy-per-cell sensitivity levels. Each target molecule is grafted onto the surface of a 5μm Dynal bead. An arbitrary but unique DNA combitag sequence is attached to a fragment of target, and the tagged library is PCR amplified. The resulting library is hybridized to beads, each of which is decorated with 100,000 identical oligonucleotide strands complementary to one of the combitags. After hybridization, each of the beads displays copies one target molecule. It has been shown (Stolovitsky, personal communication) that MPSS expression mirrors the expression distribution of Affymetrix U133A chips. SNP detection is highly precise, and quantized with a very low noise threshold and high signal-to-noise ratio.

Luminex (Riverside, CA, www.luminexcorp.com, Figure 19.2JJ) uses Dynal beads in its proprietary detection system [49,50]. With Luminex's proprietary xMAP technology, molecular reactions take place on the surface of microspheres. For each reaction in an xMAP profile, thousands of molecules are attached to the surface of internally color-coded microspheres. The assigned color code identifies the reaction throughout the test. The magnitude of the biomolecular reaction is measured using a second reporter molecule. The reporter molecule signals the extent of the reaction by attaching to the molecules on the microspheres. Because the reporter's signal is also a color, there are two sources of color, the color code inside the microsphere and the reporter color on the surface of the microsphere. To perform a test, the color-coded microspheres, reporter molecules, and sample are combined. This mixture is then injected into an instrument that uses microfluidics to align the microspheres in single file where lasers illuminate the colors inside and on the surface of each microsphere. Next, optics captures the color signals. Finally, digital signal processing translates the signals into real-time, quantitative data for each reaction. Twenty thousand microspheres can be analyzed per second, which shortens the analysis time.

THE SNP CHIP: FLOW-THROUGH METHODS

Flow-through technologies are significantly different from microfluidics systems. Typically surface hybridizations are inefficient due to the small volumes used for microarrays. Small hybridization volumes help to keep the required sample quantity to a reasonable level, but the drawback is high background, inefficient hybridizations, and extended hybridization times, often exceeding 24 h. Flow-through techniques allow a large volume of hybridization buffer wash over a probe yielding reduced background, efficient hybridization and reduced hybridization times. Companies that have developed flow-through solutions include PamGene, Metrigenix, and others.

PamGene (Hertogenbosch, Holland, www.pamgene.com, Figure 19.2KK) offers a unique approach to microarrays. The company has perfected hollow nanotubes that allow rapid back-and-forth passage of reagents across a bound probe, as opposed to the limited amount of exposure most microarray spots receive from the hybridization mixture (containing labeled target). The nanotubes can bind protein, nucleic acids, and small molecules so that any compound that can be labeled with a detectable tag can be used in the PamGene system. Direct labeling of the target takes approximately 2 h with most target molecules, RNA amplification using a linear amplification system (TYRAS) takes approximately 3 to 4 h. Hybridization takes 5 to 30 min (as opposed to 18 to 24 h for flat microarrays) due to the active pulsing of sample through the porous membrane. Reproducibility measured with CV% is of the order of 18% to 24%, depending on the complexity of the target solution and the biomolecule being used.

Metrigenix (Baltimore, MD, www.metrigenix.com, Figure 19.2LL) utilizes a patented flow-through chip technology (FTC) that serves as the technology base for the MetriGenix microarray [51]. The FTC is a state-of-the-art three-dimensional microarray technology platform that can be used to provide high-throughput, high-content assays to measure gene expression or genetic polymorphisms. The flexibility of the system is based on the detection biomolecules bound to the chip surface, but the speed of hybridization exceeds standard surface hybridization techniques by several orders of magnitude due to molecular availability. The applications to date include colon cancer biomarker detection, as well as nonsmall cell lung cancer signatures.

SNP DETECTION: SEQUENCE-BASED

Several detection technologies fall into sequence-based detection where the samples are actually shotgun sequenced in order to detect single nucleotide polymorphisms. Also included in this section are some detection methods that utilize small-molecule arrays that allow the user to screen bioactive molecules against heterogeneous protein mixtures.

Solexa (Essex, U.K., www.solexa.co.uk, Figure 19.2MM), a U.K.-based company recently merged with Lynx, has developed a nanotechnology known as a Single Molecule Array, which allows analysis of individual target molecules. This technology has been applied to genome sequencing known as Total Genotyping. Unlike conventional high-density arrays, Solexa has developed a single molecule array that randomly distributes single molecules at extremely high density (10^8 sites/cm^2) across an array substrate. Solexa's goal is to short-read sequencing bursts in order to compare individual genotypes to a reference sequence of approximately 25 bases. This allows SNP mapping of individuals in one reaction with no target amplification. Eventually an entire genome can be mapped to a single chip and an entire sequence can be SNP mapped for less than a thousand dollars.

Graffinity (Heidelberg, Germany, www.graffinity.com, Figure 19.2NN) has developed a system of immobilizing small molecules on a carrier substrate in order to screen labeled target molecules. Graffinity has pioneered the RAISE paradigm for drug discovery (Rapid Array Informed Structure Evolution) as an important application of its technology. One approach consists of binding a small molecule that has an affinity to a certain subset of proteins, and then interacting those immobilized proteins with their binding partners or a specific antibody in the detection process. These array-based "far-westerns" offer greater resolution, density, and throughput than previously possible using standard western-based hybridizations. Most importantly this approach can screen thousands of mutant proteins for target interactions and thousands of targets in one reaction, theoretically increasing drug-screening throughput several thousandfold. This approach is especially useful for structural feature detection in engineered proteins.

Febit (Mannheim, Germany, www.febit.de, Figure 19.2OO) offers automated machinery for DNA analysis and creation of custom arrays on the benchtop (Geniom). Microarray synthesis and hybridization is performed inside a three-dimensional microchannel structure that connects microfluidics and DNA synthesis components. Hybridization and detection occur immediately following the probe synthesis reactions and utilize standard fluorescent dyes. All portions of a SNP or expression microarray experiment occur within the same machine, preventing contamination or laboratory conditions from introducing external variability. Another company that specializes in the TI micromirror oligonucleotide synthesis space, Febit utilizes customization and nanofluidics to create a self-contained customizable microarray construction system.

GenoSpectra (Fremont, CA, www.genospectra.com, Figure 19.2PP) utilizes fiber-optic processes to detect RNA sequences within a still-living cell. Real-time *in vivo* expression analyses can occur as the cell is being subjected to external manipulations, the Holy Grail of systems biologists. Multiple detection events can occur over a short time span, enabling researchers to detect short-term expression level events with a degree of time resolution previously difficult to achieve. Because the detected cell remains alive, the changes in expression levels are of high quality simply because one is not overlaying a death response from killing the cell and extracting the biomolecules that one hopes accurately reflects the state of physiology due to the manipulation. The reproducibility of the platform bears this out, with repeated measures of the same cell exceeding 4% CV and less. Key products from GenoSpectra include PQB (Parallel Quantitative Biology) and XHTS (extreme high-throughput screening), both of which are still under development but are at late stages of testing. In much the same vein as a FACS-based cell sorter, GenoSpectra hopes to utilize micro fiber-optic channels to focus the detection system on single unmobilized cells using nonharmful fluorescent biomarkers.

SNP DETECTION: ALTERNATE DETECTION METHODS

Other types of detection systems include alternate approaches to binding nucleic acids other than the standard denatured single-strand hybridization. Included in this list are arrays that simply do not fit into an easily describable category.

Epoch Biosciences (Bothell, WA, www.epochbiosciences.com, Figure 19.2QQ) uses a minor-groove detection system where nucleic acids are exposed to a unique three-dimensional conformation in a nucleic acid molecule [52,53]. The detection system avoids hybridization issues that often occur during single-stranded DNA hybridization. One major advantage of this system is that there is no requirement for denaturation of nucleic acids. Large-scale and extended-sequences of DNA can be easily detected using nondenatured nucleic acids. DNA–protein interactions can be accommodated with this system as well because it is examining a tertiary structure of the DNA with any interacting binding partner. Arrays are low- to mid-density including less than 1000 features per chip, however this system can be used for extremely complex mixtures including multispecies and multi-individual forensic analysis, multi-individual polymorphism detection (microsatellites, VNTRs, and highly disrupting SNPs, that is SNPs that add large amounts of destabilization energy). Reproducibility yields CVs near 20% under standard laboratory conditions.

Exiqon, Inc. (Vedbæk, Holland, www.exiqon.com, Figure 19.2RR), has created a system of immobilization for biomolecules that uses a proprietary surface to immobilize DNA labeled with an amino-linker, protein, polypeptides, and RNA against various conjugates. Tags may include glutathione, strepavidin, nickel-chelates and other widely used biological tags from molecular biology and biochemistry. Exiqon sells blank microarray slides and custom spotted polymer arrays (using noncontact printing and capillary isolation of those full-length cDNAs directly from cell solution). Exiqon also supplies a microfluidics chip called the "Immobilizer Chip."

NimbleGen Systems (Madison, WI, www.nimblegen.com, Figure 19.2SS) creates custom arrays based on user-defined sequences (or using in-house bioinformatics expertise) and utilizes technology based on the Texas Instruments DLP, or moving mirror display products, seen in many digital projectors today. The light is used to create custom oligos without photomasking. NimbleGen recently acquired Dallas-based Light Biology in a move that parallels the considerable consolidation in the projection-based oligonucleotide synthesis field. NimbleGen designs arrays for any genome for which sequence information is available using its proprietary Maskless Array Synthesizer (MAS) technology. This unique technology provides up to 786,000 probes/array. Probe sizes range from 24 mers up to 70 mers, in any configuration, including the concept of an isothermal array where each probe is designed to hybridize ideally at one temperature, rather than fixed length oligos that hybridize at varying temperatures directly related to their sequence. NimbleGen's technology is so flexible that it has been applied to mRNA expression [54], whole genome prokaryotic resequencing [55], oligoCGH [56], and chromatin immunoprecipitation (CHiP) [57] — all techniques that will see continuing growth in the life sciences. Currently Affymetrix markets NimbleExpress Arrays for use with the GeneChip system, allowing customers to use the Affymetrix reagents, instrumentation and analysis software. The technology utilized by NimbleGen is unique and uses a computer to control movable DMD micromirrors (Digital Micromirror Device) that modify the protection status of subunits that make up the synthesized oligonucleotides. This method is distinct from the DLP-like projection system used by Xeotron and the photolithography masking system used by Affymetrix. This technology dramatically cuts the upfront cost of array design, enabling low-cost, small-scale manufacturing runs that can be scaled up when necessary. NimbleGen's subsidiary, Chemogenix, develops and markets the photochemistry technology.

Perlegen (Mountain View, CA, www.perlegen.com, Figure 19.2TT) uses Affy's *in situ* synthesis to create enormous SNP wafers that cover entire genomes with multiple redundancies, enabling scientists to assay 1.5 million SNPs at once. In contrast, the Affymetrix SNP chip will be able to assay as many as 500,000 SNPs at once per array. The advantage to the Perlegen system is that all target and buffers can be applied simultaneously onto the same large chip, avoiding the inherent prep-to-prep technical variability that occurs when using multiple smaller chips. In addition, data mining can begin as soon as a single chip is complete, enhancing the throughput of large-scale genotyping [58]. Enormous strides have been made recently in whole-genome scanning at high density, and Perlegen is applying its technology to the complex genetics of drug metabolism on an individual basis as well as diseases such as rheumatoid arthritis.

ArrayIt (Sunnyvale, CA, www.arrayit.com, Figure 19.2UU) is a basic one-stop shop/supermarket for microarray technology. Telechem International owns the ArrayIt division where products and services span a wide gamut, include robotic spotters, printing pins, slide centrifuges, protein chip spotters, clean rooms and full-service microarray services. ArrayIt has a low-density microarray service where 70 mers are used in 384-well plates allowing targeted analysis of gene subsets, but they have recently added a 25K expression chip capable of detecting 25,509 different genes in one-, two-, three-, or four-channel modes. ArrayIt also offers specialized arrays including transcription factors, cancer splice variants, the hematome, phosphodiesterases, and others.

Genomic Solutions (Ann Arbor, MI, www.genomicsolutions.com, Figure 19.2VV) is a Michigan-based biotech company that has created an automated microarrayer (the GeneMachine) to move cDNAs from clones to chip in one step. Companies that have a large collection of clones that need to be screened are able to utilize the Genomics Solutions system to quickly scan through their clones by programming the microarrayer to scroll through stacks of bacterial plates and quickly move the cloned DNA to microarray in one step. The advantage of this system is a novel method of skipping many tedious steps in the process of screening bacterial clones.

This solution makes large-scale multigenome mutagenesis screens much more viable. Beyond the standard high-throughput screening systems is the Investigator Proteomic system that offers a broad range of analysis software, hardware, and bioinformatics tools to begin mining proteomic data quickly. SynQUAD is a noncontact dispensing system that dispenses DNA onto a slide enabling highly precise SNP detection given a large library of paired perfect-match, mismatch oligonucleotides. High throughput is easily achievable — an eight-channel synQUAD system can dispense 100 nL into a 384-well microtiter plate in <10 sec [59].

Idaho Technology's RapidCycler system (www.idahotech.com, Figure 19.2WW) utilizes real-time analysis of melting curve data of prelabeled DNA to identify the temperature (and subsequently the sequence) of DNA duplexes. Many probe/amplicon heteroduplexes contain destabilizing mismatches that can be detected through melting curves [60]. Idaho still manufactures the fastest thermocycler available, using heated high-volume air rather than Peltier technology to distribute heat to reaction vessels.

Estonia has proven that SNP technology spans the globe by creating and supporting a SNP company, **Asper Biotech, Ltd.** Located in Etartu, Estonia, Asper (www.asperbio.com, Figure 19.2XX) has licensed the primer extension technology from Orchid BioSciences and develops and supplies its own SNP-detection technologies throughout the world. Asper is creating an APEX (Arrayed Primer Extension) design where up to 30,000 SNPs can be assayed on a single chip, greatly expanding previous technology in terms of both density and accuracy [61]. Primer extension using 25-mer oligonucleotides and DNA polymerase continues to form the basis of an increasing number of basic SNP-detection platforms. Sensitivity and precision are high, but the necessity of generating amplicons is often a rate-limiting step.

SNP TECHNOLOGY: ANALYSIS SOFTWARE

Scientists who are using whole-genome SNP arrays are finding significant limitations in the software that performs linkage disequilibrium, association analysis, and multilocus susceptibility studies. The limitations of current software are often obvious when hundreds of thousands or even millions of markers are loaded in the context of large families and complex pedigrees, partial penetrance and QTL phenotypes, and HW errors that occasionally occur in the data. When many of these analytical programs were written, it was unimaginable that even 1000 markers would be available for analysis, much less markers numbering in the millions. Much of the new SNP analysis programs are also sorely lacking in the statistical depth to move beyond the one marker/one disease paradigm. The simplest disease associations can be found quickly, but multiallelic susceptibility, case-control studies and disease sensitivity markers will be much harder to identify, and will require not only high-density SNP coverage, but also enormous cohorts likely exceeding thousands of samples.

Haplotype maps are relatively simple to construct and use, and enhance the inherently low information content of individual SNP markers. Although not a perfect replacement for microsatellites, which are actually highly informative, haplotypes are a robust method for data reduction in genetic linkage analysis.

HT SNPs are those markers that contribute maximal information to the definition of a single haplotype across the population that is under study. In pedigree analysis (linkage disequilibrium analysis in particular), one can filter potentially 60 to 80% of all SNPs, retaining only those highly informative markers; conversely, large-scale association studies often require the contribution of every SNP because the heterogeneity in a disease-based case-control study is substantial.

A few hundred microsatellites markers were once the limit for a genotyping experiment. When the number of markers exceeds a few thousand, linkage programs such as GeneHunter2, Linkage, Merlin, etc. use increasing amounts of memory and time to generate haplotypes and compute disequilibrium odds ratios and LOD scores. The analysis is combinatorial, iterative and recursive, therefore, time and memory is required to expand the number of markers in the analysis. SNP Assistant, Mapper, and many other software programs utilize very intuitive graphical interfaces to ease the analytical process, yet underneath these GUIs the basic algorithms are much the same from program to program. New algorithms and more efficient processing techniques are being developed and published, but data is outpacing the development of new analytical tools.

Data conversion remains a difficult task as well. Ideally, one would like a standard format from every type of SNP platform, so no interconversion would be necessary, and all software programs could read genetic information regardless of the platform. Unfortunately, manufacturers have no real interest in standardizing their output because no consensus exists among statistical geneticists on the most appropriate analytical algorithm, much less the most appropriate software for a given problem. Thus, output ranges from individual nucleotide sequence calls to calls listed as a number or nonspecific letter assignment to probability values of a heterozygote, minor or major allele call all the way to raw mass spectrometry output. Increasingly, researchers are requesting not only uniformity of data to enhance data sharing across the scientific community, but also a software tool that could integrate their data with dbSNP (www.ncbi. nlm.nih.gov/SNP) and could locate and visualize all human genes, splice variants, promoters, enhancers, and other genetic elements while at the same time supporting the visualization of the results of genetic analyses over the genome. The Affymetrix GeneChip system has developed to a point where one can buy off-the-shelf whole-genome expression and SNP chips that measure the entire human genome for mRNA expression, and up to 500K SNPs in a single experiment. Although other manufacturers may be able to manufacture higher-accuracy/higher-precision arrays for expression or SNP detection, Affymetrix is leveraging its market position to offer a simple whole-genome/whole-transcriptome one-stop shop. To this end, it has expanded the analysis and database capabilities of GCOS/GREX to encompass its genotyping products and make integration with its expression products as seamless as possible. The availability of the new high-density Affymetrix products (and the limitations of the Affymetrix software) has in fact driven rapid development of new genotyping tools, and continues to drive database development in the life sciences. Witness GPL96, a code used by GEO (gene expression omnibus, http://www.ncbi.nlm.nih.gov/geo/) to represent the Affymetrix U133A GeneChip platform. Overwhelmingly most of the data stored at GEO is of this variety, so any high volume database mechanism must take into consideration the Affymetrix format.

Silicon Genetics (www.silicongenetics.com), now part of Agilent Technologies (www.chem.agilent.com), is one of the leaders in expression software and has developed just such a tool based on its GeneSpring software package. GeneSpring GT (aka Varia for genotyping analysis) is now paired with GeneSpring GX (for expression) and is designed to handle many types of linkage and association studies, but does so in the context of visualizing the entire genome, zoomable and hotlinked. The user can display genes, cytobands, SNPs, sequence, haplotypes, linkage scores, parental contribution, GOLD plots, clustering, pedigree maps, and much more on top of a fully interactive image of all human chromosomes. Functions that do not now exist in the base version of the software can be easily added through the visual programming interface, or through R or Java connections, while still accessing the built-in chromosome viewer. GeneSpring GT handles the entire NCBI human genome in memory, and allows the user to quickly move around the chromosome while zooming in on details including SNP names, chromosomal regions, gene names, exons and introns, and other features. Most importantly, the software automatically recognizes the data format of many types of genotyping equipment and makes data import easy. There are even scripts to allow GeneSpring GT to connect directly to a relational database or data warehouse.

Insightful Corporation (www.insightful.com) has invested in genomic analysis software with its soon-to-be-released S+Gene software package, adding to its Array-Analyzer Module already in production. This new library of statistical genetics functions includes a number of analyses: QC (pedigree check, Hardy–Weinberg, etc.), linkage and association analyses, haplotype mapping and analysis models combining haplotype estimation and association, TDT, QTL, variance components, and others. Few software companies have the statistical infrastructure like S+ that can be used to build these types of sophisticated genetic algorithms, and the bonus is built-in integration with Java, C++ and relational databases as part of the core of S+. These fundamental features make it possible to add many data access and analysis routines and functions that would be extremely difficult in any other platform. S+ is based on the S language originally developed at Bell Labs and has a number of high-level statistical algorithms and visualization features as a core part of the language. Integration with external data sources (ODBC, JDBC), visualization resources (VB, C++, Java), and languages (C++, Fortran, Java) allow the user to create sophisticated multilocus analyses without undue concern about memory, speed, or platform compatibility.

DISCUSSION

Many high-quality biomolecular detection systems now exist in order to provide users with quantitative or qualitative information about interactions between biological species extracted from a cell or population of cells. The user now has a unique problem; how to choose the best platform that matches the detection requirements of the experimenter, both statistically and qualitatively, and is of appropriate density and throughput. Many biological questions must be asked with statistical limits in mind, to ensure that the inherent detection limits are not exceeded. So many alternative platforms now exist that the user must extend the experimental design beyond the standard alpha (false positive rate), beta (false negative rate), and delta (minimum detection limit) requirements, and should start to examine the experimental hypothesis to determine if the detection system chosen matches the user expectations without wasting resources. Does the experiment really require high-density SNP detection as part of an association study or would low density be adequate? Will the user look at more than a handful of genes or is the experiment designed to extract information from a whole genome scan? Will the user need quantitative protein data or will qualitative data suffice? Experimental design becomes much more important in the postgenome world, and with dozens of technological platforms to choose from, it becomes vital that the experimenter thoroughly analyze his or her experiment from the ground up if only to determine the absolute needs and expectations inherent in the experimental design.

A standard spotted array consists of an expression probe bound to a hydrophobic surface where quantitative binding of fluorescent target is measured via confocal scanning. Now much more complex tertiary interactions, such as protein–protein, or antibody–protein abundance arrays require care during development to determine the proper annealing temperatures and the best wash conditions. Now, a microarray surface can be the location of enzymatic reactions, such as single-extension priming for SNP detection or protein–protein interactions in the case of protein abundance chips. In such a case, the substrate must be enzyme-friendly and should allow free access between substrate and reactants in a near fluid-phase environment. Manufacturers have used the ongoing sophistication of microfluidic devices to ensure consistency of sample preparation, especially when the devices are used outside a biological laboratory. Increasingly, manufacturers are taking more processes traditionally left to technicians inside the MEMS devices to decrease the potential for user error and to increase precision.

A slightly more radical deviation from standard flat microarrays is electrical detection of biomolecules, especially nucleic acids. Direct detection of electrical current passing through DNA of different sequences is a sensitive and accurate method for detection of polymorphisms in defined regions of DNA. These systems are limited in the size of DNA that can be accurately detected, but using electrical charge to measure and/or capture nucleic acids has been remarkably accurate. Creating amplicons (small amplified regions of DNA that contain target SNPs) significantly reduces the burden placed on the SNP detector. Using a ferrous or other electrically conducive conjugate along with DNA amplicons effectively increases sensitivity and repeatability at the expense of more time necessary for preparation of samples. Fragmenting DNA, either randomly using physical shearing or systematically using restriction enzymes, is necessary for e-detection systems and both methods are used in many of the most popular SNP detection systems. The strength of e-detection systems is very clear when the analyst is looking at a few highly significant and/or clinically relevant SNPs, or when performing point-of-service detection, as the military is doing with its biohazard identification equipment.

Bead-based detection systems have appeared in a variety of formats. Dynal microbeads have several useful properties including magnetism, paramagnetism, the property of being easily coated with fluorescent tags, simple conjugation chemistry to DNA and RNA probes, peptides, antibodies, complex proteins, and many other biomolecules, and abundance. It is becoming obvious that microbeads are so useful that we should expect many more applications and platforms to come online in the next few years. Zero-gravity synthesis of microbeads allows even more uniform sizes and shapes and a wider variety of substrates. New physical properties have been seen in microbeads made from glass, exotic metals, and even carbon structures such as nanotubes and buckyballs.

Sequencing has always been a brute force approach to SNP and protein peptide sequence detection, but new mass spectrometry methods have increased the accuracy, speed, and multiplexing of such methods. Sequenom is leading the commercialization of SNP detection using MALDI-TOF MS and informatics-rich applications, but many other companies are leveraging the high degree of flexibility and precision of mass spectrometry in proteomics and metabolomics, such as Agilent Technologies. Mass spectrometry has always performed best when analyzing single highly purified compounds. Now mass spectrometry and the associated analytical tools can analyze highly complex biological mixtures with sensitivity and selectivity. The MS approach will be much more visible in the future, especially as portable field units are coming online.

Ever more sophisticated detection methods are developed each year. Today it is not unusual to see offshore companies developing astounding technology but at the same time wisely avoiding the associated licensing fees and potential lawsuits that hound technological development in the U.S. and U.K. Unfortunately, some very basic array-based methodologies have been patented, and patent owners continue to pursue litigation as a means to produce revenue. This situation has driven some very creative companies far from U.S. shores and in many cases has made the manufacture of array technology too expensive to complete. For example, Phalanx Biotech (www.phalanxbiotech.com) in Hsinchu, Taiwan, will market a $100 expression array to the world market but must remain ever vigilant to protect itself from lawsuits. This trend will likely continue as technology forges ever forward and prices continue to drop on experiments that increase in density and throughput. Some day a single chip will measure the genome, transcriptome, and proteome at the same time and analysis software will analyze these data in an integrated fashion yielding high-quality insights into disease, development, and basic biology. Until then companies like Agilent technologies are pushing the envelope by developing (or acquiring) technologies that enable whole cell investigation (Figure 19.2YY). The 2100 Bioanalyzer® can measure RNA, protein, cell fluorescence, and others and is highly adaptable (protein solutions utilize Caliper Life Sciences' lab-on-a-chip technology (Figure 19.2P right). Agilent also markets an HPLC chip that allows nanoflow electrospray LC/MS that encompasses all hydraulics, electrospray emitters, analytical and enrichment columns all on a single chip. New ChIP-on-Chip (chromatin immunoprecipitation) [62] arrays, oligoCGH, expression, spectroscopy and mass spectrometry, and data analysis software divisions all work together in concert to provide the potential to measure all potential signals from a single cell, analyze and interpret the data, and begin to model the underlying physiology of a living cell. This is the new biomedical paradigm, this is the direction that biotechnology leads, and this is the path we must surely follow if personalized medicine and disease eradication are ever to succeed.

REFERENCES

1. R. Shyamsundar, Y.H. Kim, J.P. Higgins, K. Montgomery, M.J., A. Sethuraman, M. van de Rijn, D. Botstein, P. Brown, and J.R. Pollack, A DNA microarray survey of gene expression in normal human tissues, *Genome Biol*, Vol. 6, r22.1–r22.9, 2005.

2. S.C. Case-Green, K.U. Mir, C.E. Pritchard, and E.M. Southern, Analyzing genetic information with DNA arrays, *Curr Opin Chem Biol*, Vol. 2, 404–410, 1998.

3. M.S. Shchepinov, S.C. Case-Green, and E.M. Southern, Steric factors influencing hybridisation of nucleic acids to oligonucleotide arrays, *Nucl Acids Res*, Vol. 25, 1155–1161, 1997.

4. S.E. Beard, S.R. Capaldi, and P. Gee, Stress responses to DNA: Damaging agents in the human colon carcinoma cell line, RKO, *Mutation Res*, Vol. 371, 1–13, 1996.

5. M.J. Lee, P. Gee, and S.E. Beard, Detection of peroxisome proliferators using a reporter construct derived from the Rat Acyl-CoA oxidase promoter in the rat liver cell line 4-II-E, *Cancer Res*, Vol. 57, 1575–1579, 1997.

6. S.B. Baylin, J.G. Herman, J.R. Graff, P.M. Vertine, and J.P. Issa, Alterations in DNA methylation: a fundamental aspect of neoplasia, *Adv Cancer Res*, Vol. 72, 141–196, 1998.

7. T.T. Nguyen, C.T. Nguyen, F.A. Gonzales, P.W. Nichols, M.C. Yu, and P.A. Jones, Analysis of cyclin-dependent kinase inhibitor expression and methylation patterns in human prostate cancers, *Prostate*, Vol. 43, 233–242, 2000.

8. P.A. Jones and P.W. Laird, Cancer epigenetics comes of age, *Nat Genet*, Vol. 21, 163–167, 1999.

9. C.M. Chen, H.L. Chen, T.H. Hsiau, A.H. Hsiau, H. Shi, G.J. Brock, S.H. Wei, C.W. Caldwell, P.S. Yan, and T.H. Huang, Methylation target array for rapid analysis of CpG island hypermethylation in multiple tissue genomes, *Am J Pathol*, Vol. 163, 37–45, 2003.

10. J.F. Costello, M.C. Fruhwald, D.J. Smiraglia, L.J. Rush, G.P. Robertson, X. Gao, F.A. Wright, J.D. Fermsco, P. Peltomäki, J.C. Lang, D.E. Schuller, L. Yu, C.D. Bloomfield, M.A. Caligiuri, A. Yates, R. Nishikawa, H.-J. S. Huang, N.J. Petrelli, X. Zhang, M.S. O'Dorisio, W.A. Held, W.K. Cavenee, and C. Plass, Aberrant CpG-island methylation has non-random and tumor-type-specific patterns, *Nat Genet*, Vol. 24, 132–138, 2000.

11. J.G. Herman, A. Umar, K. Polyak, J.R. Graff, N. Ahuja, J.P. Issa, S. Markowitz, J.K. Willson, S.R. Hamilton, K.W. Kinzler, M.F. Kane, R.D. Kolodner, B. Vogelstein, T.A. Kunkel, and S.B. Baylin, Incidence and functional consequences of hMLH1 Promoter hypermethylation in colorectal carcinoma, *Proc Natl Acad Sci U S A*, Vol. 95, 6870–6875, 1998.

12. M.F. Kane, M. Loda, G.M. Gaida, J. Lipman, R. Mishra, H. Goldman, J.M. Jessup, and R. Kolodner, Methylation of the hMLH1 promoter correlates with lack of expression of hMLH1 in sporadic colon tumors and mismatch repair-defective human tumor cell lines, *Cancer Res*, Vol. 57, 808–811, 1997.

13. R.S. Gitan, H. Shi, C.-M. Chen, P.S. Yan, and T.H.-M. Huang, Methylation-specific oligonucleotide microarray: a new potential for high-throughput methylation analysis, *Genome Res*, Vol. 12, 158–164, 2001.

14. C. Mund, V. Beier, P. Bewerunge, M. Dahms, F. Lyko, and J.D. Hoheisel, Array-based analysis of genomic DNA methylation patterns of the tumor suppressor gene p16INK4A promoter in colon carcinoma cell lines, *Nucl Acids Res*, Vol. 33, e73; doi:10.1093/nar/gni072, 2005.

15. M. Schwarz, L. Spector, A. Gargir, A. Shtevi, M. Gortler, R.T. Altstock, A.A. Dukler, and N. Dotan, A new kind of carbohydrate array, its use for profiling antiglycan

antibodies, and the discovery of a novel human cellulose-binding antibody, *Glycobiology*, Vol. 13, 749–754, 2003.

16. M. Adinolfi and J. Sherlock, Fetal cells in transcervical samples at an early stage of gestation, *J Hum Genet*, Vol. 46, 99–104, 2001.

17. E. Fung, ProteinChip clinical proteomics: computational challenges and solutions, *Biotechniques*, Vol. 34, 40–41, 2002.

18. R.R. Martel, I.W. Botros, M.P. Rounseville, J.P. Hinton, R.R. Staples, D.A. Morales, J.B. Farmer, and B.E. Seligmann, Multiplexed screening assay for mRNA: Combining nuclease protection with luminescent array detection, *ASSAY Drug Dev Technol*, Vol. 1, 62–72, 2002.

19. P. Wagner and R. Kim, Protein biochips: an emerging tool for proteomics research, *Curr Drug Discov*, May, 23–28, 2002.

20. S.R. Weinberger, T.S. Morris, and M. Pawlak, Recent trends in protein biochip technology, *Pharmacogenomics*, Vol. 1(4):395–416, 2000.

21. G.L. Duveneck and A.P. Abel, Review on fluorescence-based planar waveguide biosensors, *Proc. SPIE*, Vol. 3858, 59–71, 1999.

22. P.Y. Chan-Hui, K. Stephens, R.A. Warnock, and S. Singh, Applications of eTag assay platform to systems biology approaches in molecular oncology and toxicology studies, *J Clin Immunol*, Vol. 111, 162–174, 2004.

23. H. Tian, C. Liching, T. Yuping, S. Williams, L. Chen, T. Matray, A. Chenna, S. Moore, V. Hernandez, V. Xiao, M. Tang, and S. Singh, Multiplex mRNA assay using electrophoretic tags for high-throughput gene expression analysis, *Nucl Acids Res*, Vol. 32, e126, 2003.

24. Y. Arntz, J.D. Seelig, H.P. Lang, J. Zhang, P. Hunziker, J.P. Ramseyer, E. Meyer, M. Hegner, and C. Gerber, Label-free protein assay based on a nanomechanical cantilever array, *Nanotechnology*, Vol. 14, 86, 2003.

25. J. Fritz, M.K. Baller, H.P. Lang, H. Rothuizen, P. Vettiger, E. Meyer, H.J. Guntherodt, C. Gerber, and J.K. Gimzewski, Translating biomolecular recognition into nanomechanics, *Science*, vol. 288, 316–318, 2000.

26. X. Gao, E. LeProust, H. Zhang, O. Strivannavit, E. Gulari, P. Yu, C. Mishiguchi, Q. Xiang, and X. Zhou, A flexible light-directed DNA chip synthesis gated by deprotection using solution photogenerated acids, *Nucleic Acids Res*, Vol. 29, 2, 2001.

27. J. Philippe Pellois, X. Zhou, O. Srivannavit, T. Zhou, E. Gulari, and X. Gao, Individually addressable parallel peptide synthesis on microchips, *Nat Biotechnol*, Vol. 20, 922–926, 2002.

28. T.R. Hughes, M. Mao, A.R. Jones, J. Burchard, M.J. Marton, K.W. Shannon, S.M. Lefkowitz, M. Ziman, J.M. Schelter, M.R. Meyer, S. Kobayashi, C. Davis, H. Dai, Y.D. He, S.B. Stephaniants, G. Cavet, W.L. Walker, A. West, E. Coffey, D.D. Shoemaker, R. Stoughton, A.P. Blanchard, S.H. Friend, and P.S. Linsley, Expression profiling using microarrays fabricated by an inkjet oligonucleotide synthesizer, *Nat Biotechnol*, Vol. 19, 342–347, 2001.

29. D.J. Fu, K. Tang, A. Braun, D. Reuter, B. Darnhofer-Demar, D.P. Little, M.J. O'Donnell, C.R. Cantor, and H. Koster, Sequencing exons 5 to 8 of the p53 gene by Maldi-TOF mass spectrometry, *Nat Biotechnol*, Vol. 16, 381–384, 1998.

30. J. Leushner, MALDI-TOF mass spectrometry: an emerging platform for genomics and diagnostics, *Expert Rev Mol Diagn*, Vol. 1, 11–18, 2001.

31. N. Storm, B. Darnhofer, D. van den Boom, and C.P. Rodi, MALDI-TOF mass spectrometry-based SNP genotyping, *Methods Mol Biol.*, Vol. 212, 241–262, 2003.

32. K. Tang, D.J. Fu, D. Julien, A. Braun, C.R. Cantor, and H. Koster, Chip-based genotyping by mass spectrometry, *Proc Natl Acad Sci U S A*, Vol. 96, 10016–10020, 1999.

33. D. van den Boom, C. Jurinke, S. Higgins, T. Becker, and H. Köster, Mass spectrometric DNA diagnosis, *Nucleosides Nucleotides*, Vol. 17, 2157–2164, 1998.
34. R.L. Somorjai, B. Dolenko, and R. Baumgartner, Class prediction and discovery using gene microarray and proteomics mass spectroscopy data: curses, caveats, cautions, *Bioinformatics*, Vol. 19, 1484–1491, 2003.
35. J.N. Housby, *Mass Spectrometry and Genomic Analysis*. Dordrech, Boston, MA: Kluwer Academic Publishers, 2001.
36. A. Oliphant, D.L. Barker, J.R. Stuelpnagel, and M.S. Chee, BeadArray technology: enabling an accurate, cost-effective approach to high-throughput genotyping, *Biotechniques* (Suppl.), 56–58, 60–61, 2002.
37. S. Brenner, M. Johnson, J. Bridgham, G. Golda, D. H. Lloyd, D. Johnson, S. Luo, S. McCurdy, M. Foy, M. Ewan, R. Roth, D. George, S. Eletr, G. Albrecht, E. Vermaas, S. R. Williams, K. Moon, T. Burcham, M. Pallas, R.B. DuBridge, J. Kirchner, K. Fearon, J. Mao, and K. Corcoran, Gene expression analysis by massively parallel signature sequencing (MPSS) on microbead arrays, *Nat Biotechnol*, Vol. 18, 630–634, 2000.
38. T.M. Hsu, X. Chen, S. Duan, R.D. Miller, and P.Y. Kwok, Universal SNP genotyping assay with fluorescence polarization detection, *Biotechniques*, Vol. 31, 560–570, 2001.
39. X. Chen, L. Levine, and P.Y. Kwok, Fluorescence polarization in homogenous nucleic acid analysis, *Genome Res*, Vol. 9, 492–498, 1999.
40. P. Hardenbol, F. Yu, J. Belmont, J. Mackenzie, C. Bruckner, T. Brundage, A. Boudreau, S. Chow, J. Eberle, A. Erbilgin, M. Falkowski, R. Fitzgerald, S. Ghose, O. Iartchouk, M. Jain, G. Karlin-Neumann, X. Lu, X. Miao, B. Moore, M. Moorhead, E. Namsaraev, S. Pasternak, E. Prakash, K. Tran, Z. Wang, H.B. Jones, R.W. Davis, T.D. Willis, and R.A. Gibbs, Highly multiplexed molecular inversion probe genotyping: over 10,000 targeted SNPs genotyped in a single tube assay, *Genome Res*, Vol. 2, 269–275, 2005.
41. G.A. Thorisson and L.D. Stein, The SNP consortium Web site: past, present and Future, *Nucl Acids Res*, Vol. 31, 124–127, 2003.
42. D.G. Georganopoulou, L. Chang, J.N. Nam, C.S. Thaxton, E.J. Mufson, W.L. Klein, and C.A. Mirkin, Nanoparticle-based detection in cerebral spinal fluid of a soluble pathogenic biomarker for Alzheimer's disease, *Proc Natl Acad Sci U S A*, Vol. 102, 2263–2264, 2005.
43. M. Huber, T.F. Wei, U.R. Müller, P.A. Lefebvre, S.S. Marla, and Y.P. Bao, Homogeneous detection of unamplified genomic DNA sequences bases on colorimetric scatter of gold nanoparticle probes, *Nat Biotechnol*, Vol. 22, 883–887, 2004.
44. Y.P. Bao, M. Huber, T.F. Wei, S.S. Marla, J.J. Storhoff, and U.R. Müller, SNP identification in unamplified human genomic DNA with gold nanoparticle probes, *Nucl Acids Res*, Vol. 33, e15, 2005.
45. S.J. Rosenthal, Bar-coding biomolecules with fluorescence nanocrystals, *Nat Biotechnol*, Vol. 19, 621–622, 2001.
46. E. Tesfu, K. Maurer, S.R. Ragsdale, and K.F. Moeller, Building addressable libraries: the use of electrochemistry for generating reactive Pd(II) reagents at preselected sites in a chip, *J Am Chem Soc*, Vol. 126, 6212–6213, 2004.
47. M.A. Lyon, M.D. Musick, and M.J. Natan, Colloidal Au-enhanced surface plasmon resonance immunosensing, *Anal Chem*, Vol. 70, 5177–5183, 1998.
48. S.P. Mulvaney, M.D. Musick, C.D. Keating, and M.J. Natan, Glass-coated, analyte tagged nanoparticles: a new tagging system based on detection with surface-enhanced raman scattering, *Langmuir*, Vol. 19, 4784–4790, 2003.

49. H. Cai, P.S. White, D. Torney, A. Deshpande, Z. Wang, B. Marrone, and J.P. Nolan, Flow cytometry-based minisequencing: a new platform for high-throughput single-nucleotide polymorphism scoring, *Genomics*, Vol. 66, 135–143, 2000.

50. J. Chen, M.A. Iannone, S. Li, J.D. Taylor, P. Rivers, A.J. Nelson, K.A. Slentz-Kesler, A. Roses, and M.P. Weiner, A microsphere-based assay for multiplexed single nucleotide polymorphism analysis using single base chain extension, *Genome Res*, Vol. 10, 549–557, 2000.

51. B.J. Cheek, A.B. Steel, M.P. Torres, Y.-Y. Yu, and H. Yang, Chemiluminescence detection for hybridization assays on the flow-thru chip, a three-dimensional micro-channel biochip, *Anal Chem*, Vol. 73, 5777–5783, 2001.

52. Y. Belousov, R.A. Welch, S. Sanders, A. Mills, A. Kulchenko, R. Dempcy, I.A. Afonina, D.A. Walburger, C.L. Glaser, S. Yadavalli, N.M.J. Vermeulen, and W. Mahoney, Single nucleotide polymorphism genotyping by two colour melting curve analysis using the MGB Eclipse™ probe system in challenging sequence environment, *Hum Genomics*, Vol. 1, 209–217, 2004.

53. S. Lokhov, E. Lukhtanov, and M.W. Reed, Chemistry of minor groove binder-oligonucleotide conjugates, in *Current Protocols in Nucleic Acid Chemistry*, John Wiley & Sons, New York, 2003, pp. 8.4.1–8.4.20.

54. V. Stolc, Z. Gauhar, C. Mason, G. Halasz, M.F. van Batenburg, S.A. Rifkin, S. Hua, T. Herreman, W. Tongprasit, P.E. Barbano, H.J. Bussemaker, and K.P. White, A gene expression map for the euchromatic genome of *Drosophila melanogaster*, *Science*, Vol. 306, pp. 655–660, 2004.

55. C.W. Wong, T.J. Albert, V.B. Vega, J.E. Norton, D.J. Cutler, T.A. Richmond, L.W. Stanton, E.T. Liu, and L.D. Miller, Tracking the evolution of the SARS coronavirus using high-throughput, high-density resequencing arrays, *Genome Res*, Vol. 14, pp. 398–405, 2004.

56. R. Lucito, J. Healy, J. Alexander, A. Reiner, D. Sposito, M. Chi, L. Rodgers, A. Brady, J. Sebat, J. Troge, J.A. West, S. Rostan, K.C.Q. Nguyen, S. Powers, K.Q. Ye, A. Olshen, E. Venkatraman, L. Norton, and M. Wigler, Representational oligonucleotide microarray analysis: a high-resolution method to detect genome copy number variation, *Genome Res*, 13: 2277–2290, 2003.

57. A. Kirmizis, S.M. Bartley, A. Kuzmichev, R. Margueron, D. Reinberg, R. Green, and P.J. Farnham, Silencing of human polycomb target genes is associated with methylation of Histone H3 Lys 27, *Genes Dev*, Vol. 18, 1592–1605, 2004.

58. D.A. Hinds, L.L. Stuve, G.B. Nilsen, E. Halperin, E. Eskin, D.G. Ballinger, K.A. Frazer, and D.R. Cox, Whole-genome patterns of common DNA variation in three human populations, *Science*, Vol. 307, 1072–1079, 2005.

59. P.A.C.'t Hoen*, R. Turk, J.M. Boer, E. Sterrenburg, R.X. de Menezes, G.B. van Ommen, and J.T. den Dunnen, Intensity-based analysis of two-colour microarrays enables efficient and flexible hybridization designs, *Nucl Acids Res*, Vol. 32, e41, 2004.

60. K.M. Ririe, R.P. Rasmussen, and C.T. Wittwer, Product differentiation by analysis of DNA melting curves during the polymerase chain reaction, *Anal Biochem*, Vol. 245, 154–160, 1997.

61. K. Jaakson, J. Zernant, M. Kulm, A. Hutchinson, N. Tonisson, D. Glavaci, D. Ravnik-Glavaci, M. Hawlina, M.R. Meltzer, R.C. Caruso, F. Testa, A. Maugeri, C.B. Hoyng, P. Gouras, F. Simonelli, R.A. Lewis, J.R. Lupski, F.P.M. Cremers, and R. Allikmets, Genotyping microarray (Gene Chip) for the ABCR (ABCA4) gene, *Hum Mutation*, Vol. 22, 395–403, 2003.

62. P.M. Das, K. Ramachandran, J. VanWert, and R. Singal, Chromatin immunoprecip-itation assay, *Biotechniques*, Vol. 37, 961–969, 2004.

20 Intellectual Property Issues for DNA Chips and Microarrays

Vicki G. Norton

CONTENTS

Development of the array format for DNA hybridization assays has changed the speed and throughput that researchers can achieve, permitting researchers to study hundreds or thousands of DNA sequences in a single experiment.

DNA chips and microarrays have already proven to be valuable tools in drug discovery, leading to the filing of patent applications directed to arrays and their uses, including target identification, SNP analysis for the diagnosis of genetic diseases, gene expression, pharmacogenomics, toxicogenomics, proteomics, and bioinformatics. Recent FDA approvals of the use of microarrays as Class II medical devices should further expand commercial uses for microarrays and gene chips.

As with any valuable technology, the commercialization of DNA array and gene chip technologies has led to aggressive patenting of technologies and patent litigation as array manufacturers seek to stake out a share of the market.* Meanwhile, the wealth of genomic information available from the Human Genome Project has led to a deluge of patent applications filed at the United States Patent and Trademark Office (PTO), and initial attempts to patent raw sequencing data have resulted in changes in the guidelines used by PTO to examine biotechnology and genomic patents. This chapter surveys the impact of recent court decisions and PTO proceedings that bear on array and genomic technologies.

OVERVIEW OF PATENT LAW STATUTES AND GUIDELINES APPLIED TO GENOMIC INVENTIONS

The general rule under U.S. patent laws is that any new, useful, and nonobvious aspects of array technology may be patentable if the applicant sufficiently describes the claimed technology. Among the many types of claims found in patents issued are claims to the design itself of the arrays, methods of making and using the arrays, methods of preparing samples to be analyzed on the arrays, and methods, compositions, and devices for generating, measuring, and analyzing signals from arrays. These are further described as follows:

- Patents to the array design: Patent claims may be directed to features of the array design, such as the density, size, and arrangement of the capture oligonucleotide, polynucleotide, or cDNA; the chemical nature and spatial arrangement of the array surface; methods of making arrays, such as *in situ* synthesis, or spotting array features; and devices to make the designed arrays
- Patents to methods of using arrays and to the targets and ligands identified using the arrays: Patents can include methods of using arrays for various applications such as pharmacogenomics, toxicogenomics, and expression analysis; analysis of SNPs/mutation/polymorphism; sequencing; methods of identifying new targets and ligands; and the ligands and targets identified using the arrays. Certain aspects of new pathways discovered with the help of arrays may also be patentable.
- Patents to samples and methods of preparing samples: Claims in patents may also cover methods of preparing samples to be analyzed using the arrays, such as libraries of nucleic acids for the arrays, and robust, representative amplification methods such as whole genome amplification samples, as well as methods of labeling the samples.
- Patents to methods of hybridizing nucleic acids to the arrays: In addition, patents may cover methods of hybridization and ways to measure hybridization.
- Patents to methods and devices used in data analysis: Patents may also cover methods, devices, and software to scan, quantitate, and process data

---

* *See*, for example, "Patently Inefficient," *Scientific American*, February 2001 (setting forth in a chart a summary of U.S. and European lawsuits relating to array technology as of early 2001).

from arrays; improvements to make assays more sensitive to changes in expression level, and to determine whether differences in expression level are meaningful or result from errors in the assay; image-processing techniques; methods of overlaying array data after hybridization with two differently color-labeled samples (e.g., mRNAs from two different tissues, (or tumor vs. nontumor); and improved methods for the design and analysis of microarray-based experiments.

Patents directed to these and other aspects of array technology are subject to the same general legal requirements applicable to all patents. Those legal requirements originate in the U.S. Constitution's mandate that Congress enact laws to "promote the progress of science and the useful arts, by securing for limited times to authors and inventors the exclusive right to their respective writings and discoveries." (Art. I, § 8). The patent statutory scheme was accordingly enacted to strike a bargain between the public and the inventor, in which the inventor gains the right to exclude others from using patented technology for a limited term (generally 20 years from filing for most patents issuing under current patent laws). In return for this exclusionary right, the inventor must provide a description of the invention, which clearly informs the public what activity the patent claim excludes, permits the public to determine that the inventor was in possession of the patented technology, and provides the public with information sufficient to make and use the invention once the period of exclusivity has passed. The patent statutes therefore promote technological discoveries by offering incentives to the first discoverer of a technological innovation to commercialize the patented invention, while at the same time benefiting the public.

One important aspect of the patent law is that obtaining a patent does not give its owner the right to use the technology — a patent gives its owner the right to preclude others from using or commercializing the patented technology. However, others may own patents, which in turn may prevent the patent owner from using or commercializing a patented technology. Of course, a patent owner can license or cross-license a patent to give others the right to use a patented technology.

Because U.S. patent applications remain confidential until publication 18 months after filing, it is impossible to determine at any given time which technologies are the subject of U.S. patent applications. In addition, determining whether the use of specific DNA chip or microarray technology might be covered by a third party's patent can be a time-consuming process. *

---

* For example, a researcher who decides to use a nucleic acid sequence as a probe on an array, based on an analysis of sequences in public databases and published patents and patent applications, may risk a lawsuit for patent infringement if a private company later secures a patent on the basis of previously filed patent applications. http://www.ornl.gov/sci/techresources/Human_Genome/elsi/patents.shtml. Moreover, it might be possible for a single sequence to be the subject of more than one patent, referred to as patent "stacking." For example, depending on the information included in each patent application, it might be possible for multiple applicants to obtain a claim to a short sequence as an EST, for use as a probe, and as part of the sequence of a full-length gene. This further contributes to the difficulty in determining whether the use of specific sequences in DNA chips or microarrays might be covered by a third party's patent.

GENERAL LEGAL REQUIREMENTS FOR PATENTABILITY

Under U.S. patent laws, an invention claimed in a patent must meet the following three legal requirements (see 35 U.S.C. § 101* to 103):

1. The invention must be useful; e.g., it must have a specific, practical utility.**
2. The invention must be new (novel, in view of what was known, to one with ordinary skill in the art).
3. The invention must be nonobvious (not obvious, in view of what was known, to one with ordinary skill in the art).

The patent application must also meet certain legal requirements to ensure that the claimed invention is described sufficiently so that the public can enjoy the full benefit of its bargain once the period of the patentee's exclusive rights ends and the technology passes into the public domain for use by the public. For example, the patent application must meet the following requirements set forth in 35 U.S.C. § 112, ¶¶ 1–2:

1. The "enablement" requirement: The patent must contain an enabling description of the invention sufficient to enable those skilled in the art to make and use the invention (¶ 1).
2. The "written-description" requirement: The patent must contain a sufficient written description to inform those of skill in the art that the inventor was in possession of any claimed invention that the patent owner would seek to exclude the public from using (¶ 1).
3. The "definiteness" requirement: The patent must claim the invention in clear and definite terms, so those of skill in the art can understand what activities fall within scope of the invention (¶ 2).

In litigation, if an accused patent infringer can prove that the patent fails to meet one or more of these requirements, the patent will be invalid.

---

* 35 U. S. C. §101 provides:

"Whoever invents or discovers any new and useful process, machine, manufacture, or composition of matter, or any new and useful improvement thereof, may obtain a patent therefore, subject to the conditions and requirements of this title."

** The United States Supreme Court provided the following perspective on the origin of the "utility" requirement in the 1952 Patent Act:

The act embodied Jefferson's philosophy that "ingenuity should receive a liberal encouragement." (The V Writings of Thomas Jefferson, at 75–76. See *Graham v. John Deere Co.*, 383 U.S. 1, 7–10, 1966). Subsequent patent statutes in 1836, 1870, and 1874 employed the same broad language. In 1952, when the patent laws were recodified, Congress replaced the word "art" with "process," but otherwise left Jefferson's language intact. The Committee Reports accompanying the 1952 act inform us that Congress intended statutory subject matter to "include anything under the sun that is made by man." (S. Rep. No. 1979, 82d Cong., 2d Sess., 5 (1952); H.R. Rep. No. 1923, 82d Cong., 2d Sess., 6, 1952).

Diamond v. Chakrabarty, 447 U.S. 303, 309 (1980).

The application of the requirements for stating definiteness and written description to a patent on array technology is illustrated by the court's ruling in the patent infringement suit brought by Affymetrix against Synteni, Inc. (Synteni) and Incyte Pharmaceuticals (Incyte) in the Northern District of California. Array manufacturer Affymetrix alleged that Synteni and Incyte infringed various patents including U.S. Patent 5,800,992 (the '992 patent), directed to "two-color" methods of using a polynucleotide array to detect a substantially complementary nucleic acid sequence in two or more collections of distinguishably labeled nucleic acids, and U.S. Patent 5,744,305 (the '305 patent), directed to certain arrays of materials attached to a substrate in predefined regions.

Incyte successfully argued before the district court that the claims of the '992 patent were invalid for failing to clearly and definitely set forth what was encompassed by claims 1–3, because the meaning of the term "substantially complementary" was unclear to one skilled in the art. The court also sided with Incyte in holding that claims 4–5 of the '992 patent were invalid for failing to provide a written description of multiplex detection of a nucleic acid in two or more collections of nucleic acids labeled with distinguishable labels. In addition, the court found that Incyte's cDNA arrays did not infringe any of the claims limited to use of oligonucleotide arrays in the '934 or '305 patents (see, for example, Incyte Genomics (INCY) Receives Favorable Court Ruling Invalidating Affymetrix (AFFX) Patent, Biospace Beat, http://links.biospace.com/news_story.cfm?StoryID=6574415&full=1). The case was settled before the ruling by the district court was appealed.

In other litigation related to array technology, Oxford Gene Technology, which owns the array technology developed by Edwin M. Southern, filed lawsuits in December 2002 for infringement of U.S. Patent No. 6,054,270 against a number of companies, including BioDiscovery, Mergen, Nanogen, Genomic Solutions, Axon Instruments, BD Biosciences Clontech, Nanogen Inc., Axon Instruments Inc., Biodiscovery Inc., Mergen Ltd., and Motorola, along with PerkinElmer Life Sciences Inc. and Harvard Bioscience Inc. unit Genomic Solutions Inc. OGT has since settled the lawsuits with BioDiscovery, Mergen, Nanogen, Genomic Solutions, Axon Instruments, and Motorola. In Europe, after its array patent was challenged, OGT voluntarily narrowed its patent to cover only arrays containing oligonucleotides that are covalently attached to a smooth impermeable surface. It is not clear if a patent could be challenged in the U.S. if it extended beyond the scope of the European patent.*

In the related area of genomic inventions, which might cover, for example, the sequences attached to the surface of the array, targets identified using the arrays, or even therapeutic biologics discovered using the array, PTO and the courts have grappled most with § 101's utility requirement and § 112's written-description standard for informing one skilled in the art that the inventor is in possession of the invention. The utility requirement is usually implicated when sequence data is available but the function of the encoded protein is not known (e.g., EST sequences or homologous sequence data). Conversely, the written-description requirement is usually implicated for genomic or proteomic inventions when the function of a

---

* OGT patent 6,770,751, issued in August 2004, is directed to certain methods to detect variations in DNA sequences, e.g., SNPs or different lengths of tandem repeat regions.

protein or DNA is known (e.g., the DNA encodes a protein that inhibits a receptor) but its entire nucleotide or amino acid sequence is not known.

In the past, PTO has taken a somewhat narrow view of whether a patent application discloses an invention having sufficient utility to meet the requirement of § 101, although courts have generally taken a broader view of the type of subject matter that meets the utility requirement. It was only through the action of courts (reversing a decision by PTO) that patents were first obtained to genetically engineered organisms.

On the other hand, courts have tended to take a narrower view of the written-description requirement than PTO. Several patents issued by PTO to biomolecules described in functional terms have been held invalid by the Federal Circuit Appeals Court, which is the federal appeals court that hears all appeals in patent cases.* The Federal Circuit has characterized DNA as a large chemical compound whose structure can only be adequately described when its sequence is known. Accordingly, the Federal Circuit has held that claims to biochemicals stated in terms of biological function lack sufficient written description unless a structure (i.e., the sequence or a deposited cell line containing the cloned gene) is associated by those of skill in the art with that function.

THE UTILITY REQUIREMENT: AT WHAT STAGE OF DISCOVERY CAN A BIOCHEMICAL BE PATENTED?

Federal courts began applying patent law requirements to products obtained from living systems in 1911 when a court sustained the validity of claims to an extracted and concentrated form of adrenaline. (*Parke Davis & Co. v. H.K. Mulford Co.,* 189 Fed. 95, S.D.N.Y. 1911, *aff'd*, 196 F. 496, 2d Cir. 1912).** The Federal Appeals Court held that although powdered adrenal gland tissue had been used for medicinal purposes, purified adrenaline was free of the disadvantages of using the unpurified form and was therefore a new commercial therapeutic entitled to patent protection. Thus, although hormones, genes, proteins, and other biochemicals are not patentable in their naturally occurring form, purified or isolated hormones, genes, gene fragments and proteins may be patentable if they meet the criteria for patentability, including utility, novelty, and nonobviousness, and they are properly described.

After molecular biology techniques enabled the genetic engineering of living organisms, courts faced new questions of how to apply the legal requirements of patentability to living systems. Prior to 1980, PTO had refused to grant patents to living organisms, which were considered products of nature lacking sufficient utility to meet the requirement of section 101.

In 1980, Dr. Chakrabarty appealed PTO's rejection of his patent application to a bacterium genetically engineered to consume oil spills, and the case was finally

---

* Parties who lose before the Federal Circuit can request review of the case by the Supreme Court of the United States; however, it is relatively uncommon for the Supreme Court to agree to review a patent case.

** Prior to the establishment of the Federal Circuit on October 1, 1982, by the Federal Courts Improvement Act of 1982, Pub. L. No. 97-164, 96 Stat. 25., appeals from patent cases were heard by various regional circuit courts of appeal, such as the Second Circuit.

heard by the Supreme Court. In *Diamond v. Chakrabarty*, 447 U.S. 303, 309 (1980), the Supreme Court construed § 101 broadly, holding that Dr. Chakrabarty's bacterium was patentable; the court cited the legislative history of § 101 to support its conclusion that, "Congress intended statutory subject matter to 'include anything under the sun that is made by man.'" See *Diamond v. Chakrabarty*, 447 U.S. 303, 309 (1980) (quoting S. Rep. No. 82-1979, at 5 (1952); H.R. Rep. No. 82-1923, at 6 (1952)). Later cases reaffirmed the basic rule that under section 101 products or processes made by man are patentable, whereas products and processes of nature are not. See *SmithKline Beecham Corp. v. Apotex Corp.*, 365 F3d 1306 (Fed. Cir. 2004) (citing *Chakrabarty* at 313; *J.E.M. Ag Supply v. Pioneer Hi-Bred Int'l*, 534 U.S. 124, 130 (2001)); see also *Gottschalk v. Benson*, 409 U.S. 63, 67 (1972) ("Phenomena of nature, though just discovered, mental processes, and abstract intellectual concepts are not patentable, as they are the basic tools of scientific and technological work.")*. In 2001, PTO promulgated utility guidelines to address concerns raised over the examination of genomic-related inventions.

Early projects relating to the Human Genome Project brought many issues relating to patenting of human DNA sequences to the forefront. When the Human Genome Project began in 1990, scientists had discovered fewer than 100 human disease genes; today, more than 1400 disease genes have been identified. http://www.genome.gov/11006929.**

The question, who owns the human genome, was first raised in 1991 when the NIH applied for a patent on brain cDNA discovered in an EST*** project led by Dr. Craig Venter. (Tom Strachan and Andrew P. Read, *Human Molecular Genetics 2*, BIOS Scientific Publishers Ltd. 1999). PTO rejected the applications for lack of utility. Venter later left the NIH to set up a commercially backed Institute of Genome Research, and eventually joined forces with Applied Biosystems/Celera Genomics to sequence the human genome.

As the private sector vied with the public sector to be the first to sequence the human genome, the race was on to file thousands of applications directed to DNA sequences, gene fragment sequences, or expressed DNA sequences before any further information was ascertained about the function of the encoded protein. Although Celera Genomics, the winner of the race to sequence the human genome initially filed over 6500 patent applications on gene sequences, its then president, Craig Venter, said in a 2000 interview that the company intended to pursue 500 or so patents on specific genes that might be significant for drug development. (Kristen Philipkoski, "Celera Wins Genome Race," *Wired*, 4/6/2000.) Over 3 million genome-related patent applications have been

---

* In 1987, in *Ex parte Allen*, 2 USPQ2d 1425, 1428 (Bd. Pat. App. & Inter. 1987).[3] The Board of Patent Appeals and Interferences found that subject matter patentable under §101 extended to man-made life forms. One year later, the USPTO issued the famous "Harvard mouse" patent to a transgenic mouse.

** In 2003, the Human Genome Project Consortium announced that the goal to sequence the human genome had been achieved at a cost under the $3 billion estimate, and 2 years ahead of schedule. The wealth of genomic information made available from sequencing of the entire human genome, combined with high-throughput array technology, has permitted scientists in industry and academia to quickly explore the genomic data.

*** ESTs, or expressed sequence tags are 300- to 500-bp gene fragments, which represent only 10 to 30% of the length of the average cDNA, often 10 to 20 times smaller than the corresponding genomic gene.

filed.* (Human Genome Project information at http://www.ornl.gov/sci/
techresources/Human _Genome/elsi/patents.shtml).

The number of patent applications filed for gene fragments such as ESTs, and other
partial gene sequences raised concerns among researchers in both academia and private
industry, as well as commentators, over whether PTO should grant patents based on
early-stage genome sequence data, and if so, what standards should be applied.

Some researchers and industry commentators maintained that patenting gene fragments was inappropriate because the methods used to obtain raw sequence data or to
identify ESTs were routine, and the amount of work to obtain the sequence data was
small compared to the amount of work spent developing a drug.** In other words,
permitting patents on ESTs would reward the smallest up-front contribution, but not
the harder effort of isolating and sequencing a gene, finding out what the gene product
does, and developing a commercial product based on it once the target gene and protein
product were known. Allowing applicants to obtain "gatekeeper" patents on raw
sequences would permit them to control the use and commercialization of further
genomic research. Proponents of patenting ESTs, however, argued that the gene fragments were useful as molecular probes to search for complete cDNA sequences.

In 1999, to address concerns raised in debates between the academic scientific
community, the private sector, industry commentators, and bioethicists, USPTO
promulgated guidelines for examiners to determine whether applications for patents
to genomic inventions meet the legal standard for utility under § 101 and to address
questions and policy concerns raised over the patentability of sequence information
obtained in the Human Genome Project.

The PTO's Revised Interim Utility Examination Guidelines issued in January
2001 addressed comments from both sides and appeared to resolve competing
concerns by maintaining the patentability of gene sequences, although tightening
the utility requirement for patents to genes and gene fragments.

With its systematic consideration of a wide range of comments, the 2001 Utility
Guidelines drew praise from those associated with genomic companies. William Haseltine,

---

* Many biotech companies have applied for provisional patents, which provides a 1-year grace period
for determining whether the company wishes to proceed with filing a utility application. This means that
companies filing the provisional patent application have up to 1 year to file their actual patent claims in
a utility patent application. The 1-year grace period does not count as one of the 20 years that the patent
is issued for.

** For example, in a comment in *Nature Biotechnology*, John H. Barton proposed three further limitations
(1) whether data obtained from automated gene sequencing is really something "made by man"*;
(2) the weighing of economic policy considerations in favor of rewarding the development of research
tools with a patent monopoly with the need to split the benefits of the monopoly rent on a new
pharmaceutical drug with all tool makers who helped in development of the drug dictates that a tool such
as an SNP or an EST should be patentable only if the benefit of such a patent in strengthening incentives
to develop genomic information is greater than the costs of the patent in foreclosing others' ability to
use the information about the genome; (3) determining whether bioinformatics data is patentable as a
description of a property of a physical chemical, or if it is not patentable subject matter because it merely
represents naturally occurring information (and a law of nature cannot be patented); if rulings in court
cases finding software applications patentable are extended to bioinformatics, it could blur the distinction
between patentable subject matter and a principle of nature and may spur the filing of patent claims
designed to exclude others from using genomic information. (John H. Barton, Commentary, 18:9 *Nature
Biotechnology* 804, August 2000).

then the chairman and CEO of Human Genome Sciences (Rockville, MD), was quoted as saying, "I think this could be the Magna Carta of biotechnology." (David Holzman, "Magna Carta" of biotechnology, *Genetic Engineering News*, February 1, 2001. "This is probably the most systematic review of this field that has ever been done. It acknowledges every comment and lays out a rational, clear response to each one.")

The 2001 Utility Guidelines clarified to patent examiners that a DNA fragment such as an EST would not meet the utility requirement if the patent application failed to disclose at least one practical utility. In the past, patenting of a gene sequence was allowed based on general claims such as its use as a probe for locating the full-length gene; now, such a general claim would be insufficient. To meet the requirement, either (1) the application had to disclose a specific, substantial, and credible utility or (2) those skilled in the art would have to readily recognize that the invention was useful based on a well-established utility (Utility Guidelines at 1094). The burden of disclosure, or of demonstrating that those skilled in the art are aware of a well-established utility, would fall on the patent applicant (Utility Guidelines at 1096–1097).

Under the revised guidelines, a specific, substantial utility is provided only if the applicant has disclosed that the invention is useful for a particular, practical purpose. A specific, substantial, and credible utility cannot be "throwaway," "insubstantial," or "nonspecific" such as use of a composition as landfill (Utility Guidelines at 1098). Credibility is measured from the perspective of one of ordinary skill in the art who has viewed the disclosure and any other evidence; e.g., test data, affidavits, or declarations from experts in the art, patents, or printed publications (Utility Guidelines at 1098). The PTO further elaborated that "[a] claimed DNA may have a specific and substantial utility because, e.g., it hybridizes near a disease-associated gene or it has a gene-regulating activity." (January 5, 2001 Utility Examination Guidelines, 66(4) Fed. Reg. 1092, 1095).

The PTO guideline requirement for a specific, credible, substantial utility appeared to track the utility requirement formulated by the Supreme Court in *Brenner v. Manson*, 383 U.S. 519 (1966). In *Brenner*, the Supreme Court emphasized the need for the disclosure of a present, real-world utility in ruling that utility was not established for a process of making a chemical compound where no specific utility had been demonstrated for the compound itself except as the object of further testing.

> This is not to say that we mean to disparage the importance of contributions to the fund of scientific information short of the invention of something "useful," or that we are blind to the prospect that what now seems without "use" may tomorrow command the grateful attention of the public. But a patent is not a hunting license. It is not a reward for the search, but compensation for its successful conclusion. "[A] patent system must be related to the world of commerce rather than to the realm of philosophy…" (*Brenner v. Manson,* 383 U.S. 534–535 [1966]*).

Thus, the PTO's stricter standard guidelines appeared to draw from the United States Supreme Court's recognition in *Brenner* that a patent is not "a reward for the search"

---

* In so ruling, the Supreme Court noted it was not relying on reference to an article in the November 1956 issue of the *Journal of Organic Chemistry*, 21 J. Org. Chem. 1333–1335, which revealed that steroids of a class that included the compound in question were undergoing screening for possible tumor-inhibiting effects in mice, and that a homologue 3 adjacent to Manson's steroid had proven effective for that purpose. Id.

for an invention's utility but, rather, "compensation" for successfully finding its utility.

In adopting the requirement for a specific, credible, substantial utility, the PTO recognized that under traditional patent law principles a patent on a composition, such as DNA, granted exclusive rights to any use of the composition. The PTO also addressed many policy-based concerns relating to the patentability of genomic inventions and whether a patentee should be permitted to assert claims against speculative uses of DNA that were unforeseeable at time of filing. The PTO responded to some of the issues as follows:

- The PTO rejected the suggestion that genes ought to be patentable only when the complete sequence of the gene is disclosed and a function for the gene product was determined; the PTO noted that a partial gene sequence might have use, for example, if it hybridizes near a disease-associated gene or has a gene regulating activity (Utility Guidelines at 1095). Thus, recitation of use of a DNA sequence in detecting or regulating a particular disease state may satisfy the utility requirement.
- The PTO also rejected the suggestion that the DNA sequences should be freely available for research or that certain raw DNA sequence data might be unworthy of any patent protection — instead, the PTO determined that DNA patents were subject to the same statutory interpretation governing the scope of patentable subject matter (Utility Guidelines at 1095).
- In addition, the PTO rejected the suggestion that because methods of DNA sequencing had become so routine, determining the DNA sequence was not inventive. The PTO quoted from the statute governing obviousness in rejecting this argument (Utility Guidelines at 1095, quoting § 103: "Patentability shall not be negatived by the manner in which the invention was made").
- Another rejected comment suggested that the use of computer-based analysis of nucleic acids to assign a function based on homology to prior art nucleic acids was unpredictable and should therefore not form a sufficient basis for assigning a function to a putatively encoded protein (Utility Guidelines at 1096). The PTO indicated that such findings should be made on a case-by-case basis and that it would take into account the nature and degree of the homology (e.g., whether the class of proteins was defined and highly conserved) in deciding whether a specific, substantial and credible utility had been asserted (Utility Guidelines at 1096).
- The PTO also responded to one comment asserting that utility based on homology data should correspondingly render the sequence obvious by noting that even where a homology-based utility was established, a complete inquiry into whether the sequence would be obvious under § 103 should be made on a case-by-case basis, but it should not result in a *per se* finding of obviousness (Utility Guidelines at 1096).

In the absence of a disclosed specific, substantial, and credible utility, a well-established utility would have to be readily apparent to one of skill in the art as of the time the application was filed, not based on a later-discovered utility (Utility Guidelines at 1096). The well-established utility is a specific, substantial, and credible utility that must be readily apparent to one skilled in the art (Utility Guidelines at 1097). Moreover, if the examiner does not readily perceive a well-established utility, the guidelines provide that the examiner may issue a rejection for lack of utility, thereby placing the burden on the applicant to establish that such a utility exist (Utility Guidelines at 1097).

It will remain to be seen if separate utility guidelines are the proper way to alter the balance struck between the discovery of bioinformatic information and the researcher's ability to use downstream in applied research for new drug innovation bioinformatics information, including information gleaned from the use of gene chips and microarrays.

REQUIREMENTS OF WRITTEN DESCRIPTION UNDER 35 U.S.C. SECTION 112

Several appeals court decisions have discussed the legal requirements for written description in the context of genomic inventions. "The written-description requirement serves a teaching function, as a *quid pro quo* in which the public is given meaningful disclosure in exchange for being excluded from practicing the invention for a limited period of time." *University of Rochester v. G.D. Searle & Co., Inc.*, 358 F.3d 916, 919 (Fed. Cir. 2004) (quoting *Enzo*, 323 F.3d at 970). Thus, the written-description requirement "ensure[s] that the scope of the right to exclude, as set forth in the claims, does not overreach the scope of the inventor's contribution to the field of art as described in the patent specification." *Reiffin v. Microsoft Corp.*, 214 F.3d 1342, 1345 (Fed. Cir. 2000).

The legal standard for written description under §112, ¶1,* requires that the patent specification clearly "describe the claimed invention so that one skilled in the art can recognize what is claimed." *University of Rochester v. G.D. Searle & Co., Inc.*, 358 F.3d 916, 924 (Fed. Cir. 2004) (*quoting Enzo v. Gen-Probe*, 323 F.3d at 968). That is, the patent must contain enough detail "to allow a person of ordinary skill in the art to understand what is claimed and to recognize that the inventor invented what is claimed." *University of Rochester*, 358 F.3d at 929; *Regents of the Univ. of Cal. v. Eli Lilly & Co.*, 119 F.3d 1559, 1568 (Fed. Cir. 1997).

The Federal Circuit Court of Appeals has held that a patent claiming a DNA-related invention in functional terms must also describe a structure associated with that function, unless those skilled in the art are aware of a structure associated with that function. See, for example, *Enzo Biochem Inc. v. Gen-Probe Inc.*, 323 F.3d 956, 964 (Fed. Cir. 2002); *Regents of the Univ. of Cal. v. Eli Lilly & Co.*, 119 F.3d 1559, 1568 (Fed. Cir. 1997). To describe a class of cDNAs, such as mammalian cDNA for insulin, a patent must describe the sequence of a sufficient number of sequences so that one skilled in the art can visualize or recognize the members of the genus;

---

* The first paragraph of 35 U.S.C. § 112 requires as follows:
"[t]he specification shall contain a written description of the invention, and of the manner and process of making and using it, in such full, clear, concise, and exact terms as to enable any person skilled in the art to which it pertains, or is most nearly connected, to make and use the same, and shall set forth the best mode contemplated by the inventor of carrying out his invention."

thus, it is not a sufficient description of a mammalian cDNA to provide the sequence for a rat insulin cDNA and the amino acid sequence for human insulin:

> In claims to genetic material, however, a generic statement such as "vertebrate insulin cDNA" or "mammalian insulin cDNA," without more, is not an adequate written description of the genus because it does not distinguish the claimed genus from others, except by function. It does not specifically define any of the genes that fall within its definition. It does not define any structural features commonly possessed by members of the genus that distinguish them from others. One skilled in the art therefore cannot, as one can do with a fully described genus, visualize or recognize the identity of the members of the genus... The description requirement of the patent statute requires a description of an invention, not an indication of a result that one might achieve if one made that invention. [citation omitted]. Accordingly, naming a type of material generally known to exist, in the absence of knowledge as to what that material consists of, is not a description of that material.

> ...A description of a genus of cDNAs may be achieved by means of a recitation of a representative number of cDNAs, defined by nucleotide sequence, falling within the scope of the genus or of a recitation of structural features common to the members of the genus, which features constitute a substantial portion of the genus. (Regents of the Univ. of Cal. v. Eli Lilly & Co., 119 F.3d 1559, 1568 [Fed. Cir. 1997]).

In *Enzo*, a later case involving claims to DNA sequences, the Federal Circuit Court of Appeals held that sufficient written description of a claim to a full-length DNA sequence with recited functional characteristics may not constitute adequate written description of subsequences having those functional characteristics. (*Enzo v. Gen-Probe*, 323 F.3d at 964–966). The Federal Circuit appeared to relax the rule in *Lilly* requiring description of the sequence of a cloned DNA by holding that a cloned-DNA sequence could be sufficiently described by referring to a publicly available ATCC culture deposit containing the sequence, instead of sequencing the cloned DNA and describing that sequence. Id.

However, the Federal Circuit determined that factual issues remained concerning whether the patent description was sufficient to demonstrate that the inventors were in possession of claimed subject matter to sequences, subsequences, and mutated sequences that preferentially hybridized to *Neisseria gonorrhoeae*. Id. The Federal Circuit therefore sent the case back to the district court for a determination of whether a person of skill in the art would glean from the written description, subsequences, mutated variants, and mixtures sufficient to demonstrate possession of the class of sequences covered by the claims. Id. The appeal court also adopted the standard set forth in the PTO guidelines for written description:

> In its guidelines, PTO has determined that the written description requirement can be met by "showing that an invention is complete by disclosure of sufficiently detailed, relevant identifying characteristics..., i.e., complete or partial structure, other physical and/or chemical properties, functional characteristics when coupled with a known or disclosed correlation between function and structure or some combination of such characteristics." (Guidelines, 66 Fed. Reg. at 1106).... Thus, under the guidelines, the written-description requirement would be met for all of the claims of the '659 patent

if the functional characteristic of preferential binding to *N. gonorrhoeae* over *N. meningitidis* were coupled with a disclosed correlation between that function and a structure that is sufficiently known or disclosed. We are persuaded by the guidelines on this point and adopt PTO's applicable standard for determining compliance with the written-description requirement.

Id.

More recently, the Federal Circuit applied this test in affirming the PTO's rejection of claims to DNA encoding a specific protein isolated from human urine that selectively inhibits the cytotoxic effect of tumor necrosis factor (TNF), where the only data in the specification was a partial amino acid sequence of the N-terminal portion of the protein, and a determination of the molecular weight of the intact, isolated protein. *In re Wallach*, 378 F.3d 1330 (Fed. Cir. 2004), the Federal Circuit concluded that the decision of the Board of Patent Appeals and Interferences that the claims lacked written description was consistent with the PTO's policy set forth in its Manual of Patent Examining Procedure, which advises that "disclosure of a partial structure without additional characterization of the product may not be sufficient to evidence possession of the claimed invention." Id. (quoting MPEP § 2163.II.A.3.a.i.). The Federal Circuit cited the PTO's written-description guidelines in rejecting the patent applicant's argument that possession of the isolated protein, its molecular weight and a partial sequence provided possession of the protein's full amino acid sequence, which the Federal Circuit acknowledged would have been sufficient to give possession of any DNA sequence that could code for that protein sequence:

Appellants have provided no evidence that there is any known or disclosed correlation between the combination of a partial structure of a protein, the protein's biological activity, and the protein's molecular weight, on the one hand, and the structure of the DNA encoding the protein on the other.

Whether Appellants were in possession of the protein says nothing about whether they were in possession of the protein's amino acid sequence. Although Appellants correctly point out that a protein's amino acid sequence is an inherent property of the protein, the fact that Appellants may have isolated and thus physically possessed TBP-II does not amount to knowledge of that protein's sequence or possession of any of its other descriptive properties. Appellants have not provided any evidence that the full amino acid sequence of a protein can be deduced from a partial sequence and the limited additional physical characteristics that they have identified. Without that full sequence, we cannot agree with Appellants that they were in possession of the claimed nucleic acid sequences.

A gene is a chemical compound, albeit a complex one, and it is well established in our law that conception of a chemical compound requires that the inventor be able to define it so as to distinguish it from other materials, and to describe how to obtain it…. Until Appellants obtained the complete amino acid sequence of TBP-II, they had no more than a wish to know the identity of the DNA encoding it. (*In re Wallach*, 378 F.3d 1330).

Finally, in a decision that may have implications for the extent to which newly discovered target pathways must be described to obtain claims to therapeutic uses of ligands to the targets, the Federal Circuit ruled that an inventor must describe more than a target gene and a compound that binds to that gene *in vitro* in order to sufficiently describe a method of inhibiting the target gene's activity in a human by administering a compound that selectively inhibits the target. See *University of Rochester v. G.D. Searle & Co., Inc.*, 358 F.3d 916, 919 (Fed. Cir. 2004). In *University of Rochester*, the Federal Appeals Court affirmed the District Court's finding of lack of written description, where the patent claimed a method of selectively inhibiting PGHS-2 activity in a human host" by "administering a nonsteroidal compound that "selectively inhibits activity of the PGHS-2 gene to a human to [or in] a human host in need of such treatment," but the specification did not describe any compound capable of achieving the claimed effect of selectively inhibiting PGHS-2 activity. Id. at 924; see also id. at 928. Without that disclosure, the Federal Circuit held that "the claimed methods cannot be said to have been described." Id. Accordingly, the court found the claims invalid for lack of written description.

Thus, the determination of whether a patent specification contains a sufficient description of a claim to a genomic invention includes case-specific factual determinations of whether one of ordinary skill in the art would glean from the particular patent specification at issue a structure associated with a claimed biological function.

IP IMPLICATIONS FOR DNA CHIPS AND MICROARRAYS

NONUNIFORM APPLICATION OF EVOLVING GUIDELINES FOR EXAMINING BIOTECHNOLOGY INVENTIONS

Since 1991, when the NIH filed its patent application claiming thousands of EST sequences, controversy and disagreement about the patenting of such sequences culminated in the promulgation of the 2001 Utility Guidelines by PTO. Critics argued that patents on gene fragments, especially uncharacterized cDNA sequences may reward activity too early in the drug discovery process — it will benefit those who contributed early by sequencing DNA using routine technologies, but penalize those who perform the more difficult task of discovering biological function or pathways.

Because of the evolving standards for examining of biotech inventions, the standards may not have been uniformly applied to all issued patents. For example, patents may be issued that were examined under older, more lax utility guidelines. On October 6, 1998, the PTO issued to Incyte Pharmaceuticals, Inc., U.S. Patent No. 5,817,479, one of the first patents known to include claims to purified polynucleotide EST sequences for "Human Kinase Homologs." Similarly, PTO issued U.S. Patent 6,025,154 to Human Genome Sciences with claims directed to an isolated polynucleotide encoding CCR5, a receptor that binds chemokines, the CCR5 protein and host cells containing the gene. Human Genome Sciences obtained the sequence while sequencing the entire human genome. The patent specification did not disclose any experimentally determined function for the CCR5 protein, but deduced from the protein's homology to other chemokine receptors that the new protein would bind chemokines. Other researchers later independently discovered that the CCR5 protein is a coreceptor for HIV binding. However, Human Genome Sciences' patent is a potential block for researchers using the CCR5 gene or protein as a target for AIDS research, even the researchers who were the first to discover the function of the protein. See, for example, J. Madeline Nash, "Who Owns The Genome? Battle Pending," CNN.com, at http://cgi.cnn.com/ALLPOLITICS/time/2000/04/10/ genome.html (April 10, 2000).

The issues of whether the standard set forth in the 2001 guidelines should apply to patents issued before their implementation has not yet been litigated, although recent litigation over whether the description in the patent provides a specific, substantial, credible utility meeting the standard of the 2001 utility guidelines has occurred. Despite the earlier issuance of U.S. patents with claims to ESTs, the Federal Circuit recently affirmed rejection of claims to EST sequences based on lack of utility, approving the Board of Appeals' reliance on the Supreme Court's analysis in *Brenner*. *In re Fisher*, 421 F.3d 1365, 1371–74 (Fed. Cir. 2005); *see also Ex parte Fisher*, 72 U.S.P.Q.2d (BNA) 1020 (unpublished decision).

In *Fisher*, the Board of Appeals relied on the Supreme Court's ruling in *Brenner* to reach its decision that the application for EST sequences did not describe a specific, substantial, credible utility. *Ex parte Fisher*, 72 U.S.P.Q.2d at 1020, 1023–26. The Federal Circuit affirmed, and noted that the PTO's 2001 Utility guidelines were consistent with its interpretation of the utility requirement. *In re Fisher*, 421 F.3d at 1372. Thus, as with the guidelines for written description (*see Enzo v. Gen-Probe*, 323 F.3d at 964–966), the Federal Circuit approved the PTO's guidelines for utility. The Federal Circuit's affirmance of the Board's rejection of the EST sequence claims provides some useful guidance on the description an Applicant for a patent on EST sequences must provide in order to meet the requirements for the patentability of DNA sequences.

The EST sequences in *Fisher* were raw EST sequences with no description of its function, no confirmed characteristic based on sequence homology, or any other feature that would establish a specific use of the ESTs. Although *Fisher* had described several catch all uses in the application including the identification and detection of polymorphisms, use as probes in hybridization assays, and as primers for amplification, without further information such as the proteins encoded by the genes the ESTs are derived from, the Federal Circuit concluded that the stated uses were general uses for nucleic acids and thus did not satisfy the requirement for specific utility. *In re Fisher*, 421 F.3d at 1374; *see also ex parte Fisher*, 72 U.S.P.Q.2d at 1029.

PATENTING ON GENES AS AFFECTING DIAGNOSTIC RESEARCH

Patents on genes and gene sequences could impact the development of diagnostics, including array diagnostics, because of the costs associated with licensing and using patented sequences.

Patents issued in the U.S. and Europe have recently brought into focus the debate over how the patenting of genes affects patient care and research, and how private companies that own certain patents can monopolize certain genetic testing markets. For example, after Myriad Genetics Laboratories received a patent to the BRCA1 gene and BRCA2 mutations associated with a risk for breast cancer, the company sent a letter to the Genetics Diagnostics Center at the University of

Pennsylvania seeking licensing fees for use of a BRCA1 screening test. The letter informed researchers performing diagnostic screens for the gene that they would have to pay a fee for each test performed, raising concerns over the cost and availability of genetic diagnostic testing. (see Charles Schmidt, Cashing in on gene sequences, *Money Matters Corporate*, http://pubs.acs.org/subscribe/journals/mdd/ v04/i05/html/ 05money.html.). In Europe, a challenge to Myriad's patents resulted in nullification of one patent and narrowing of a second patent to the one mutation/ sequence correctly identified in the original patent application. Myriad's European BRCA2 patent had been earlier successfully opposed because the charity Cancer Research UK had filed a patent on the gene first. (Breast Cancer Gene Patent Revoked in Europe, Science and Intellectual Property in the Public Interest, May 19, 2004 at http://sippi.aaas.org/ipissues/updates/?res_id=312.).

Restrictions on use of a gene sequence might also apply to use of DNA microarrays or gene chips to carry out diagnostic testing.

MULTIPLE PATENTS: COST IMPACT ON DRUG PRODUCT DEVELOPMENT

Stacking of multiple patents on the same sequence might also discourage the development of diagnostics and drug products as a result of multiple royalty payments.

Under current laws, patent stacking is permitted, for example, where a single genomic sequence is first patented as an EST, then a gene, then an SNP, or where different patents cover both the composition of an array surface and a method of attaching oligonucleotides to the surface. In addition, as noted previously, as researchers use array technology to identify the genes involved in genetic diseases or predisposition to genetic diseases, holders of earlier patents to genes or gene fragments may require a royalty payment and/or prohibit the development or use of hybridization assays to screen for individuals at risk for developing the disease. Stacking may therefore discourage the development of diagnostic and drug product development because payment of multiple royalty costs owed to each of the patent owners of that sequence may be prohibitive.

Allowing multiple patents on the same sequence, or different aspects of the same genome region, also creates a related concern — that the costs to determine whether patents are blocking freedom to use any given nucleic acid sequence may, in itself, become prohibitive. Moreover, DNA chips and diagnostic devices may contain tens of thousands of DNA sequences. The lag in publication of patent applications adds to the complexity because of the difficulty of determining, at any given time, whether a third party has filed a patent on a given sequence. Additional patents on other aspects of array technology may add further complexity to the analysis of freedom to operate for use of a given sequence on an array.

CLAIMS TO SUBFRAGMENTS OF CLONED DNA WITH UNIQUE FUNCTIONAL PROPERTIES

A catch-22 might arise for patent applications directed to subsequences with unusual functional properties. In a recent Federal Circuit decision, the appeals court held that a claim to a purified oligonucleotide comprising a promoter having activity for the human involucrin gene lacked novelty in view of the disclosure by the inventor's publication of a plasmid containing a cloned gene (including the promoter region) and the method of obtaining the plasmid, even though the promoter region had not yet been sequenced. (see *In re Crish*, 393 F. 3d 1253).

On the other hand, the written-description standard dictates that a DNA cannot be patented unless its sequence is known or unless it is cloned in a cell line deposited in a public cell depository. Thus, if claims are not carefully drafted, the information required to satisfy the written-description requirement for the gene sequence may eliminate novelty for a subfragment of the cloned DNA later shown to have unusual functional properties.

OTHER SOLUTIONS TO PROMOTE INNOVATION: FREE SHARING OF SNP INFORMATION

Several genetic diseases have already been shown to be the product of SNPs, sequence variations that occur when a single nucleotide in a sequence is altered. It is believed that SNPs could be partially responsible for genetic conditions, predispose individuals to disease, or influence metabolism of, and responses to, drugs, toxins, or infectious agents. Using array methodology to study SNPs should permit scientists to establish additional correlations and associations of multiple genes that jointly contribute to these conditions. SNP analysis may also help to predict the action of drug candidates, and the effect of genetic makeup on the mode of drug action. This makes the use of arrays to study SNPs of great value in new drug discovery, and many arrays are designed for genotyping of single nucleotide polymorphisms.

In April 1999, fearful that patenting of SNPs would severely impact drug discovery research, ten pharmaceutical, biotech, array and/or bioinformatics companies and the U.K. Wellcome Trust announced the establishment of a nonprofit foundation, the SNP Consortium Ltd., to find and make publicly available without patent restrictions a highly reliable map of 300,000 common SNPs distributed evenly throughout the human genome. The SNP Consortium planned to patent all the SNPs found as part of the project, but to enforce the patents only to prevent others from patenting the same information. The consortium has made its information freely available to the public.

During the project, which began in April 1999 and continued until the end of 2001, the SNP Consortium identified nearly 1.8 million SNPs, many more than originally planned. See notes on the last data release (Sept 2001) for more information. Various member laboratories are now in the process of genotyping a subset of those SNPs as a part of the Allele Frequency/Genotype Project.

Whether the consortium's unique approach to solving the debate over patenting SNPs is successful in eliminating freedom to operate issues relating to SNP sequences remains to be seen. In addition, processes for analyzing SNPs using array technology have been separately patented.

CONCLUSION

Microarray and gene chip technology has proven to be a valuable tool in explaining the genetic basis of disease, identifying new targets in the search for new therapeutics and the diagnosis of genetic predisposition to disease. The increasing market for array and gene chip technologies will likely lead to the enforcement and challenge of patents to array and gene chip technologies, and genomic discoveries made possible by those enabling technologies. As courts continue to grapple with the application of patent law requirements and PTO guidelines to the patentability of array, gene chip, and related technologies, new court rulings may determine and impact the balance between rewarding early discovery and encouraging long-term innovation.

21 Biochips: Market Drivers and Commercial Prospects

Jing Xu

CONTENTS

Since their conception in the late 1980s, DNA microarrays and biochips have not only proven their value in genomic research but also created a multibillion dollar industry with diverse applications. From DNA microarray to protein arrays, microarrays have made their impact in many aspects of life science and forever changed the way scientists in life science approach their research. The microarray industry has been one of the fastest growing fields since the mid-1990s and will remain so in years to come.

DNA MICROARRAYS: AN ESTABLISHED AND EXPANDING BUSINESS

By all accounts the research community, either academia or pharmaceutical industry, has embraced DNA microarrays. Since the publication of the first microarray study in *Science* in 1995 [1], the DNA microarray burst into a $48 million industry in 1997 [2]. As DNA microarrays quickly became the standard in genomic research, the market expanded to $232 million in 1999 [2,3]. As the genomic research further expanded, the DNA microarray market further doubled to $596 million in 2003 [4].

The extremely fast growth of the DNA microarray industry from 1997 to 2000 had led to overoptimistic projections for the industry. Market research reports published in late 2000 and early 2001 predicted a compounded annual growth rate of more than 50% and over a $2 billion market in 2004 [3]. Such predictions did not come true. After a take-off period from 1997 to 2000, the DNA microarray market continued to grow, but at a marked slower pace from 2001 to 2003. According to a report published by Frost & Sullivan in 2004, the global market for DNA microarrays is poised to grow at an annual average of 6.7% from $596 million in 2003 to $937 million in 2010 [4].

From 2000 to 2005, applications of DNA microarray have expanded from now standard gene expression profiling and genotyping research and development in the pharmaceutical industry and academia to new markets in agriculture, business, environmental analysis, forensic analysis, and clinical diagnostics. If successfully adopted in molecular diagnostics, the DNA microarray and related reagents will experience an even more explosive surge in market growth. Much of this growth will be from new growth areas for microarray, such as SNP analysis, Comparative Genome Hybridization (CGH), and the study of myriad epigenetic factors with whole genome or tiling arrays.

Geographically, the U.S. is still the leader as well as the biggest market in DNA microarray, followed by Europe and Japan. According to a report by Frost and Sullivan at the end of 2005, the total U.S. DNA microarray revenue was approximately $447 million in 2005, and projected to grow at a compounded annual growth rate of 10.9% to $532 million by the end of 2012 [5].

THE EVOLVING INDUSTRY LANDSCAPE

The DNA microarray industry emerged in the mid-1990s. In the past decade, the industry has matured, and the landscape has changed significantly from its early days. Affymetrix of Santa Clara, CA, was one of the very first companies to commercialize microarrays. The company pioneered the *in situ* synthesis of oligonucleotides on glass chips. Affymetrix's proprietary processes combine solid-phase chemical synthesis with photolithography. It was the first to offer commercial human genome microarrays, as well as the first to offer human "whole-genome" arrays. Its GeneChip — an Affymetrix trademark — now contains over 1 million different oligonucleotides, representing more than 33,000 of the best-characterized human genes. The price of GeneChips has come down significantly, bringing them within the reach of at least some academic researchers. On a cost per feature basis, they have come down an order of magnitude. Affymetrix's strategy has been to offer more per chip as compared to reducing individual chip price, partly because the company has command over the high-density market. Affymetrix had an estimated 70% market share in 2003.

There have been major shakeups in the industry. Incyte Genomics of Palo Alto, California, one of the leading suppliers of microarrays in the 1990s, quit the chip-making business in 2001, deciding to refocus on its core business of bioinformatics. Incyte shifted its focus again from informatics to drug discovery a year later. In part, Incyte had hoped or hopes to utilize revenues gained from its informatic sales and milestone payments resulting from utilization of this information to leverage itself into a full-fledged pharmaceutical company, but this remains to be seen.

Motorola of Northbrook, IL, intending to enter the already crowded market, launched CodeLink bioarray in summer 2001. Motorola's move into the DNA microarray industry turned out to be unsuccessful. In summer 2002, Motorola sold its CodeLink™ prearrayed slides business to British biochip maker Amersham in an effort to downsize the company and refocus on its core business. To further mark transition for this company, its new parent company was sold in January 2004 to General Electric. Concurrent with this the CodeLink platform offered full arrays for human, rat and mouse genomes in 2004. It is expected that General Electric will breathe new life into this once ailing business, in part with an emphasis of utilization of this technology in the diagnostic market.

Motorola and Incyte may be gone, but there are more competitors entering the market, including the No. 2 rank holder, Agilent Technologies in Palo Alto, California, and General Electric (GE). Other recent entrants, such as Illumina of San Diego, California and Applied Biosystems of (Foster City, California), are vying to grab chunks of Affymetrix's market share as well.

In contrast to Affymetrix's methods, Agilent Technologies uses proprietary Sure-Print inkjet technology and offers human, mouse, and rat cDNA arrays and custom oligonucleotide arrays. As a recent spun-off subsidiary of Hewlett-Packard, Agilent still has access to considerable expertise in inkjet printing methods and high-end analytical instrumentation, principally high-performance liquid chromatography and mass spectrometry.

Much of the new demand for chips is for more customized chips than Affymetrix makes. Once the first draft of the human genome was completed in 2001, researchers quickly turned up a vast number of possible targets for potential drugs. The need thereafter is to narrow the field by validating which targets deserve further study. This new mission requires chips whose focus is small groups of specific genes, potentially run in higher throughput, or the genetic makeup of obscure organisms.

Agilent is benefiting from this trend. Its inkjet manufacturing process seems to be much more adaptable to customization than the photolithography process that Affymetrix uses. Agilent estimates that its share of the biochip market rose to 15% in 2003, up from what analysts say was single digits previously. That growth came from sales of "catalog chips," which contain the genetic maps of organisms widely used in research, such as human, rat, and mouse, and from sales of customized chips.

Other players are trying to expand the limits to how much can be put on a single chip. Illumina launched a six-genome chip in mid-January 2004 that will let customers run six of the same experiments on a single chip, saving time and money. Illumina hopes that this product will take some market share from Affymetrix. At the press release price of $160 per genome chip this will be a very compelling offering compared to the current GeneChip. Unfortunately a lawsuit alleging infringement of various Affymetrix patents quickly ensued following this press release. Illumina fought back with a countersuit against Affymetrix for unfair competition. The patent war saga continued when Illumina challenged Affymetrix' right of a different patent in October 2005. It remains to see the final outcome of these lawsuits and how it will affect the industry. Recent press releases in the industry are suggestive of Illumina's favorable position in that even the more conservative pharmaceutical customers are now willing to work with Illumina, with public statement, even with these pending lawsuits. It is also interesting to note that in 2006 the total market cap of Illumina, once dwarfed by mighty Affymetrix, has now for the first time, exceeded that of its rival. Much of this growth for Illumina has been largely based upon its expansion into the new SNP markets, after Affymetrix had significant setbacks with its 500K chip-set in early 2006.

Along with actively protecting its voluminous patent position, Affymetrix has stepped up its efforts to push the GeneChip envelope. The company in the fall of 2003 was the first large player to put the entire human genome on a single chip, and it continues to add to its large library of chips containing whole genomes of heavily researched organisms. The company also strengthened its custom array offerings by initiating a partnership with NimbleGen, a DNA custom array maker, in mid-2004. Under the agreement, Affymetrix will market NimbleGen's made-to-order Nimble-Express arrays for use on Affymetrix's GeneChip instrument systems.

Yet such dominance has not stopped new entrants into the market. Undaunted, Invitrogen of Carlsbad, California, made a foray into the DNA custom array market through its acquisition of Xeotron in 2004. Still, Affymetrix enjoys a comfortable lead in the DNA microarray market. A survey in late 2005 shows that 75% of researchers use Affymetrix arrays [6]. A recent report by BioCompare Inc. of South San Francisco in mid-2006 shows that Affymetrix's microarray sales bring in well over $350 million revenue annually, while the revenue of Agilent, GE Healthcare, and Illumina combined only account for just over one-third of that amount [7]. Indeed, Affymetrix's competitors find it the biggest challenge to convert the Big Pharma customers who have been entrenched with Affymetrix. Once again you don't lose your job by buying IBM conceptually.

CONTINUED MARKET DRIVERS AND COMMERCIAL PROSPECTS

Suppliers of DNA microarray products and services will face major challenges in a market environment characterized by rapidly changing technology, extensive government regulation, downward pricing pressures and ongoing intellectual property battles. Genomic research will continue to be the leading application of DNA microarrays in the foreseeable future. Genomic research is moving well beyond deciphering genetic code sequences, rather focusing on analyzing specific actions of genes and gene-encoded proteins. As a result, for DNA microarrays, gene expression profiling is projected to also comprise the leading application over the next several years.

After a rapid embrace of DNA microarray technology, the research community soon realized the problems associated with a lack of consensus on how to analyze microarray data. In response to standardizing data collection from different microarray platforms so that they can be accurately compared, the Microarray Gene Expression Data Society (www.mged.org) has developed guidelines for the publication of DNA microarray data. The guidelines, Minimal Information about a Microarray Experiment (MIAME), are meant as provisional solutions to the broader problem of standardizing microarray data annotation and interpretation. Some journals — including *Nature*, *Cell*, and *The Lancet* — have adopted MIAME. The desire to standardize may lead more researchers to choose commercial arrays over the home-grown variety, and will inspire investigators to purchase complete array systems to ensure the reproducibility of their data. As prices of DNA microarrays continue to drop, more researchers are expected to be able to buy into the commercial arrays. In fact the economy of scale, in part, dictates that it will be a cost-effective, and therefore a wise business decision to choose manufactured arrays over self-spotted. Self-spotted arrays pose myriad problems for the spotter, such as the aforementioned standardization and quality control issues inherent in self-spotting.

Pharmaceutical companies and academic laboratories will remain the leading markets for DNA microarray products and services. DNA microarrays have been used in almost all aspects of drug discovery and development, from target discovery and identification to toxicogenomics and pharmacogenomics. Pharmacogenomics, for example, stands to benefit from DNA chips. Defects in certain genes are associated with adverse reactions to commonly prescribed drugs. Current tests for those genes are expensive. If microarrays become inexpensive enough, they will present an effective alternative means of testing. In fact with the advent of pharmacogenomic screens we may see future drugs that will have as prerequisite to their dispensation a chip-based prescreen. Ongoing efforts to improve the efficiency of drug discovery and screening processes will boost growth opportunities among pharmaceutical manufacturers.

However, there has been increasing pressure from the public for drug price control in the U.S., as well as demand for approval of drug reimportation to the U.S. (It can be argued that this is due to a misguided insurance industry and the lack of a proper safety web for the poor and elderly). Such changes may cap the growth of the U.S. pharmaceutical market, which is the single largest pharmaceutical market. With patents for several blockbuster drugs expected to expire in the next 5 years, big pharmas are facing fierce competition from generic drug makers, who do not have the significant R&D overhead. If big pharmas, the most lucrative customer base for DNA microarray products, fail to sustain their sales, it is likely that they will cut the hefty R&D cost, leading to less demand for DNA microarrays. The real downside to this of course will be that development of new drugs will be thwarted, and likely be a death spiral for pharma, driving further consolidation. Alternative viewpoints suggest that if DNA microarrays can be used to fail poor compounds earlier in the discovery process, cost savings would dictate increased expenditure.

Demand in the academic community will increase as life science researchers seek powerful and versatile tools for basic research. On the other hand, college and university laboratories will remain price sensitive due to ongoing funding constraints. The academic community will remain more likely to use "homemade" biochip products whenever feasible. This trend will impose downward pricing pressures on industry suppliers. Again economy of scale and desire for standardization may be the ultimate drivers.

Though genomic research will remain the leading application of DNA microarrays, the market is reaching saturation due to constraints on private sector's R&D costs and public sector's funding. Therefore, it is imperative for the microarray market leaders to move aggressively into other markets for further growth. Rapid growing niche opportunities will emerge in *in vitro* diagnostic testing, forensic medicine, agriculture, and food diagnosis. DNA fingerprinting has become widely accepted in forensic medicine, plant genetics, and food diagnosis. DNA microarrays are very effective test systems for nucleic acids and will likely become the new standard tool for these tests, if the price becomes affordable. With razor-thin margins in the diagnostic industry, it is likely that these costs will have to be driven down an order of magnitude or more.

Still, the move into diagnostics has the greatest potential, in part because of the scale, opening the doors to a market that could exceed $1 billion [4,8]. Affymetrix and Roche Diagnostics have launched in June 2003 the AmpliChip CYP450 array (an Affymetrix chip), which measures genetic variations in drug-metabolizing enzymes. Agilent and Agendia are in collaboration to develop what could become the first gene-expression-based test used in the clinic. Amersham's 2004 acquisition by GE was facilitated by their mutual interest in pharmacoge-nomics and diagnostics.

As mentioned, diagnostic chips will have to be priced significantly lower to be competitive in a market saturated with low-priced qualitative diagnostic kits. Currently, microarrays range from a few hundred dollars to $1000 or more. The good news is that diagnostic chips will carry far fewer genes than the latest research arrays, making it possible to be in smaller size and priced in the double digits. Time will tell how much share of the diagnostic market DNA chips can grab.

PROTEIN ARRAYS: AN INDUSTRY IN TAKE-OFF

Though not yet at DNA microarray status, protein arrays are on their way to become a key technology. Because the size of the protein field is much larger than that of the DNA field, many expect the protein microarray potential to be much larger than the DNA microarray field.

In their simplest iterations the principle that underlies protein microarrays are similar to DNA microarrays. When the protein microarray is exposed to a mixture of other proteins, molecules that naturally interact with the proteins fixed on a slide bind to the protein probes. The proteins bind to the probes and can be labeled and visualized similar to the way the gene sequences in DNA microarrays are. However, many more potential applications and specific detection methods than can be mentioned here are possible.

Making a protein biochip is technically much more challenging than making a DNA chip. Proteins are far less stable than DNA. They tend to be active only when in their native conformation. Changes in pH, temperature, or the ionic strength of a solution can cause native proteins to change conformation and denature, rendering them inactive. Therefore, the research community was much more cautious to accept this technology than DNA microarray.

Many of these initial technical problems are now solved or being solved and scientists can now choose among various ready-made antibody- or protein-profiling arrays. For those that prefer to make their own protein chips, tools and kits for do-it-yourself arrays are also available. As researchers overcome initial skepticism, protein arrays are on their way to become a mainstream tool. Still, protein array developers have to continue to address problems such as manufacturing cost and sensitivity.

EMERGING INDUSTRY LANDSCAPE

Protein microarrays have only just begun to make an impact on the life science community. Although not as significant as its DNA counterpart yet, the industry is maturing from its emerging state. Many companies are betting that proteomics will emerge as the next wave of research in basic life science and drug discovery. Early entrants into the protein chip business from the late 1990s to the early 2000s included Ciphergen of Fremont, California, Large Scale Proteomics, a division of Large Scale Biology Co. of Vacaville, California, Packard BioScience of in Meriden, Connecticut, and Phylos of Lexington, Massachusetts. There have been major changes of these early entrants in the protein microarray market. Packard was acquired by PerkinElmer in 2001. Phylos was acquired in 2004 by Compound Therapeutics, the now Adnexus Therapeutics of Waltham Massachusetts, and refocused on its pursuit of protein therapeutics. Large Scale Biology filed bankruptcy in January 2006. Ciphergen, while formerly a major player, eventually decided to focus on its diagnostics business in 2006. The company recently announced the sale of its life science research business, including its protein chip business, to Bio-Rad of Hercules, California.

Several companies offering DNA microarray technologies have also begun migrating into the field. For example, Amersham Bioscience of GE Healthcare, No. 3 in the DNA microarray industry, has been actively working on high-density protein chips and hopes to sell them in 2005. Likewise, Boston-based PerkinElmer, provider of MICROMAX™ line of cDNA arrays, now offers a complete proteomics package from slides, printing and scanning equipment to labeling and amplification reagents for the production of protein microarrays.

Major life science reagent companies are eyeing the field as well. For example, Invitrogen has strengthened its position in the proteomics research tool market through a series of acquisition and licensing deals since 2003. In summer 2004, the company launched ProtoArray Yeast Proteome Microarray, the world's first proteome microarray, soon after its acquisition of Protometrix of Branford. Connecticut, the developer of the pioneering ProtoArray technology. Invitrogen has since launched several ProtoArray human protein microarrays. Clontech of Mountain View, California, now a Takara Bio company, offers Clontech Ab Microarray 500 Slides, which contain 512 well-characterized antibodies. The two companies emerge as top suppliers of protein microarrays, as indicated by a market survey in fall 2005 [9].

Although the content of most protein arrays is proteins, ranging from antibodies and antigens to recombinant proteins and peptides, some companies are currently developing alternatives. SomaLogic of Boulder, Colorado, for example, employs photoaptamers, single-stranded nucleic acids selected to bind specifically to target proteins. These aptamers form covalent bonds with their targets when irradiated with ultraviolet light, allowing more stringent washing and better signal-to-noise ratios. This approach also seems to be more cost-efficient and less time-consuming than developing specific monoclonal antibodies.

It is too early to predict whether aptamer arrays, a nucleic acid-based technology, will outshine antibody arrays and become the dominant protein chip. Only time will tell. Neither is it absolutely clear at this current stage which company, if any, will become the dominant leader in the protein chip industry. Several companies have invested heavily in protein chips. Like many other emerging industries, there will be a consolidation in the protein chip industry in the next 10 to 15 years. Whatever the fate of individual protein microarray companies, it is clear that the era of microarray-based proteomics has arrived.

MARKET DRIVERS AND COMMERCIAL PROSPECTS

Proteomics has been the most imminent driver for protein microarrays and will remain so in the next 5 years. The proteomics market has reportedly grown from $1.1 billion in 2001 to $1.4 billion in 2002 [10]. Although DNA microarrays can measure mRNA expression levels in a cell, they cannot yield direct measures of the proteins produced. Protein chips on the other hand can directly measure both the relative level of proteins and their interactions with other molecules. Recognizing that mRNA expression levels do not always accurately reflect the expression of corresponding proteins or secondary processing/modification events researchers are increasingly combining gene expression and proteomic data. As genomic research is moving toward analyzing specific actions of genes and gene-encoded proteins, protein arrays' potential in the proteomics field appears tremendous.

Furthermore, protein microarrays will permit researchers routinely to study families of proteins in their efforts to understand the complex interactions of protein systems inside cells. Such studies can illustrate protein interactions and signaling pathways, providing more definite answers to whether a potential target is indeed involved in the disease etiology and how it affects the disease development. Scientists can address more complicated questions about gene function and drug interactions than they otherwise might.

Protein microarrays have great potential for use in drug development as well. One such application is for toxicology studies. Currently, drug developers screen compounds for toxicity only in animal trials. With effective toxicity assays, protein microarrays could identify toxic compounds well before the animal trials, saving pharmaceutical developers the costs of further developing drug candidates that will prove fruitless — in other words, taking the fail-early strategy.

Beyond proteomics research, protein microarrays have great potential in clinical trials and diagnostics. Yet it is not likely that commercial products in this arena will be widely available in the next 5 years. As in DNA microarrays, challenges to develop clinical products are higher than research products. One cannot sacrifice accuracy or ease of use in clinical products. With that said, protein microarrays can be adapted into many different applications. Market potential for protein chips is likely to be much bigger than that of DNA microarrays, which already reached $596 million in 2003 [4].

BIOCHIP INDUSTRY: THE NEXT 5 YEARS

The biochip industry has grown beyond DNA and protein microarrays to tissue arrays, cell arrays and other arrays. The fast adoption of biochips also drives the growth of related services such as instruments, software, reagents, and other consumables. According to a recent market report, demand for biochip products and services in the U.S. will expand more than 25% annually to nearly $1.8 billion in

2006 [2]. Sales of biochips will reach $720 million, while combined sales and revenues of related products and services will reach $1 billion.

Different microarray industries will see different growth projections in the next 5 years. The DNA microarray industry will continue to grow, but the market has matured. The past 5 years saw the fastest growth of DNA microarrays in genomic research. Although DNA microarrays will be more widely used in the next 5 years, the total revenue for DNA microarrays in genomic research will grow at a much slower pace as the price of DNA chips fall due to competitive pressures. Diagnostic DNA chips will likely become available in the next 5 years and open a whole new market for DNA microarrays.

Protein microarrays have overcome initial technical challenges and started making an impact in the life science community. Protein chips will see the strongest growth owing to demand for functional genomics studies by pharmaceutical and other life science researchers. A market report projected the protein microarray market to grow at an annual growth rate of 85% to 2006 [11].

Tissue arrays, cell arrays, and other arrays are in an emerging stage. The next 5 years may see the emergence and maturation of commercialization of these novel arrays. Microarrays are such a versatile technology that can be adapted into a spectrum of different applications. In many ways, this ensures microarray remains one of the fastest growing fields for years to come.

REFERENCES

1. Schena, M., Shalon, D., Davis, R.W., Brown, P.O. Quantitative monitoring of gene expression patterns with a complementary DNA microarray. *Science* 270: 467–470, 1995.
2. Biochips to 2006. Cleveland, OH: The Freedonia Group, Inc. 2002.
3. Frost and Sullivan. World Biochip Market. 2001.
4. Frost and Sullivan. Strategic Analysis of World DNA Microarray Market. 2004.
5. Frost and Sullivan. U.S. DNA Microarray Markets. 2005.
6. The Market for DNA Microarrays: Core Labs & End-Users Perspectives. Arlington, VA: BioInformatics LLC. 2005.
7. DNA Microarrays: A Market Update. South San Francisco. CA: BioCompare, Inc. 2006.
8. Castrodale, B., Pollard, J., and Branca, M.A. Outlook for DNA Microarrays: Emerging Applications and Insights on Optimizing Microarray Studies. Needham, MA: Cambridge Healthtech Institute, 2002.
9. Protein Microarrays: Technology Adoption and Utilization. Arlington, VA: BioInformatics LLC. 2005.
10. Bodovitz, S. *Proteomics*. Sudbury, U.K.: Select Bioscience, Ltd. 2003.
11. Nabhan, A. Protein Arrays: Commercial Opportunities, Legal Trends, and Technology Analyses Westborough, MA: Drug and Market Development Publications. 2003.

22 A Pharmaceutical Perspective for Microarrays and Biochips: Current-Market Overview and Future Trends

Lisa Milne

CONTENTS

MICROARRAYS: CURRENT-MARKET OVERVIEW

Since the completion of the Human Genome Project, DNA microarrays have become a significant laboratory research tool. The microarray industry is now poised for explosive growth into other areas including clinical diagnostics/toxicogenomics, environmental analysis, antiterrorism monitoring, comparative genomics, and forensic testing. Most industry analysts would agree with a recent Frost & Sullivan report that puts the current size of the DNA microarray market at nearly $600 million [1]. In their report, Frost & Sullivan predict that total DNA microarray revenue will reach $937 million by the end of the decade with a compound annual growth rate of 6.7%. Estimates regarding the DNA microarray industry could go higher if one considers that Affymetrix alone posted $222 million in product revenue in 2003 and if it were to continue to post growth at 10% compounded annually for the next 6 years (a low rate when compared with its previous 5 years) Affymetrix could anticipate product revenues of $430 million by the end of the decade [2]. When one considers that close to 50% of the array market, by volume, was done by do-it-yourselfers in 2003, shrinking to less than 25% in 2006, and not with commercially purchased arrays, and that Affymetrix accounts for 80% of commercial array sales, the total market for DNA arrays could be well over $1 billion by 2010 [3]. The continued conversion from the home-grown variety of arrays to the commercial arrays, as economy of scale and qualitative considerations takes hold, could greatly drive these numbers. Additionally, the array industry as a whole, including DNA, protein, cell, and tissue arrays could exceed $5 billion by 2010 [4].

Though these estimates are impressive, they still fail to take into account the explosive industry growth that could occur, should microarrays take off in any of the four emerging growth areas, namely clinical diagnostics, toxicogenomics, environmental monitoring, and bioterrorism. Current estimates put the DNA-based (nonmicroarray) diagnostics industry at more than $1 billion (2002) with expectations that it will reach $3 billion by 2008 [5]. Toxicogenomics, with a current market of only $120 million (2003), is expected to grow to almost $259 million by 2008 [6]. But the real financial incentives are coming from the U.S. government's commitment to combating terrorism. There is an expectation that DNA microarrays will be used in every aspect of counterterrorism from surveillance to exposure diagnosis to treatment monitoring. The stakes are high and so are the funding opportunities for microarray technologies. For example, funding at the CDC for initiatives regarding public health preparedness and response to bioterrorism topped $49 billion in 2004 [7]. To add to these numbers, the budget proposed by President Bush for 2005 includes a 10% increase to Homeland Security funding and an increase in the Health and Human Services budget for bioterrorism preparedness to over $5 billion from just $300 million in 2001 [8]. Included in the 2005 number is money for Project Bioshield with a $2.5 billion budget proposal before Congress (up from $900 million in 2004) and an estimated $5.5 billion secured for the project over the next 10 years and a new $129 million biosurveillance initiative. The proposed funding for the NIH tops $28 billion in 2005 with $1.7 billion set aside for bioterrorism research. There is also a $567 million allocation for defending food and agriculture systems. In many of these areas DNA microarray technology could prove useful pending further development. It is clear that federal initiatives offer great financial incentives for those in the DNA microarray industry to push this technology forward.

Of these "new frontiers" for DNA microarrays, early commercial applications and market adoption is expected in the area of molecular diagnostics. Diagnostic arrays could be used to profile tumor subtypes to identify patients who would respond to a particular drug treatment *a priori*, for treatment decisions or for inclusion in clinical trials, and might also be used to follow a patient through treatment to see if the tumor is responding [9–13]. One recent study using DNA microarrays to diagnose non-Hodgkin's lymphoma (NHL) illustrates this point [14]. In 2000, there were 22,553 deaths from NHL subtypes, and 54,900 patients were newly diagnosed. Of the various subtypes of NHL, diffuse large B-cell lymphoma (DLBCL) subtype is the most common. Within this subtype, only 35–40% of patients can be cured with chemotherapy, the remaining 60–65% with DLBCL will die from the disease. Current techniques, such as histopathology and molecular markers, are unable to distinguish between responsive and nonresponsive DLBCL tumors. However, using DNA microarrays to profile gene expression, researchers were able to classify DLBCL tumors into several subclasses and to further link those profiles with patient prognosis.

The application of microarrays to the study of toxicogenomics is already moving into the commercial realm. Affymetrix and Roche have teamed up with a $70 million multiyear partnership to develop commercial array applications. Their first product is the Amplichip™ CYP450 microarray launched in June 2003 in the U.S. and, in fall 2004, received its CE mark for launch in Europe [15]. The CYP450 chip is used to identify variations in two genes (CYP2D6 and CYP2C19) that play a major role in the metabolism of various drugs. Revenue expectations are high, and corporate sources are hoping for $100 million annually by 2008 from this single product, though the pair expects the P450 array will be only one of several pharmacogenomic tests in the platform [16]. However, the FDA has already begun to question Roche's classification of the Amplichip as an ASR (analyte-specific reagent), which did not require review prior to launch [17]. The FDA is investigating whether the claims made by the company for the chip would require it to be designated as a medical device that would then be subject to FDA review and likely array platform standardization requirements for both test procedures and data analysis. Currently, the device is approved for research use only, and is not indicated for clinical diagnosis, though Roche is pursuing IVD (*in vitro* diagnostic device) status for the Amplichip. Regardless of its designation, toxicogenomic chips are in demand not only for clinical patient profiling but also from drug industry researchers to enable them to identify toxic compounds earlier in the R&D timeline, saving time and money. Meanwhile, there are those in market research who anticipate that over the next year, the use of microarrays for toxicogenomics could grow by approximately 33% among existing users [18].

The combination of molecular diagnostics and toxicogenomics studies are leading the way toward personalized medicine where treatment plans would be tailored to an individual's response and potential for side-effects based on the personal genetic expression pattern. There is a growing eagerness for such personalized medicine. In a recent, though small, survey, 43 participants were asked how much more they would be willing to pay for tailored drugs that better matched their body type and have fewer adverse reactions [19]. Of the 43 questioned, 27 said they would pay 10% or more for such personalized medicines. There are, however, significant barriers to the widespread adoption of microarrays in clinical diagnostics such as reimbursement, education, awareness of the test's utility, nonuniform testing procedures, data interpretation, and intellectual property rights covering specific genes that limit their inclusion in such tests [20].

Federal funding of environmental and antiterrorism surveillance has helped to stimulate the development of commercially viable microarray-based detection products. In order to be competitive, biosurveillance devices must become more accurate, faster and cheaper as outlined by panelists at a bioterrorism session at the Biotechnology Industry Organization (BIO) Convention in Washington, D.C. (June, 2003). For example, current air monitoring tests cost about $40 and are truck-mounted but will need to be made both cheaper and smaller according to Cindy Bruckner-Lea (staff scientist, U.S. DoE Pacific Northwest National Laboratory). It is in this context that DNA microarrays could prove useful if the analysis device could be miniaturized. Recently, BioLog Corporation has been awarded an NIAID grant to develop their Phenotype Microarray™ for bioterrorism agents, and a group at Lawrence Livermore National Laboratory has constructed a Multi-Pathogen Identification array that proved to be 91% accurate and able to detect as little as 10fg of *B. anthracis* DNA in a sample [21]. Microarrays can also be used to aid in determining if there has been a terrorist attack at the patient level. Because an outbreak of flu might present symptoms similar to an outbreak caused by exposure to a biowarfare agent, there is a real need to be able to quickly determine the source of a patient's symptoms. One can imagine a scenario at local emergency rooms during such an outbreak where physicians would have to make a rapid assessment about public threat levels and will need diagnostic tools to do so. With this in mind, Dr. Maria Salvato at the Institute of Human Virology, University of Maryland Biotechnology Institute is employing DNA microarrays to identify gene expression profiles that correspond to a flu infection vs. infection from a more serious threat [22]. There are many other potential applications for microarrays in combating terrorism that have not been discussed here.

DNA sequencing is also becoming an important growth area for microarray technology and has applications in both terrorism surveillance and public health monitoring. For example, the recent severe acute respiratory syndrome (SARS) outbreak highlighted the need for rapid genetic level identification of the virus, as well as an immediate need for an easy-to-use, accurate screening tool for patient diagnosis. There was really no time to develop a traditional antibody-based diagnostic test in light of the rapid spread of the virus from Asia to North America and Europe. The ability to quickly identify the strain of the outbreak through comparative sequencing could lead to a better understanding of how the virus spreads and how best to treat patients. Monitoring mutation patterns and rates among other viruses such as HIV, polio, and rhinoviruses might enable researchers to create better vaccine or drugs that target specific variants.

In summary, there are many areas in which DNA microarrays will have a significant impact. From bioterrorism, cancer profiling, or to monitoring global flu outbreaks, there are a wide range of opportunities and significant increases in both public and private funding to stimulate microarray technology development.

MICROARRAYS: COMPANY PROFILES

The following is a list of companies that are involved in the microarray arena. The list is by no means exhaustive, nor is it a complete description of the offerings of an individual company. The following is meant to provide an overview of the kinds of business involved in microarray development and include companies that provide: sample preparation, microarray construction, do-it-yourself providers, robotics, data analysis, reagents and full-service array outsourcing.

Affymetrix (Santa Clara, CA) generates oligonucleotide arrays by synthesizing short oligonucleotide probes (20 to 25 bases) directly on the surface of chips via photolithography and combinatorial chemistry using photomasks akin to the semiconductor industry. Although their DNA density is quite high (500,000 features per chip), Affymetrix continues to work on decreasing feature size to realize significant reductions in manufacturing cost, especially as measured on a per feature basis. Affymetrix expects to get to features that are orders of magnitude smaller than currently produced. Several new high-density array products have recently been launched including CustomSeq, for large-scale resequencing projects and Mapping 10K, Centurion 100K, and reports of a release of 500K SNP genotyping chips in 2005, which allows genotyping of 10,000, 100,000 or 500,000 SNPs, respectively. At 500K density, this will give a serious competitive advantage to Affymetrix for genetic association studies that predict the need for greater than 300K SNPs needed for most screens. Affymetrix also has prelaunched a 96-well-based system for expression profiling geared toward large pharmaceutical users; this has the potential to provide a much lower cost alternative to the cartridge arrays. This system is based upon a modified CaliperLS SciClone robotic platform. Whole plate scanning is achieved via an Axon OEMed whole plate scanner.

Agencourt Bioscience Corporation (Beverly, MA) provides DNA sequencing, SNP discovery and library construction services.

Agilent (Palo Alto, CA) provides Printed Microarray Solutions, which are a line of catalog oligo microarrays printed with 60-mer probes including a new whole mouse genome microarray kit and which has been a widely used source for custom array design and printing services.

AGOWA (Berlin, Germany) provides DNA sequencing and genomics services.

Agilix Corporation (New Haven, CT) has designed a set of universal probes and a procedure called rolling circle amplification (RCA) to provide full transcriptome analysis that can be done without the need for prior consent?

Applied Biosystems (Norwalk, CT) offers the Applied Biosystems Expression Array System that includes the Human Genome Survey Microarray. The current version of Human Genome Survey Microarray contains 31,077 probes that cover 53,977 individual transcripts and target a complete, annotated, and fully curated set of 27,868 human genes from the public and Celera databases and is used with the Applied Biosystems 1700 Chemiluminescent Microarray Analyzer. They also offer a macroconfocal system (8200 Cellular Detection System) that employs the company's FMAT® platform [3] for assaying cells or beads in standard microtiter plates.

Applied Precision, LLC (Issaquah, WA) is a provider of imaging, measurement, and analysis systems for life science research and of outsourced microarray analysis services.

Avalon Pharmaceuticals (Germantown, MD) is a drug discovery company that measures changes in gene expression patterns in response to an applied drug candidate.

Axon Instruments (now part of Molecular Devices) (Union City, CA) produces the GenePix® 4000B microarray scanner equipment, one of the most well adopted scanners for smaller scale microarray, as well as a line of specialized imaging systems.

Bangs Laboratories (Fisher, IN) is a supplier of uniform microsphere products including its QuantumPlex™ product line for flow cytometry-based multiplexed analyses. The company offers bead sets preconjugated with streptavidin or goat-anti-mouse IgG, or beads bearing a carboxylate group.

BD Biosciences Clontech (Franklin Lakes, NJ) manufactures the Atlas™ gene expression microarrays made with spotted 70-mer probes and also supplies antibody microarrays and Cytometric Bead Arrays.

BioArray Solutions (Warren, NJ) provides custom bead arrays for DNA, protein, and cell analysis.

Biocept Laboratories (Carlsbad, CA) employs the 3D HydroArray Technology platform using a 'hydrogel' substrate to generate two-dimensional DNA or protein microarrays with a focus on prenatal health.

Bioforce Nanosciences, Inc. (Ames, IA) is applying its nanoarray technology to proteomics and diagnostics uses and had developed the ViriChip™ pathogen screening system.

BioLog (Hayward, CA) offers array sets, testing 2000 phenotypes of microbial cells called the Phenotype MicroArray™.

Bio-Rad Laboratories (Hercules, CA) offers the VersArray™ microarray hybridization chamber and offers Luminex's instrumentation integrated with Bio-Rad's Bio-Plex™ software for simplified system setup, data analysis/management, and automated hardware validation/calibration.

BioSource International (Camarillo, CA) offers antibody bead array kits using the Luminex platform.

Caliper Life Sciences (Hopkinton, MA) is a microfluidics company that offers the LabChip 90 for DNA analysis of multiplex PCR and digest fragments as well as protein analysis. It also now offers a new version, the LabChip 3000, with a 16-sipper design that may prove to be a fundamental part of higher throughput systems requiring high-density plate formats. It would be intriguing if they combine an RNA analysis for high-throughput applications and QC. They also have had a limited foray into automation cassettes and instrumentation for higher throughput slide-based applications, as well as newer joint ventures with Affymetrix to provide instrumentation for high-density plate arrays. Perhaps this company is well poised to take part in the next wave of higher density microarray.

Compugen (Tel Aviv, Israel) is an informatics company that has agreements with Sigma-Genosys, Pfizer, and Novartis. Microarray manufacturers are increasingly forming partnerships with bioinformatics companies to develop probes that more accurately reflect mRNA variations.

Combimatrix (Mukilteo, WA) sells a semiconductor-like custom array with the potential for electrochemical detection as well as standard fluorescent detection. It has had collaborations with Roche for diagnostic systems and with the Department of Defense.

Corning (Corning, NY) sells GAPS™ II and Ultra-GAPS™ coated ultraflat glass slides for self-spotted arrays as well as many reagents and plasticware for microarray. They had made an earlier unsuccessful foray into spotting of microarrays with limited acceptance, using a fairly unique process utilizing a ceramic spotting head system for which they still hold the patent.

Decode Genetics (Reykjavik, Iceland) is a biopharmaceutical company that employs a population-based approach to gene discovery using genetic and genealogical data gathered from more than half the adult population of Iceland. Decode currently has lead programs in heart attack, arthrosclerosis, asthma, stroke, schizophrenia, diabetes and obesity.

EraGen Biosciences (Madison, WI) uses novel DNA chemistry, incorporating synthetic base pairs into the natural DNA code to develop "universal DNA microarrays" that are not specific to a single organism.

Fluidigm Corporation (San Francisco, CA) produces integrated fluidics circuits with the TOPAZ™ System that are currently used to identify conditions for protein crystallization. Microfluidics techniques such as those offered by Fluidigm may be used 1 d to create integrated systems for both sample preparation and analysis of DNA and protein microarrays.

Furuno Electronic, Ltd. (Japan) has recently announced that it will be offering DNA microarray synthesizers through a deal with Combimatrix. The pair will codevelop a commercial bench-top synthesizer (BTS) instrument to be launched in 2005.

Future Diagnostics (Netherlands) specializes in the contract development of assays for other companies.

GE Healthcare (former Amersham BioScience, former Motorola) (Piscataway, NJ) offers CodeLink™ prearrayed oligonucleotide bioarrays for gene expression and SNP analysis as well as CyScribe™ labeling kits. CodeLink™ is a new type of array that reportedly eliminates hybridization problems and reduces background levels by putting the oligo probes into a gel matrix that more closely resembles a solution-phase environment. The CodeLink™ brand was acquired from Motorola Life Sciences. They also offer the Typhoon™ series imager as well as reagents and slides for microarray applications.

Genomic Solutions (Ann Arbor, MI) produces the GeneMachines® line of array instrumentation.

Genomics USA (Chicago, IL) is a private company working on a Human ID chip, a Bacterial/Viral ID chip and an HLA (human leukocyte antigen typing) chip for which they received almost $500K to develop.

Genisphere, Inc. (Hatfield, PA) offers the 3DNA™ microarray detection kits.

GenoSpectra (Fremont, CA) provides gene expression profiling using the QuantiGene® Reagent System based on branched DNA signal amplification for RNA quantification. GenoSpectra is developing a multiplex method for analyzing the activity of up to 30 genes from a single sample simultaneously.

Illumina (San Diego, CA) pioneered the BeadArray® multiplex array technology comprised of addressable beads that self-assemble into microwells etched into an array substrate. The company has developed two different platforms based on BeadArray: the Sentrix™ 96 Array Matrix for high-throughput applications, and the Sentrix BeadChip for lower-throughput applications. Though primarily noted for its work in the SNP arena, the company also recently announced a foray into the whole genome expression array at a low price point utilizing its bead technology.

Invitrogen (Carlsbad, CA) is the leader in custom oligonucleotide synthesis, enzymes and reagents for genomic expression profiling with new products focused on proteomic arrays including the Yeast Protoarray PPI (protein–protein interaction) kit.

Luminex (Austin, TX) developed the widely popular xMAP platform enabling up to 100 assays to be performed simultaneously in a single multiplex reaction. Tm Biosciences (Toronto, Canada) offers beads with its Universal Array™ oligonucleotide adapters to create DNA multiplexed arrays for use on the xMAP platform.

The Tm 100 Universal Sequence Set™, used with Luminex's FlexMAP™ beads, consists of 100 unique 24-mer tags designed to work in a single reaction with minimal cross-hybridization. The tags can be integrated into virtually any primer for genomic screens and SNP detection.

Marligen Biosciences (Ijamsville, MD) offers two preconfigured multiplexed systems in its Signet™ line of xMAP-based genotyping assays: one for detecting 43 male-specific SNPs on the Y chromosome, and one for distinguishing 30 genetic variations in the hypervariable regions I and II of mitochondrial DNA. The company also will work with customers to custom-design multiplexed SNP assays.

MicroFab Technologies (Plano, TX) uses inkjet technology to generate DNA and protein microarrays.

MiraiBio (Alameda, CA) offers software for streamlining data analysis including its MasterPlex GT program for genotyping applications.

Motorola Life Sciences (Schaumburg, IL) acquired Clinical Microsensors in June 2000 and now offers their **eSensor™** Cytochrome P450 DNA Detection System that simultaneously detects ten well-characterized mutations belonging to the 2D6, 2C9 and 2C19 genes of the Cytochrome P450 superfamily. CodeLink systems also originated from this group, later sold to Amersham, now part of GE.

MWG Biotech (Germany) offers catalog arrays and oligo sets using its HPSF oligonucleotides. The MWG human array comprises 40,000 genes on two slides.

Nanoplex Technologies (Mountain View, CA) has developed novel particles that are an alternative to traditional microsphere beads. Nanoplex has developed a technology for manufacturing tiny, cylindrical particles that serve as the nanoscale equivalents of bar codes. Fashioned from inert metals such as gold, nickel, platinum, or silver, these Nanobarcodes™ are encoded with a series of submicron stripes that can be machine-read. The rods can be complexed directly to biological molecules for various applications, including multiplexed assays

Nanosphere (Northbrook, IL) offers the Verigene™ platform that can perform several tests including a SNP-based test for hypercoagulation disorder and is developing assays for infectious diseases, cancer and CNS disorders.

NimbleGen (Madison, WI) offers versatile low-volume, but high-density custom arrays synthesized using DLP technology vs. photomasks. This product is now offered via Affymetrix as NimbleExpress.

Orchid Diagnostics (Stamford, CT) offers the SNP-IT tag array for genotyping capabilities and provides paternity testing through its GeneScreen product.

OriGene Technologies, Inc (Rockville, MD) specializes in low-density, focused arrays to permit study on one specific biological pathway or the study of a limited number of genes.

Perkin Elmer (Boston, MA) provides an Integrated Microarray Laboratory including the ScanArray scanners and MICROMAX™ line of cDNA arrays and reagents and the Piezorray microdispensing system to generate spotted arrays.

Pharmaseq (Monmouth Junction, NJ) applies its patented microtransponder technology to DNA probe diagnostics, single-nucleotide polymorphism (SNP) detection, and proteomics.

Qiagen (Netherlands) is the leader in oligonucleotide isolation kits and now sells SensiChip spotted arrays. Each SensiChip Bar contains six arrays separated by microfluidic hybridization chambers permitting the researcher the flexibility to run six low-density experiments at once

Quantum Dot (Hayward, CA) has developed a system based on its Qdot™ quantum dot conjugates with the following advantages over traditional labeling methods cited: increased probe brightness, photostability, broad excitation ranges, and narrow-emission spectra. QDC is currently developing a product line featuring QDot-labeled beads for multiplexed assays, with applications for SNP detection and immunoassays that may enable the simultaneous analysis of up to thousands of analytes at once. This technology is now owned by Invitrogen.

Radix BioSolutions (Georgetown, TX) specializes in custom-developed assays using the client's chosen platform. A large portion of Radix's business is currently devoted to developing xMAP (Luminex) assays for protein and DNA analytes.

Roche (Basel, Switzerland) launched the first commercially available pharmacogenomic microarray, the AmpliChip CYP450, in June 2003 for metabolic profiling and is working on approval of a clinical diagnostic version of the test in 2004. Roche also offers the MagNA Pure Compact for benchtop nucleic acid purification and the LightTyper for single nucleotide polymorphism (SNP) analysis.

Sequenom (Boston, MA) offers the MassARRAY™ genetic analysis system for SNP detection primarily and quantitative gene expression determination both based upon a high-throughput MALDI-MS technology.

SuperArray Bioscience Corporation (Frederick, MD) is a privately held company that offers its own GEArray™ line of pathway-specific DNA microarray products as well as other expression products and services.

Sigma-Genosys (Haverhill, UK) is well known for providing custom oligonucleotides and oligo kits in addition to offering its Panorama™ expression arrays and ORFmers.

SmartBead Technologies (Cambridge, UK) employs an alternate to bead arrays by using a barcoding strategy called The UltraPlex™ system. This system employs microscale aluminum particles that have a series of holes that make up the barcode. UltraPlex™ barcodes can be used to tag biomolecules in multiplexed assays.

SuperArray Bioscience Corporation (Frederick, MD) specializes in low-density, focused arrays for researchers who want to study a limited number of genes involved in a specific biological pathway using GEArray® focused microarrays.

SurModics (Eden Prairie, MN) offers reagents to stabilize proteins bound to microarrays or to microspheres for maintenance of protein integrity during storage. They are the supplier to GE for the CodeLink microarrays substrate.

TeleChem International (Sunnyvale, CA) offers the ArrayIt™ platform for colorimetric protein microarray analysis with several new applications anticipated in low- and high-throughput nucleic acid and protein array analysis using fluorescent, colorimetric, chemiluminescent and surface plasmon resonance detection. Telechem also offers microspotting devices as well as array slides and labeling kits.

3D Molecular Sciences (Cambridge, UK) has created "optically readable" microparticles that are identified by both their shape and the holes contained in the particle. The particles can be coupled to biomolecules for multiplex array analysis in research, clinical diagnostics, and environmental monitoring and antiterrorism applications.

MICROARRAYS: FUTURE MARKET DIRECTIONS AND HURDLES

Many exciting advances are being made in the field of microarray technology, but there continue to be several major hurdles for the technology to evolve fully.

1. Regulatory: As demonstrated by the Roche launch of the AmpliChip, there are many FDA regulatory problems that need to be resolved. These include not only the designation as an IVD vs. ASR, but also how to standardize the sample preparation, and the interpretation of results. What would happen to isolated failures? Would failure of one gene or one section of the array invalidate the entire assay? What controls would be required? Many other similar questions would need to be resolved.

2. Quality control: Other FDA issues include quality control in both the design and the manufacturing process of the array itself. One can hope for improvements in microfabrication to minimize defects (and to reduce costs).

3. Cost: The cost per datapoint and cost per assay will need to be comparable to current assays, and in the context of increasing overall medical costs, the array data will likely need to be even more cost-efficient than current tests in order to be competitive.

4. Standardization: Currently genomic expression data is reported in many different ways in the literature. SuperArray (Frederick, MD). is developing a method of reporting expression levels in "copies per microgram of total RNA per cell." This would help to allow researchers in different labs to compare data; however, it does not address the role of alternate splicing of mRNA. This lab-to-lab and platform-to-platform standardization is a major hurdle to widespread routine use of DNA microarrays. The creation of the Genetic Analysis Technology Consortium (GATC), initiated by Affymetrix and Molecular Dynamics, is working to standardize array-based genetic analysis.

5. Sensitivity: Improvements in detection level of RNA to identify rare transcripts is critical to array technology.

6. Ease of use: For DNA microarrays to become universally useful, especially in a large clinical diagnostics lab, the integration of sample preparation with detection and analysis in one machine is fundamental to moving into the diagnostics area.

7. Reimbursement: The problem of who will pay for the adoption of DNA microarray technologies in the clinical setting has yet to be tested. But the ability of microarrays to generate diagnoses in hours rather than days may help to push insurers to cover the cost of the new technologies.

Now that the Human Genome Project is completed, researchers have a nearly complete catalog of all human genes. Because the practice of genomics relies on large-scale, comprehensive analyses of genes, the ability to parallel process expression information using DNA microarrays has become invaluable in genomics research. Advances in our understanding of genomewide expression patterns will foster many uses for microarray technology including: determining the potential for disease, predicting drug response, disease diagnosis and monitoring, bioterrorism surveillance, identifying infectious disease outbreaks, as well as forensic and paternity identification.

Furthermore, the U.S. government has indicated its inclination toward acceptance of genomic data in clinical settings. In 2003, the NIH issued a vision statement describing how genomics can contribute to the future of the practice of medicine and the FDA issued draft guidelines for the submission of genomic data in drug applications [23,24]. In September 2003, J. Craig Venter announced at the 15th Annual Genome Sequencing and Analysis Conference that his Science Foundation would establish a $500,000 cash prize to the person(s) who can develop a whole human genome sequencing technique at a cost of $1,000 per genome. The use of microarrays for DNA sequencing holds great promise for reaching Dr. Venter's goal. The recent increases in funding sources, the creation of new corporate partnerships, the launch of new commercial products and the tantalizing potential for wide-use applications combine to make the future of DNA microarray technology very bright.

REFERENCES

1. Frost and Sullivan, *Strategic Analysis of World DNA Microarray Markets*, March 2004.
2. Affymetrics, SEC filing 10-K, 2003.
3. John L. Sullivan at Stephens, Inc. (Little Rock, AR) estimates Affymetrix garners $240 M of the total $300 M array market. Rhonda Ascierto, Affymetrix Sequences Gene Profits, Silicon Valley Biz Inc., July 2003.
4. Kalorama Information, February 12, 2003.
5. *Lucy Sannes, Sannes and Associates,* Molecular Diagnostics: Technological Advances Fueling Market Expansion.
6. Global Information, Inc., Toxicogenomics: A Strategic Market Analysis, September 2003.
7. CDC Web site.
8. Bush Proposed Budget 2005, White House, Office of Management and Budget.
9. Sakamoto, M. et al., Analysis of gene expression profiles associated with cisplatin resistance in human ovarian cancer cell lines and tissues using cdna microarray, *Hum Cell*, 14: 305–315, 2001.
10. Luker, K.E. et al., Overexpression of IRF9 confers resistance to antimicrotubule agents in breast cancer cells, *Cancer Res*, 61: 6540–6547, 2001.
11. Certa, U. et al., High density oligonucleotide array analysis of interferon-alpha2a sensitivity and transcriptional response in melanoma cells, *Br J Cancer* 85: 107–114, 2001.
12. Chang, J.C. et al., Gene expression profiling for the prediction of therapeutic response to docetaxel in patients with breast cancer, *Lancet* 362: 362–369, 2003.
13. Buchholz, T.A. et al., Global gene expression changes during neoadjuvant chemotherapy for human breast cancer, *Cancer J*, 8: 461–468, 2002.
14. McShea, A., Arjormand, A., Kumar, A., Schmechel, S., *IVD Technologies*, 2004.
15. Roche Web site.

16. Heino von Prondzynski, member of Roche Executive Committee and head of Roche Diagnostics.
17. Filmore, D., Arrays and the FDA, *Mod Drug Discov*, June 2004.
18. Robin Rothrock, director of market research at BioInformatics, Microarray technology drives toxicogenomics market, *Bioinformatics*, Arlington, VA, May 2003.
19. Advances in Life Science Reports, Successful Pharmacogenomics Business Models, Sam Tetlow, Clearview Limited.
20. Luch Sannes, Sannes and Associates, Molecular Diagnostics: Technological Advances Fueling Market Expansion.
21. Wilson, W.J., Strout, C.L., DeSantis, T.Z., Stilwell, J.L., Carrano, A.V., Andersen. G.L., Sequence-specific identification of 18 pathogenic microorganisms using microarray technology, *Mol Cell Probes*, 16(2): 119–127, April 2002.
22. Common Flu or Bioterrorism? August 2002.
23. Medicines for You, NIH NIGMS brochure.
24. FDA, Guidance for Industry: Pharmacogenomic Data Submissions and Speech before PhRMA, Remarks by Mark B. McClellan, Commissioner, Food and Drug Administration, March 28, 2003.

Index